Springer Proceedings in Physics 12

Springer Proceedings in Physics

Managing Editor: H.K.V.Lotsch

Springer Proceedings in Physics is a new series dedicated to the publication of conference proceedings. Each volume is produced on the basis of camera-ready manuscripts prepared by conference contributors. In this way, publication can be achieved very soon after the conference and costs are kept low; the quality of visual presentation is, nevertheless, very high. We believe that such a series is preferable to the method of publishing conference proceedings in journals, where the typesetting requires time and considerable expense, and results in a longer publication period. Springer Proceedings in Physics can be considered as a journal in every other way: it should be cited in publications of research papers as *Springer Proc.Phys.*, followed by the respective volume number, page number and year.

Quantum Optics IV

Proceedings of the
Fourth International Symposium,
Hamilton, New Zealand
February 10–15, 1986

Editors: J. D. Harvey and D. F. Walls

With 101 Figures

Springer-Verlag Berlin Heidelberg New York
London Paris Tokyo

Professor John D. Harvey
University of Auckland, Private Bag, Auckland, New Zealand

Professor Daniel F. Walls
Department of Physics, University of Waikato, Hamilton, New Zealand

ISBN-13: 978-3-642-71409-2 e-ISBN-13: 978-3-642-71407-8
DOI: 10.1007/978-3-642-71407-8

Softcover reprint of the hardcover 1st edition 1986

Preface

This volume contains notes based on the lectures delivered at the fourth New Zealand Symposium in Laser Physics, held at the University of Waikato, Hamilton, February 10–15, 1986. At this meeting, about 80 physicists working in many parts of the world met to discuss topics of current interest in contemporary laser physics and quantum optics. These symposia, which have been held triennially since 1977, have evolved into an important meeting ground for experimentalists and theoreticians working in a very rapidly developing field. As the format has evolved, the number of participants, including the number from overseas, has grown steadily, and this year a poster session was included for the first time, enabling a far greater range of topics to be discussed than was possible in the limited lecture time available.

At this meeting the major interest of the participants concerned the theoretical investigation of squeezed states of the radiation field and the very recently reported experimental observations of such states. Other related areas of work reported here include bistability and chaotic behaviour of optical systems, the quantum theory of measurements, optical tests of general relativity, and the current technological limitations governing the stabilization of lasers.

The editors would like to thank the participants for providing detailed notes for publication shortly after the meeting, and the various organisations that have provided financial support. These include the New Zealand Institute of Physics, the Royal Society of New Zealand, the International Union of Pure and Applied Physics, the British Council, the University of Waikato, and the following companies: Coherent Radiation, Quentron Optics Pty, Burleigh Instruments, and Melles Griot.

The continuing success of these meetings has established them as triennial events, and we look forward to another stimulating meeting in 1989.

Hamilton, New Zealand
March 1986

D.F. Walls J.D. Harvey

Preface

[illegible]

Contents

Part I

Squeezed States of Light

Penetrating the Quantum Limit: The Experimentalists' View

M.D. Levenson and R.M. Shelby

IBM Almaden Research Laboratory, 650 Harry Road, San Jose, CA 95120, USA

0. Introduction

For someone concerned with the sensitivity or precision of a measurement, achieving the standard quantum limit has been considered the ultimate challenge. Around 1975, several theoreticians suggested that the standard quantum limit was not in fact a fundamental consequence of the uncertainty principle, and that quantum states existed which allowed more accurate measurements than had previously been thought possible [1]. They even suggested some plausible methods for generating certain of these states, now generally termed "squeezed states," and defined detection techniques which would demonstrate the noise reduction.

Experimentalists reacted to these early insights with apathy, cynicism and outright disbelief, but the need to make ever more subtle measurements and the hope for new discoveries would not go away [2]. After about seven years, some of us began exploring what could be done in the laboratory - rather than what could be done in principle [3]. As in all new endeavors, the experimentalists were confronted by a number of difficulties which could (usually) be overcome or circumvented. In this paper, we will discuss these difficulties and our conquest of them or lack there of.

1. First Difficulty: Understanding What The Theoreticians Are Talking About

I am not sure how theorists visualize light. Their results in quantum optics tend to be phrased in terms of correlation functions and to look like [4]:

$$g^{(n)}(r_1, t_1, ...r_n, t_n; r_{n+1}, t_{n+1}....r_{2n}, t_{2n}) = \frac{<\hat{E}^-(r_1,t_1)...\hat{E}^-(r_n,t_n)\hat{E}^+(r_{n+1},t_{n+1})....\hat{E}^+(r_{2n},t_{2n})>}{\left\{<\hat{E}^-(r_1,t_1)\hat{E}^+(r_1,t_1)> ..... <\hat{E}^-(r_{2n},t_{2n})\hat{E}^+(r_{2n},t_{2n})>\right\}^{1/2}} . \quad (1)$$

To an experimentalist, a light wave is a sinusoidal electric field: $E = E_1 \cos \omega t + E_2 \sin \omega t$ where the quantities E_1 and E_2 are called quadrature amplitudes. These "quadrature amplitudes" can be portrayed as axes of a two-dimensional map or "phasor diagram" and any state of the field can be specified on such a map.

In classical physics, both quadrature amplitudes can be exactly specified, and thus, a state of the field corresponds to a single point. Figure 1 shows the classical state of Athens on such a map. The amplitude of a classical state is the vector from the origin to the point specifying the state. The phase is the angle of that vector with respect to the E_1 axis. The two quadrature amplitudes are the projections of the vector on the two axes.

In quantum mechanics, the quadrature amplitudes are conjugate operators, and thus the Heisenberg Uncertainty principle prevents them both being specified exactly. Thus a modern quantum state must be portrayed as a region on the map with an uncertain border. Figure 2 shows the quantum state of Texas along with a theoretician's impression of that state. Such a state still has an average amplitude, average quadrature amplitudes and an average phase, but all of these quantities also have non-zero mean square deviations from the average values.

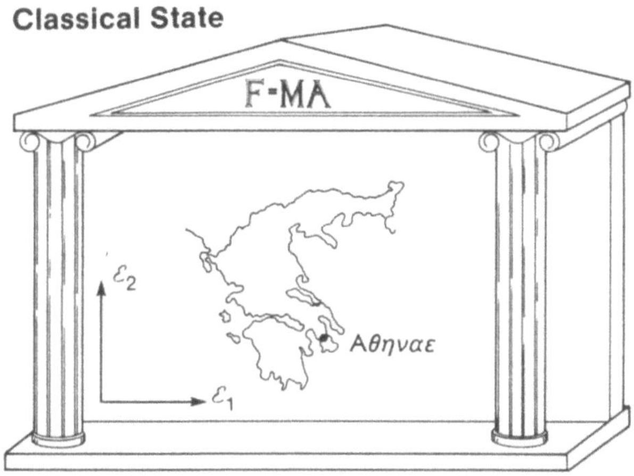

Fig. 1. A classical state is represented as a single point on the phasor map. This is the state of classical Athens. The outlined region contains other classical states.

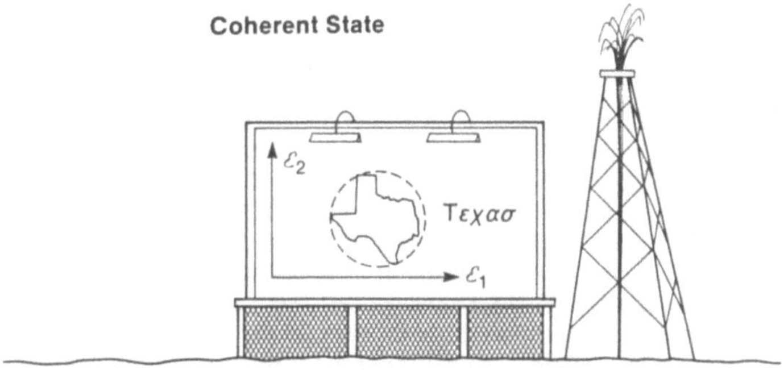

Fig. 2. In modern quantum physics, a single state covers a region in the map. A coherent state such as Texas (shown here) has circular probability contours.

The quantum states most similar to the classical states are called "coherent states" and have the additional property that the mean square deviations of the two quadrature amplitudes are equal [5]. On the phasor diagram, these states are represented as circular probability profiles. They are minimum uncertainty states; the probability profiles for all coherent states are the same. Only the average amplitudes can differ. The vacuum state is a coherent state with zero average amplitude, and all coherent states can be constructed by adding a classical amplitude to the fluctuation operators describing the vacuum state. Since there is no phase reference in the vacuum, the fluctuations for the vacuum state must be symmetric on the phasor diagram. These vacuum fluctuations are the source of the quantum noise measured in experiments [1,2].

In a squeezed state, the fluctuations in one quadrature amplitude are less than in the other and less than those of a coherent state [6]. Since the uncertainty principle still applies, the region of the phasor diagram corresponding to such a state has the same area as the circle of a coherent state. Figure 3 shows a typical squeezed state: Tennessee, along with the theorist's elliptical probability contour. Clearly the uncertainty in one quadrature amplitude has been reduced; however, the orientation of that quadrature with respect to the axes of measurement need not be so simple.

When the major axis of the squeezed state ellipse is along the vector of the average amplitude, the state has a poorly defined amplitude and a well-defined phase. Such states might be useful in interferometry. Conversely, when the major axis is perpendicular to the average amplitude, the magnitude of the amplitude is well defined. This might be useful in communication. Since light detectors measure only the square of the amplitude and are insensitive to phase, one must vary the

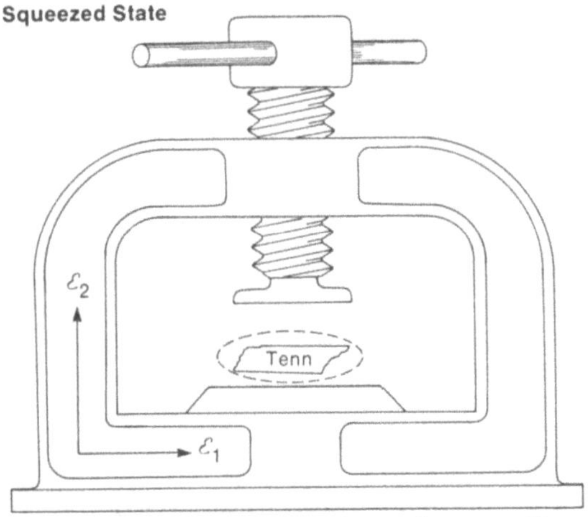

Fig. 3. Tennessee, a squeezed state, has a smaller extent along one quadrature axis than along the orthogonal axis. The area of a minimum uncertainty squeezed state is equal to that of a coherent state. The uncertainty of the E_2 coordinate of Tennessee is less than that of a coherent state.

relative phase of the average amplitude with respect to the ellipse axes to verify the generation of a squeezed state. The techniques for doing so are known in optical homodyne detection [7].

Having understood what the theorists are talking about, the experimentalist then has the problem of believing it. We had all been taught that quantum noise resulted from the fundamental uncertainty of the time at which a photon would reach the detector. Calculations based on that model give quantitatively correct results (for coherent states). To understand squeezed states we must reject that familiar particle model and learn a new (and more complex) field theory of noise. We also are asked to believe that common and relatively simple nonlinear optical processes can compress the vacuum fluctuations to a measurable extent. And then one of us actually has to do it.

2. Second Experimentalists' Difficulty: Being Taken Seriously

One who has just read the previous section might not be surprised at this difficulty, but it is serious nonetheless. There are two aspects of the problem: making theorists accept one's insights on the real issues in a feasible experiment and making supervisors, contract monitors and colleagues accept such a speculative project instead of what one had been doing previously. Both aspects become more difficult if commitments of time or money are required.

3. Third Difficulty: Measuring a Quadrature Amplitude

Simple light detection measures the energy flux which is proportional to the sum of the squares of the quadrature amplitudes. Optical heterodyne and homodyne detection can measure single quadratures, but special procedures are necessary to penetrate the quantum limit.

What one actually measures is the electronic noise from a light detector. Figure 4 shows the simplest experiment of that type. Light is incident upon a detector with unity quantum efficiency. The electrical current produced is amplified with a device with time constant τ. A meter then measures the current. The intensity at the detector is proportional to the square of the total field amplitude:

$$I = |E|^2 = |E_0|^2 + \delta E^* E_0 + E_0^* \delta E + |\delta E|^2 \qquad (2)$$

where E_0 is the average amplitude and δE is the fluctuation operator. The last term on the right is

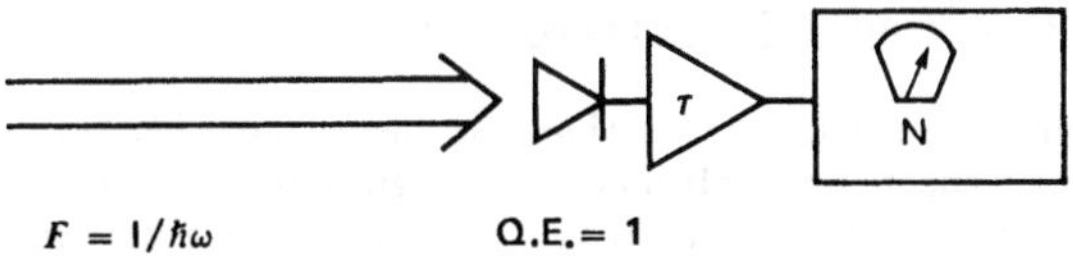

Fig. 4. Schematic of the photodetection process. A beam with intensity I and photon flux F is incident on a detector with unit quantum efficiency (Q.E.=1). The electrical signal is amplified by a device with time constant τ. The number of electrons in time τ is recorded as N on a meter.

the energy content of the fluctuations and is negligible. The operator describing the current is proportional to this intensity operator. The photon flux is $F = I/\hbar\omega$ and the average number of detected photoelectrons is $N = <F> \tau = |E_0|^2\tau/\hbar\omega$.

The meter does not read a constant value, however. The mean square fluctuation in the number of electrons is related to the mean of the fluctuation operators:

$$< \Delta N^2 > = 4|E_0|^2\{< \delta E_1^2 > \cos^2\Phi + < \delta E_2^2 > \sin^2\Phi\} \frac{\tau}{\hbar\omega} \tag{3}$$

where $< \delta E_i^2 > = < E_i^2 > - < E_i >^2 = V(E_i)$ is the mean square fluctuation in quadrature i and is often called the variance [8]. For coherent states in our notation $V(E_i)=1/4$. A little algebra then gives the familiar Poisson statistics result $< \Delta N^2 > = N$.

For a squeezed state, the variances are unequal:

$$V(E_1) = \exp(-2r)/4, \qquad V(E_2) = \exp(2r)/4 \tag{4}$$

where r is the squeeze parameter [2]. The mean square fluctuation then depends upon the phase angle Φ between the average amplitude and the ellipse axis. The electronic noise falls below the coherent state limit when Φ is correct.

The simplest way to vary the phase angle without introducing loss which destroys squeezing is the Yuen-Shapiro test for squeezing shown in Figure 5 [9]. Light that might be squeezed is incident on a low reflection beam splitter. A strong coherent wave which acts as the local oscillator is incident from the other side. The amplitude reflection coefficient is ε. The dominant fluctuations on the combined beam come from the potentially squeezed signal wave. The dominant average amplitude comes from the local oscillator. Thus when the phase of the local oscillator is varied (for example by moving a mirror), the mean square fluctuation of the detected current should vary as well.

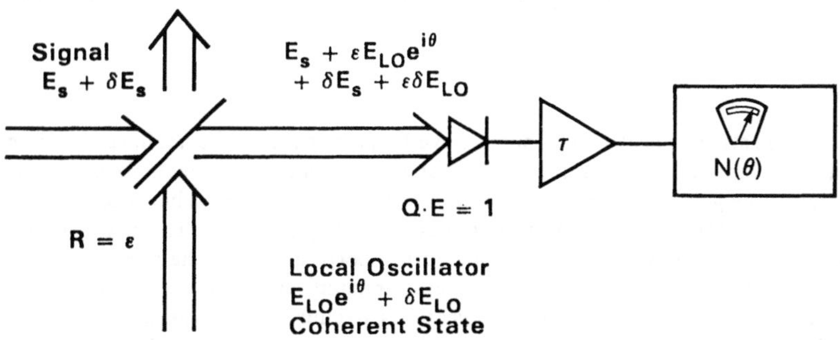

Fig. 5. The Yuen Shapiro Test for Squeezing. The signal (which might be squeezed) is incident on a beam splitter with a amplitude reflection coefficient ε. A strong local oscillator is incident from the other direction. The wave at the detector is the superposition of these two. The dominant fluctuations come from the signal, but the dominant average amplifier comes from the local oscillator.

4. Fourth Difficulty: Laser Power Fluctuations Contribute 80dB of Excess Noise

This can be overcome by looking at fluctuation frequencies higher than the fastest possible laser fluctuation, typically 5 MHz. Radio frequency spectrum analyzers are designed to make such measurements, so there is no electronic difficulty. There is an optical problem, however; the electronic noise at a nonzero frequency results from two bands - not one band - of optical frequencies [7].

Imagine a local oscillator at the optical frequency ω. Then a (fluctuating) wave at frequency $\omega+\nu$ can beat with this local oscillator on the detector surface to produce electronic signals and noise at frequency ν. However, an independent wave at frequency $\omega - \nu$ also beats with the local oscillator to produce an electronic response at frequency ν. If only one of these bands is squeezed, the potential noise reduction is halved. To realize the full reduction, one must squeeze both the "signal" band at $\omega+\nu$ and the "image" band at $\omega-\nu$, which otherwise will contain quantum noise even if the average amplitude is zero.

5. Fifth Difficulty: Learning to Squeeze Two Bands of Vacuum States

The Hamiltonians that produce squeezed states are proportional to the square of photon creation operators or to the product of two creation operators [2]. The proportionality constant is a functional on atomic operators. In the simplest approximation the coefficient is proportional to a nonlinear susceptibility times a pump amplitude to some power. The (three wave) parametric amplification process employs a second order optical nonlinearity. The overall process splits pump photons into signal and idler photons. In four wave mixing, the third order nonlinearity creates signal and idler photons from two pump photons. In each case there is parametric gain, and the phase reference is the pump field.

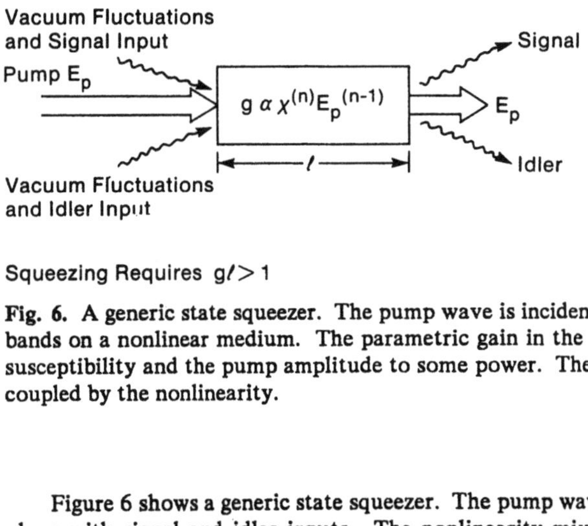

Fig. 6. A generic state squeezer. The pump wave is incident along with inputs at the signal and idler bands on a nonlinear medium. The parametric gain in the medium is proportional to the nonlinear susceptibility and the pump amplitude to some power. The output consists of signal and idler bands coupled by the nonlinearity.

Figure 6 shows a generic state squeezer. The pump wave E_p is incident on the nonlinear medium along with signal and idler inputs. The nonlinearity mixes these together. Detectable squeezing requires that the product of the gain and length be of order unity. Figure 7 shows schematically what happens to the signal and idler. Each of them has an amplitude in each quadrature. The mixing is best described by the eigenvectors of the mixing process which are superpositions of the signal and idler quadratures. One eigenvector is amplified while the other is attenuated. According to squeezed state theory, even the quantum noise of the deamplified eigenvector is attenuated [8]. The amplified noise is familiar as parametric fluorescence. To detect the deamplified noise, one must add a local oscillator before the detector. When the local oscillator phase is correct, the field at the detector has sufficient average amplitude to show that the noise is reduced. The number of detected electrons in the squeezed frequency band varies little around its average value. The essential insight is that homodyne detection can access the deamplified eigenvector of the parametric mixing process.

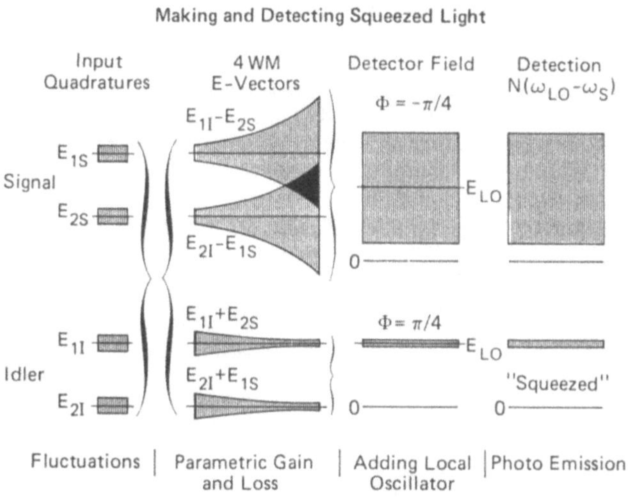

Fig. 7. What goes on in a generic state squeezer. The fluctuations in the two quadratures of the signal and idler bands are shown as input. The eigenmodes of the parametric mixer are superpositions of the inputs. One is amplified and the other attenuated. The average amplitude with noise inputs remain zero. The local oscillator adds a nonzero average amplitude with a definite phase with respect to the pump and eigenmodes. At the detector, one phase of local oscillator projects out a state with smaller fluctuations in the amplitude and the detector current - the squeezed state.

6. Sixth Difficulty: No Nonabsorbing Material is Nonlinear Enough to Produce Detectable Squeezing. with C.W. Laser Pumping in the Length of a Single Beam Waist

Absorption ruins squeezing by randomly destroying photons that had been created in highly correlated pairs. Detectable squeezing implies at least a 10% reduction in the noise; sufficient to be measured unequivocally. Pulsed lasers are simply too unreproduceable for such subtle measurements, and really big C.W. lasers are very expensive.

One solution of this problem is to enclose the nonlinear medium in a "single port optical cavity," thus allowing many beam waists worth of interaction [10]. Another paper in this volume discusses the success of a technique of that type [11]. Our own research has focussed on optical fibers where the light is tightly confined and the length can be equivalent to millions of beam waists. All else being equal, a 4 micron core fused silica fiber will give 10dB squeezing with 1W of pump power and 50 m of interaction length.

7. Seventh Difficulty: One Watt of Light Will Not Go Through the Fiber

The stimulated Brillouin effect limits the power that can be transmitted through a single mode fiber. For monochromatic light, the threshold for SBS in 50m of fiber is about 100mW. Only the threshold power can be transmitted, the rest is reflected. One solution is to impose a temperature gradient along the fiber, thus broadening the SBS gain line [8]. Another solution is to broaden the pump light by sending it through a phase modulator. In a fiber optic ring resonator, SBS can be suppressed by a Faraday rotator optical diode [12].

8. Eighth Difficulty: Excess Noise From The Nonlinear Medium

We were surprised to find that optical fibers themselves generate noise by means of forward light scattering from the acoustic eigenmodes of the fiber and from motion of atoms in the glass [13]. The Guided Acoustic Wave Brillouin Scattering (GAWBS) produces a complex structured spectrum of phase noise well above the quantum noise level. The widths of the peaks depend on the acoustical properties of the jacket material used to protect the silica fiber. Some fibers seem to show a noise floor between the GAWBS peaks. At first we thought that this noise was at the quantum level, but it now seems that there is excess noise at those frequencies as well, apparently due to motion of individual atoms or small groups of atoms. The level of this noise depends on the chemistry of the glass. There has long been a prediction that light scattering should result from such motion at two-level-tunneling centers in glass [14]. We may have been the first to detect it directly.

The problem is that the light scattering from such tunnelling centers is quantum mechanical in origin and should not disappear even at zero degrees Kelvin. The temperature dependence that we measure suggests that such is the case. Even so, one might imagine that a fiber could be designed with a low ratio of light scattering noise to squeezing, or that some small squeezing could be detected even in the presence of the noise. As an experimental problem this noise from the nonlinear medium is similar to that caused by fluorescence in the atomic experiments. Those experiments appear to have succeeded.

9. Ninth Difficulty: Establishing Plausibility

The nonlinear interactions that suppress quantum noise also should reduce classical noise on the input beam. We have used the apparatus of Fig. 8 to test this hypothesis and prove that our proposed squeezed state generation and detection system is likely to work [15]. The noise is generated by electro-optical amplitude and phase modulators driven by an r.f. noise source. The modulator drives are carefully balanced to simulate the phase independence of coherent state fluctuations, but the noise level is 100 times larger. The noise and a strong pump wave are coupled into a fiber and the fiber output is reflected from a ring interferometer onto a photodiode. The interferometer phase shifts and attenuates the pump, and the reflected pump frequency wave acts as a local oscillator at the detector.

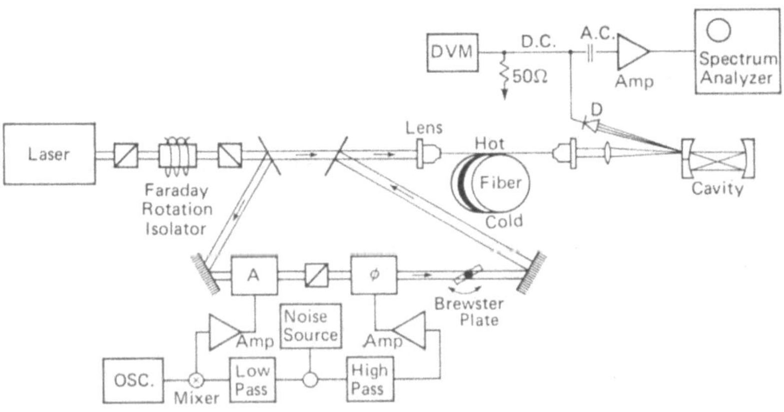

Fig. 8. The apparatus for detecting squeezing of classical noise. Noise sidebands, generated by modulators, are coupled into the fiber along with the pump wave. The output is reflected from a cavity. The pump field is phase shifted by the external cavity. The reflected beam falls on a photodetector. The average detector current is measured using a digital voltmeter while the noise signals at about 50 MHz are amplified and displayed on the spectrum analyzer.

The interferometer resonance is locked at some offset from the pump frequency and the phase is varied by changing this offset. The average power and the noise level at the detector are both measured as a function of the interferometer offset. Typical data appears in Fig. 8. For a short fiber and low pump power, the noise and detector current are proportional at all phases as shown in Fig. 8a. With 180 mW pump in a 100m fiber, the noise becomes phase dependent, with a minimum on one side of the minimum detector current. The high power in the fiber clearly squeezes the externally generated noise by a factor of 3, just as theory would predict.

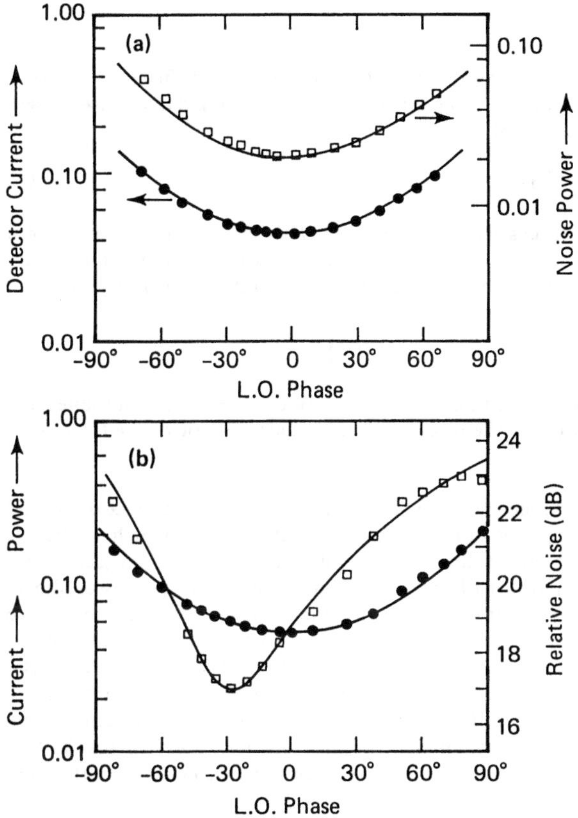

Fig. 9. Plots of the detector current and noise level for a short fiber and low power (top) and a 100m fiber and 180 mW pump (bottom). Without squeezing, the noise level is proportional to the current and independent of phase. Squeezing deamplifies the noise for negative phase shifts, making it phase dependent.

The noise minimum is at -30° phase shift, not -45° as predicted by the simple parametric theory. This seemingly unimportant shift is one result of a major new effect, nonlinear dispersion. The nonlinear susceptibility which couples the waves parametrically also causes an intensity-dependent phase shift between the pump and signal and idler [15]. The phase matching then gets worse as the nonlinearity gets stronger. These experimental results require a revision of the standard theory.

The correct form for the nonlinear interaction Hamiltonian becomes:

$$\mathcal{H} = \hbar\kappa\left[E_p^2 a_s^\dagger a_I^\dagger + E_p^{*2} a_s a_I + 2|E_p|^2\left(a_s^\dagger a_s + a_I^\dagger a_I\right)\right] \tag{5}$$

where $a_s^\dagger$ and $a_I^\dagger$ are the creation operators for the signal and image (idler) photon modes, respectively. The ordinary parametric interaction Hamiltonian lacks the final term. The parameter coupling coefficient is $\kappa = \frac{12\pi\omega}{nc}\chi_{eff}^{(3)}$. This Hamiltonian applies to forward near-degenerate four wave mixing in general, not just in fibers. The corresponding matrix for the propagation (Liouville) equation is singular.

The correct theoretical treatment including the nonlinear dispersion shows that there is no exponential parametric gain; the actual form for the variances of the unsqueezed and squeezed quadratures are:

$$V(E_\pm) = \frac{1}{4}\left(1 + 2r^2 \pm 2r\sqrt{1 + r^2}\right) \tag{6}$$

where $r = \kappa\ell\,|E_p|^2$ is our squeeze parameter. Another result is that phase noise of the sort generated by the fiber is immune to the squeezing interaction. The Hamiltonian describing phase noise caused by light scattering in the medium has the form [16]:

$$\mathcal{H}_N = \left(a_s^\dagger E_p + a_I E_p^*\right)B + \left(a_I^\dagger E_p + a_s E_p^*\right)B^\dagger \tag{7}$$

where B and $B^\dagger$ are delta correlated noise forces when $kT/\hbar\nu >> 1$. This Hamiltonian adds a noise term to the variance proportional to the intensity and to $\sin^2 \Phi$. If such a term is present at all, its level constitutes the noise floor. As a function of the phase angle Φ, the total noise variance is:

$$V(\Phi) = \frac{1}{4}\left[1 + 2r\sin 2\Phi + 2r^2(1 - \cos 2\Phi) + \rho r(1 - \cos 2\Phi)\right] \tag{8}$$

where ρ parameterizes the ratio of the excess noise to the squeezing. Both excess noise and squeezing increase linearly with pump intensity, the ratio is the most important characteristic of the fiber. Our best fibers have $\rho = 5$.

Since the excess noise is small for small values of Φ, one might imagine that some squeezing could be detected for a certain range of Φ. Verifying that squeezing has been detected under such conditions requires that the coherent state noise level be accurately determined. One might also imagine that a thermal light source, where the occupation number of any mode is much less than unity, would give an accurate level for quantum noise at any detector current. Unfortunately, real detectors are not so simple. Low-energy photons from the thermal source penetrate more deeply into the active region than laser photons. Such photons contribute less to the noise level measured at 50 MHz than do laser photons. Thus one must filter the thermal radiation. The noise level from a filtered thermal source agrees with that from an attenuated laser beam within our precision. Even so, detectors can heat up and saturate, amplifiers can drift, and noise is always difficult to measure accurately. Thus the best expected precision is only 5% or so, and 5% squeezing in the presence of phase noise from the fiber requires large pump powers.

We have done experiments like those in Fig. 8, but without the electronic modulators. The ratio of noise to detector current for our best result to date appears in Fig. 10. The minimum noise level

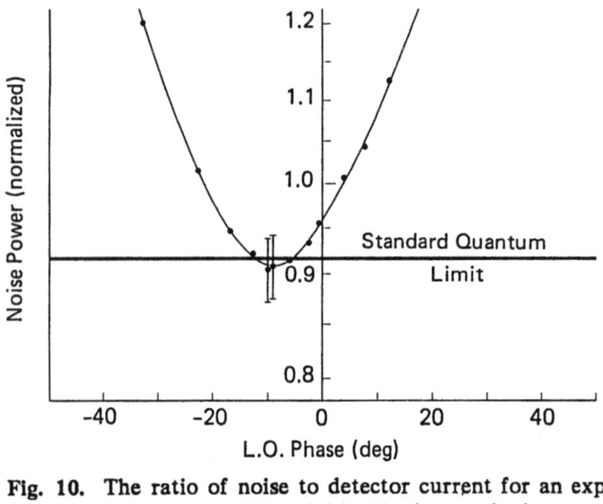

Fig. 10. The ratio of noise to detector current for an experiment similar to Fig. 8, but without electronic modulation. The solid line is the standard quantum limit measured using a filtered black body source. The noise level from the fiber shows squeezing-related asymmetry, but does not fall significantly below the standard quantum limit. The fact that the minimum noise is at an angle less than the -30 degrees indicates that excess phase noise makes up for the suppression of quantum noise.

is quite close to the quantum limit as we define it operationally. But is it significantly below? The shape of the curve can be fit by Eq. (8) with r=.3 and ρ=8. That would imply that quantum noise is being deamplified by ~3 dB, but the squeezing is being masked by excess phase noise. More pump power increases both the squeezing and the excess noise, but does not lower the minimum level significantly. At some pump level, the stimulated Brillouin effect turns on and creates more noise.

10. Tenth Difficulty: Removing Excess Phase Noise

In all of our best results, the ratio of excess noise to squeezing is 5-8. If the excess noise could be reduced - even by a factor of 2 - the minimum noise level would fall a few dB below the standard quantum limit. One can imagine various ways to accomplish this.

The excess noise level does fall with decreasing temperature. Cooling a fiber that gives ρ=5 at room temperature should yield $\rho<1$ at liquid helium temperature. Unfortunately, the best room temperature fibers become lossy at low temperatures. Figure 10 was taken with a fiber at 1.7K. Loss was minimal, but the phase noise did not vanish.

The fiber used in our ring fiber resonator can be cooled, but some of the optical elements in that device cannot be. Even so, the enhancement of squeezing in the resonator is so great that considerable noise reduction can be expected even if only half of the fiber is at cryogenic temperature [12].

The excess noise is classical; that means that two detectors would see the same noise signals, which could then be subtracted electronically in a dual differential detector scheme [7]. The problem is that most such schemes also cancel the squeezing from a fiber. One scheme that might not requires two pump wavelengths that can be spectrally separated after the fiber. If the two pump intensities are each sufficient to cause squeezing, subtracting the signals from two detectors - one for each wavelength - should reveal the suppression of the quantum noise. The amount of squeezing would be less than that obtained for a single pump frequency with the total power in both wavelengths used in the differential experiment, but the quantum level would be no higher. Thus a two wavelength differential detector scheme should give substantial noise reduction.

Neither of the two waves produced in this way need have a total noise level below the quantum limit, but the optical interaction in the fiber would trade uncorrelated quantum noise for correlated classical noise. The differential detector then eliminates the classical noise to reveal the squeezing. The experiment has extra complexity compared to those requiring one pump wavelength and it involves one unknown - the extent to which the classical noise fields probed by different wavelengths actually are correlated. Still, the risk seems worthwhile.

11. A Final Observation

It is clear now that several systems provide sufficient optical nonlinearity to get observable squeezing [8,11]. Neither absorption nor laser noise pose insuperable problems. The difficulties in all experiments to date arise from excess noise contributed by the nonlinear medium and proportional to the squeezing itself. While fluorescence from two level atoms was known, the noise in other experiments result from processes that were unknown when the experiments were conceived. The next generation of experiments will be designed to avoid these noise sources. The future will show whether the result will be many dB of noise suppression or the discovery of new sources of optical noise.

12. References

1. H. P. Yuen: *Phys. Rev.* **A13**, 2226 (1976).
2. C. M. Caves: *Phys. Rev.* **23D**, 1693 (1981).
3. For a reasonably current review, see the papers in session FM 1985 Annual Meeting, Optical Society in America, Washington, D. C. October 14 - 18, 1985, to be published in *J. Opt. Soc. Am.* (December, 1985).

4. R. Loudon: *The Quantum Theory of Light* (2nd Edition) (Clarendon Press, Oxford, U.K. (1983)), p. 219.
5. *Ibid*, p. 46 FF.
6. D. F. Walls, *Nature* **306**, 141 (1983).
7. B. C. Schumaker, *Opt. Lett.* **9**, 189-191 (1984), and references therein.
8. M. D. Levenson, R. M. Shelby, M. D. Reid, D. F. Walls and A. Aspect, *Phys. Rev.* **A32**, 1550 (1985).
9. H. P. Yuen and J.H. Shapiro, *Opt. Lett.* **4**, 334 (1979).
10. B. Yuerke, *Phys. Rev.* **A29**, 408 (1984).
11. Slusher, This Volume.
12. R. M. Shelby and M. D. Levenson paper WJ16, 1985 Annual Meeting of the Optical Society of America, October 14 - 18, 1985, to be published, *J. Opt. Soc. Am.* (December, 1985).
13. R. M. Shelby, M. D. Levenson and P. W. Bayer, *Phys. Rev.* **B31**, 5244-5252 (1985).
14. P. W. Anderson, B. I. Halperin and C.M. Varma, *Phil. Mag.* **25**, 1 (1972).
15. M. D. Levenson, R. M. Shelby and S. H. Perlmutter, *Optics. Lett.* **10**, 514-516.
16. R. M. Shelby, M. A. Levenson, D. F. Walls, A. Aspect and G. J. Milburn, *Phys. Rev. A* (submitted).

Inhibition of Atomic Phase Decays by Squeezed Light: A Direct Effect of Squeezing

C.W. Gardiner

Physics Department, University of Waikato, Hamilton, New Zealand

ABSTRACT

It is shown that by reducing the electromagnetic field fluctuations in one quadrature phase, squeezed light can inhibit the phase decay of an atom. The effect is to give three relaxation times; the usual longitudinal relaxation time, and two different transverse relaxation times, which are inversely proportional to the variances of the two quadrature phases of the incident light.

By reducing the variance of one quadrature phase sufficiently, the corresponding transverse relaxation time can be made as long as one wishes.

The two transverse decay times are observable in the spectrum of the fluorescent light, thus providing measure of the squeezing in the incident light.

The fundamental property of squeezed light is that of reduced quantum fluctuations in one quadrature phase. The early works of Yuen and Shapiro [1], and Caves [2], in which squeezed states were first investigated, have been followed by much activity, which has been reviewed by Walls [3]. Squeezing has recently been observed by Slusher et al. in 4-wave mixing [4] in confirmation of the predictions of Reid and Walls [5]. It is of interest to consider how physical effects of these reduced fluctuations might be detected, and in this paper we shall show that, under the influence of squeezed light, atomic phase decays can be partially suppressed, and the consequent lengthened lifetime of a decaying atom can be observed in the spectrum of light fluorescing from such an atom.

One considers the effect of an incident squeezed vacuum on a two-level atom. Such a state can (in principle) be produced by a degenerate parametric amplifier, whose input is vacuum fluctuations. A detailed treatment of the output field of such a system has been carried out in Ref.6 and in Ref.7.

Squeezing is in practice described by several parameters, and in a multimode situation, which we must use in the case of a travelling wave input, one of the principal parameters is the bandwidth over which the light is squeezed. From eq. 47 of [7] we can obtain the correlation functions of the output light from an ideal degenerate parametric amplifier. We assume (a) the amplifier is in a single-ended cavity, (b) that it operates around a frequency Ω, (c) the damping constant of the cavity is γ, (d) the amplification constant is a real $\epsilon > 0$, (e) the input is the vacuum fluctuations. Fourier transforming eq. 47, we find the correlation functions for $b(t)$, $b^+(t)$, the positive and negative frequency parts of the travelling wave output field, evaluated at the location of the atom are stationary, and satisfy

$$\langle b(t)\rangle = \langle b^+(t)\rangle = 0$$

$$\langle b^+(t)b(t')\rangle = \frac{\lambda^2-\mu^2}{8\mu\lambda}\left(\lambda e^{-\mu|t-t'|} - \mu e^{-\lambda|t-t'|}\right) \tag{1}$$

$$\langle b(t)b(t')\rangle = \frac{\lambda^2-\mu^2}{8\mu\lambda}\left(\lambda e^{-\mu|t-t'|} + \mu e^{-\lambda|t-t'|}\right)e^{-i\Omega(t+t')} \quad \text{where}$$

$$\lambda = \frac{1}{2}\gamma+\epsilon, \quad \mu = \frac{1}{2}\gamma-\epsilon \tag{2}$$

If we consider the case of strong amplification and large damping, the exponentials can be replaced by delta functions, and we obtain

$$\left.\begin{aligned}
&\langle b(t)\rangle = \langle b^+(t)\rangle = 0 \\
&\langle b^+(t)b(t')\rangle = N\delta(t - t') \\
&\langle b(t)b^+(t')\rangle = (N + 1)\delta(t - t') \\
&\langle b(t)b(t')\rangle = Me^{-i\Omega(t + t')}\delta(t - t') \\
&\langle b^+(t)b^+(t')\rangle = M^*e^{i\Omega(t + t')}\delta(t - t')
\end{aligned}\right] \tag{3}$$

where

$$\left.\begin{aligned}
N &= \frac{\lambda^2-\mu^2}{4}\left(\frac{1}{\mu^2} - \frac{1}{\lambda^2}\right) \\
M &= \frac{\lambda^2-\mu^2}{4}\left(\frac{1}{\mu^2} + \frac{1}{\lambda^2}\right)
\end{aligned}\right] \tag{4}$$

(The commutation relation $[b(t),b^+(t')] = \delta(t-t')$ is implicit in (3): this is not exactly true, but is a reasonable approximation when only frequencies in a bandwidth about Ω which is small compared with Ω are considered. In order to avoid confusion, we emphasise that $b(t)$ is a field operator evaluated at the position of the atom. The Fourier components of $b(t)$ each correspond to a mode of the incoming field. See refs. 6 and 7 for a detailed explanation).

This is what we have called squeezed white noise in [7]. It will be valid to use (3) instead of (1) as long as the time constants of any system driven by the squeezed light are significantly longer than both $1/\lambda$ and $1/\mu$. In general, squeezed white noise does not require $M = M^*$, but a phase choice can always be made to fix M to be real and positive, and we shall do so in the remainder of this paper. By noting that $\langle (b(t) + \lambda b^+(t))(b^+(t)+\lambda^* b(t))\rangle$ is positive for all λ, one readily deduces that

$$|M|^2 \leq N(N + 1) \tag{5}$$

The equality is in fact achieved from the output of an ideal degenerate parametric amplifier, as can be deduced from eq. 4, and indeed it is possible simultaneously to have $N \to \infty$ (how well these can be realised in practice is a matter of current investigation).

Choosing M to be real and positive in this case, the quadrature phases are most conveniently chosen as

$$\left.\begin{aligned}
X(t) &= \frac{b^+(t)e^{-i\Omega t} + b(t)e^{i\Omega t}}{2} \\
Y(t) &= \frac{b^+(t)e^{-i\Omega t} - b(t)e^{i\Omega t}}{2i}
\end{aligned}\right] \tag{6}$$

and the correlation functions are

$$\left.\begin{aligned} \langle X(t)X(t')\rangle &= \frac{1}{2}\left(N + M + \frac{1}{2}\right)\delta(t - t') \\ &\sim \left(N + \frac{1}{2}\right)\delta(t - t') \quad \text{if } N \to \infty \text{ and } M = \sqrt{N(N+1)} \\ \langle Y(t)Y(t')\rangle &= \frac{1}{2}\left(N + \frac{1}{2} - M\right)\delta(t - t') \\ &\sim 1/16N \quad \text{if } N \to \infty \text{ and } M = \sqrt{N(N+1)} \end{aligned}\right\} \tag{7}$$

Thus, in the limit considered in eq. 7, we see that the fluctuations in the quadrature phase $Y(t)$ can be made as small as one pleases.

If one shines such light on a two-level atom whose transition frequency is Ω, the effect of the fluctuations on the Bloch vector will be felt only by that component of the Bloch vector in phase with $X(t)$; the component in phase with $Y(t)$ will feel fluctuations which can be arbitrarily small.

The decay of a two-level atom consists of two parts; Polarisation decay and Inversion decay. The polarisation decays as the result of field fluctuations gradually randomising the phase of the Bloch vector, whereas the inversion decay is related to the radiation of photons into space.

If we squeeze the input light we can reduce the fluctuations in one quadrature phase, and that component of the polarization which is in phase with the low noise quadrature phase will feel reduced fluctuations, and not decay as fast as the other, which will feel increased fluctuations.

The formalism developed by Gardiner and Collett [8] is well adapted to the treatment of this problem, since the properties (2) are those of Quantum White Noise, for which a formalism is developed in their paper. We consider a two level system described by spin operators

$$S_z = \begin{pmatrix} 1 & 0 \\ 0 & -1 \end{pmatrix}, \quad S^+ = \begin{pmatrix} 0 & 1 \\ 0 & 0 \end{pmatrix}, \quad S^- = \begin{pmatrix} 0 & 0 \\ 1 & 0 \end{pmatrix} \tag{8}$$

coupled to the electromagnetic field. The Hamiltonian is

$$\left.\begin{aligned} H &= H_{SYS} + H_{EM} + H_{INT} \\ {}_{EM} &= h\int_0^\infty d\omega\ \omega\ b^+(\omega)b(\omega) \\ {}_{NT} &= ih\int_0^\infty K(\omega)[b^+(\omega)S^- - b(\omega)S^+]d\omega \\ {}_{YS} &= \frac{1}{2}\ h\Omega S_z \end{aligned}\right\} \tag{9}$$

In writing (9) we make the electric dipole approximation and the rotating wave approximation. If $K(\omega)$, the dipole moment is assumed smooth around $\omega = \Omega$, and Ω is sufficiently large that the lower limit of the integrals in (9) can be considered to be essentially $-\infty$. Collett and Gardiner derive a master equation for the atomic density operator ρ for the situation in which the incoming light is squeezed quantum white noise: this is (in a frame rotating at frequency Ω)

$$\left.\begin{aligned} \frac{d\rho}{dt} &= \frac{\gamma}{2}(N + 1)(2S^-\rho S^+ - S^+S^-\rho - \rho S^+S^-) \\ &+ \frac{\gamma}{2}N(2S^+\rho S^- - S^-S^+\rho - \rho S^-S^+) \\ &- \frac{\gamma}{2}M(2S^+\rho S^+ - S^+S^+\rho - \rho S^+S^+) \\ &- \frac{\gamma}{2}M^*(2S^-\rho S^- - S^-S^-\rho - \rho S^-S^-) \end{aligned}\right\} \tag{10}$$

Here $\gamma = 2\pi K(\Omega)^2$ (11)

This equation is valid provided the bandwidth of the squeezed white noise is substantially larger than γ. If we define

$$S^{\pm} = \frac{1}{2}(S_x \pm iS_y) \tag{12}$$

then we find equations of motion for the means of S_x, S_y, S_z to be

$$\begin{aligned}
\langle\dot{S}_x\rangle &= -\gamma\left(N + M + \frac{1}{2}\right)\langle S_x\rangle \equiv -\gamma_x\langle S_x\rangle \\
\langle\dot{S}_y\rangle &= -\gamma\left(N - M + \frac{1}{2}\right)\langle S_y\rangle \equiv -\gamma_y\langle S_y\rangle \\
\langle\dot{S}_z\rangle &= -\gamma(2N + 1)\langle S_z\rangle - \gamma \equiv -\gamma_z\langle S_z\rangle - \gamma
\end{aligned} \tag{13}$$

The quantities N and M can be varied independently in principle, though for the ideal squeezed input, (such as is produced from a DPA) N and M are connected by the relation

$$M = \sqrt{N(N+1)}$$

We can see therefore that for large squeezing both γ_z and γ_x become large. In fact γ_x and γ_y are proportional to the coefficients of intensity of the respective quadrature phases of the input, as defined in eq. 7. Since this noise can approach zero in the Y quadrature phase, it is clear that the decay constant γ_y can be as small as we please, while at the same time γ_x becomes extremely large. Thus in a time scale short compared with γ_y^{-1}; but substantially larger than γ_x^{-1} and γ_z^{-1}; we find

$$\begin{aligned}
\langle S_x\rangle &\to 0 \\
\langle S_z\rangle &\to -1/(2N + 1)
\end{aligned} \tag{14}$$

while $\langle S_y\rangle$ is essentially unchanged. Thus the projection of the original orientation of the Bloch vector on the direction of the low-noise quadrature phase is preserved - nevertheless the inversion $\langle S_z\rangle$ decays rapidly to its stationary value.

To detect this effect one can look at the spectrum of the fluorescent light. Using the quantum regression theorem, and standard methods connected with it [9] the stationary correlation functions of the atom are:

$$\begin{aligned}
\langle S^+(t)S^-(0)\rangle &= \frac{1}{2}\,\frac{N}{2N+1}\left[\exp(-\gamma_x t) + \exp(-\gamma_y t)\right] \\
\langle S^-(t)S^+(0)\rangle &= \frac{1}{2}\,\frac{N+1}{2N+1}\left[\exp(-\gamma_x t) + \exp(-\gamma_y t)\right] \\
\langle S^+(t)S^+(0)\rangle &= \frac{1}{2}\,\frac{N+1}{2N+1}\left[\exp(-\gamma_x t) - \exp(-\gamma_y t)\right] \\
\langle S^-(t)S^-(0)\rangle &= \frac{1}{2}\,\frac{N}{2N+1}\left[\exp(-\gamma_x t) - \exp(-\gamma_y t)\right]
\end{aligned} \tag{15}$$

The above has been formulated in the case that there is one input and one output channel. In the case of interaction of light with an atom, this is realistic if we view the channels as corresponding to the various partial waves, since only the electric dipole partial wave, (corresponding to the electric dipole approximation) interacts appreciably with the atom. If the experiment can be carried out with a squeezed electric dipole wave, then the output would be of the same kind. We will not attempt a consideration of how this might be achieved in practise, though schemes involving parabolic mirrors come to mind immediately.

In this case the amplitude correlation function of the fluorescent light (whose Fourier transform is the spectrum) is computed from the boundary condition.

$$b_{out}(t) = \sqrt{\gamma}S^-(t) + b(t) \tag{16}$$

which is derived in [6]. We find

$$\begin{aligned}\langle b_{out}^+(t)b_{out}(0)\rangle = N\delta(t) &+ \gamma u(t)\langle[S^+(t),NS^-(0) - MS^+(0)]\rangle \\ &+ \gamma u(-t)\langle[NS^+(t) - MS^-(t),S^-(0)]\rangle \\ &+ \gamma\ \langle S^+(t)S^-(0)\rangle\end{aligned} \tag{17}$$

(where $u(t) = 0\ t<0;\ = 1\ t>0$)

This correlation function contains elements arising from the input correlation function ($N\delta(t)$) and elements from the atom itself - arising because the fluorescent light consists of a radiated part and a reflected part. Using (11) and (13), the correlation function becomes

$$\langle b_{out}^+(t)_{out}b(0)\rangle = N\delta(t) + \frac{1}{2}\frac{\gamma M}{2N+1}\left[\exp(-\gamma_y t) - \exp(-\gamma_x t)\right] \tag{18}$$

which displays the characteristic decay constants.

The spectrum obtained by the Fourier transform of (18) consists of a flat background, from $\delta(t)$, plus a negative peak of width γ_x (which will be very broad), and finally a positive peak of width γ_y, which can be as narrow as we please. Such narrowing is an effect which can only be produced by some reduction of noise, i.e. squeezing. The narrowed width is in fact directly proportional to the squeezing, and thus provides a measurement of squeezing.

One might prefer to use plane waves as an input, but in fact most of the squeezing effect is then lost. For, we characterise the state of the 3-dimensional input field by operators $b_i(\underset{\sim}{n},t)$ such that

$$\left.\begin{aligned}[b_i(\underset{\sim}{n},t),\ b_j(\underset{\sim}{n}',t')] &= \hat{\delta}(\underset{\sim}{n},\underset{\sim}{n}')\delta(t-t')\delta_{ij} \\ \langle b_i^\dagger(\underset{\sim}{n},t)b_j(\underset{\sim}{n}',t')\rangle &= \bar{N}_I(\underset{\sim}{n},\underset{\sim}{n}')\delta(t-t')\delta_{iI}\delta_{jI} \\ \langle b_i(\underset{\sim}{n},t)b_j(\underset{\sim}{n}',t')\rangle &= \bar{M}_I(\underset{\sim}{n},\underset{\sim}{n}')\delta(t-t')\delta_{iI}\delta_{jI}\end{aligned}\right\} \tag{19}$$

Here $\underset{\sim}{n}$ and $\underset{\sim}{n}'$ are unit vectors, $\hat{\delta}(\underset{\sim}{n},\underset{\sim}{n}')$ is delta function on the unit sphere, and the operators $b_i(\underset{\sim}{n},t)$ are Fourier transform wrt frequency of free-wave input destruction operators with a given direction of propagation $\underset{\sim}{n}$. The indices i, j, are polarisation indices, and I is the particular polarisation which is squeezed.

The functions $\bar{N}_I$ and $\bar{M}_I$ characterise the squeezing, and are in principle arbitrary. The annihilation operator corresponding to a square normalised wave-function $f_I(\underset{\sim}{n})$ is

$$A_I(t) = \int d\underset{\sim}{n} f_I^*(n) b_I(\underset{\sim}{w},t) \tag{20}$$

The case in which only this direction is squeezed is given by

$$\left.\begin{aligned} \bar{N}_I(\underset{\sim}{n},\underset{\sim}{n}') &= f_I^*(\underset{\sim}{n}) f_I(\underset{\sim}{n}') N_I \\ \bar{M}_I(\underset{\sim}{n},\underset{\sim}{n}') &= f_I^*(\underset{\sim}{n}) f_I(\underset{\sim}{n}') M_I \end{aligned}\right] \tag{21}$$

since in this case

$$\left.\begin{aligned} \langle A_I^\dagger(t) A_I(t')\rangle &= \delta(t-t') N_I \\ \langle A_I(t) A_I(t')\rangle &= \delta(t-t') M_I \end{aligned}\right] \tag{22}$$

and similar averages corresponding to all orthogonal functions vanish.

The corresponding parameters to N_I and M_I in (22) which would arise from the operator $B(t)$ corresponding to a wavefunction $g(n)$ are obtained by multiplying N and M in (22) by $|\int dn f_I^*(n) g(n)|^2$.

This can be a considerable reduction. For example if $f_I(n)$ corresponds to an almost plane wave which includes only directions n confined to a solid angle Ω with equal amplitude, while $g(n)$ corresponds on electric dipole wave, i.e. which is of order of magnitude $1/\sqrt{4\pi}$ on the unit sphere, then this overlap factor is of order of magnitude $\frac{\Omega}{4\pi}$ which is vanishingly small as $\Omega \to 0$, corresponding to a plane wave. In order therefore to utilise squeezing effectively, Ω must be as close to 4π as possible.

If the experiment is done with an incoming wave that is not a perfect electric dipole, then the effect of multiplying M and N by the factor $\epsilon = |\int dn f_I^*(n) g(n)|^2$ is to yield

$$\left.\begin{aligned} \gamma_x &\to \gamma \left[\epsilon(N+M+\tfrac{1}{2}) + (1-\epsilon)\tfrac{1}{2}\right] \\ \gamma_y &\to \gamma \left[\epsilon(N-M+\tfrac{1}{2}) + (1-\epsilon)\tfrac{1}{2}\right] \end{aligned}\right. \tag{23}$$

The term proportional to $(1-\epsilon)$ in both these expressions represents the input of vacuum fluctuations from those modes in the electric dipole wave other than the input mode. Unless ϵ is very close to one, these can completely mask the effect. The production of a squeezed electric dipole incident wave will therefore be an important component of the experimental verification of this effect, unless it is found possible to set up a genuine one-dimensional experiment. This would involve the manufacture of an appropriate waveguide, at whose termination one would locate the 2 level atom. This might be difficult at visible wavelengths, though it might be possible at the wavelengths corresponding to transitions in Rydberg atoms.

In practice the light will not be perfectly delta correlated, but the analysis will be valid provided the input squeezing bandwidth is larger than γ_x, γ_y and γ_z. This can be achieved in principle in 4 wave mixing [10]. It will therefore be relevant to find ways of computing the effects of finite squeezing bandwidth.

ACKNOWLEDGMENTS

I would like to thank Craig Savage for conversations which led me to consider this problem, and Dan Walls for his enlightenment on certain points.

REFERENCES

1) H.P. Yuen, Phys. Rev. A13, 2226, (1976); H.P. Yuen and J.H. Shapiro, IEEE, Trans. Inform. Theory IT 24, 657 (1978); IT 26, 78, (1980)
2) C.M. Caves, Phys. Rev. D26, 1817, (1982)
3) D.F. Walls, Nature 306, 141, (1983)
4) R.E. Slusher, L.W. Hollberg, B. Yurke, J.C. Mertz and J.F. Valley Bell Labs. Preprint.
5) M.D. Reid and D.F. Walls, Phys. Rev. A31, 1622 (1985)
6) C.W. Gardiner and C.M. Savage, Opt. Comm. 50, 173, (1984)
7) M.J. Collett and C.W. Gardiner, Phys. Rev. A30, 1386, (1984)
8) C.W. Gardiner and M.J. Collett, Phys. Rev. 31A, 3761, (1985)
9) C. W. Gardiner, "Handbook of Stochastic Methods", Springer, Berlin-Heidelberg-New York - 1983, Ch. 10
10) D.F. Walls and A.S. Lane, private communication.

Broadband Squeezing*

C.M. Caves and B.L. Schumaker

Theoretical Astrophysics, California Institute of Technology,
Pasadena, CA 91125, USA

1. Introduction

The phenomenon of squeezed states of light has led to exciting theoretical and experimental work in quantum optics. Squeezed states might have important applications, in optical communications [1-3] or high-precision interferometric measurements [4-6], or perhaps in other areas not yet apparent. Just as important, squeezed states offer a fundamental probe into the quantum nature of light. The recent experiment at AT&T Bell Laboratories [7], which successfully generated and detected squeezed-state light, gives impetus to further theoretical and experimental efforts.

In this lecture we review the theoretical description of broadband squeezing (Sec. 2). This description has been developed in greatest detail independently by YURKE [8,9] and by us [10,11]. It describes broadband squeezing in terms of the quadrature components – the quadrature phases – of a plane electromagnetic wave and, more specifically, in terms of the Fourier components – the quadrature-phase amplitudes – of the quadrature phases. In addition, we go beyond the usual description of broadband squeezing to consider another phenomenon recently investigated by SCHUMAKER [12,13] – a phenomenon which might be called "quadrature squeezing" or "squeezing of the squeezing" (Sec. 4).

Our aim in this lecture is to tie the description of squeezing as closely as possible to detection techniques used in real experiments. Squeezing refers to a non-random distribution of noise within the phase of some sinusoidal oscillation. Thus, squeezing shows up when one uses phase-sensitive detection.

At optical frequencies the only phase-sensitive technique available is optical heterodyning [3]: the signal to be detected is combined at a beamsplitter with a powerful, nearly monochromatic beam at frequency Ω, generated by a local oscillator; the combined beam is then directed onto a photodetector. The output photocurrent contains information about the signal quadrature that is in phase with the local oscillator. The noise in this signal quadrature can be characterized by the spectral density of the photocurrent. Hence, defined operationally, broadband squeezing is simply a variation of the current spectral density with local-oscillator phase.

If one examines only a narrow bandwidth of the photocurrent at zero frequency (homodyning), then one is looking for degenerate (single-mode) squeezing (Sec. 3). If one examines a narrow bandwidth about a nonzero frequency ε, then one is looking for nondegenerate (two-mode) squeezing (Sec. 3), a result of a special correlation between signal modes at frequencies $\Omega \pm \varepsilon$. These two types of squeezing are aspects of the same phenomenon; both are incorporated in a broadband description.

Quadrature squeezing is different. In detecting two-mode squeezing at frequency ε, one is indifferent to how the photocurrent noise is distributed in phase. Quadrature squeezing *is* the variation in phase of the photocurrent noise at frequency ε (Sec. 4). To detect that variation, one uses a second local oscillator to homodyne the photocurrent at frequency ε down to dc. Defined operationally, quadrature squeezing is a variation of the resulting noise at dc with the phase of the second local oscillator.

* Supported in part by the US Office of Naval Research [Contract No. N00014-85-K-0005].

Ordinary broadband squeezing can be regarded as squeezing of the electromagnetic field; it shows up when the field is decomposed into its quadrature components, the quadrature phases. Quadrature squeezing is squeezing of one of the quadrature phases; it shows up when the quadrature phase is decomposed into its quadrature components.

Our goal in this lecture is to give a pedagogical account of broadband and quadrature squeezing. To that end, we aim to provide enough detail so that the lecture can serve as an accessible, self-contained reference to the basic concepts.

2. Broadband Squeezing

Consider a beam of light — the signal — emerging from some nonlinear device capable of generating squeezed states. Idealize the signal as a plane wave with a single polarization propagating in the $+x$-direction. We are interested only in a finite bandwidth B symmetrically located about frequency Ω; B consists of frequencies ω such that $\Omega - \varepsilon_{max} \leqq \omega \leqq \Omega + \varepsilon_{max}$. The electric-field operator for the signal can be written in the form

$$E(t,x) = E^{(+)}(t,x) + E^{(-)}(t,x)\,, \tag{2.1}$$

where the positive-frequency part $E^{(+)} = E^{(-)\dagger}$ can be expanded in terms of the Fourier modes $e^{-i\omega(t-x/c)}$ within B:

$$E^{(+)}(t,x) = \int_B \frac{d\omega}{2\pi}(\hbar\,\omega/2)^{1/2}a(\omega)e^{-i\omega(t-x/c)}\,. \tag{2.2}$$

The annihilation and creation operators, $a(\omega)$ and $a^\dagger(\omega)$, obey the continuum commutation relations

$$[a(\omega),a^\dagger(\omega')] = 2\pi\delta(\omega-\omega')\,. \tag{2.3}$$

With these choices, $a^\dagger(\omega)a(\omega)d\omega/2\pi$ is the number of photons within bandwidth $d\omega$, and the units of the electric field are such that $2E^{(-)}E^{(+)}$ is the intensity (energy per sec) of the plane wave.

The preferred method of heterodyning for detecting squeezing, called balanced heterodyning (or a double-balanced mixer) [14,15], is illustrated in Fig. 1. The signal beam and a local-oscillator (LO) beam are combined at a 50-50 beamsplitter, both of whose outputs are directed onto identical photodetectors, and the photocurrents

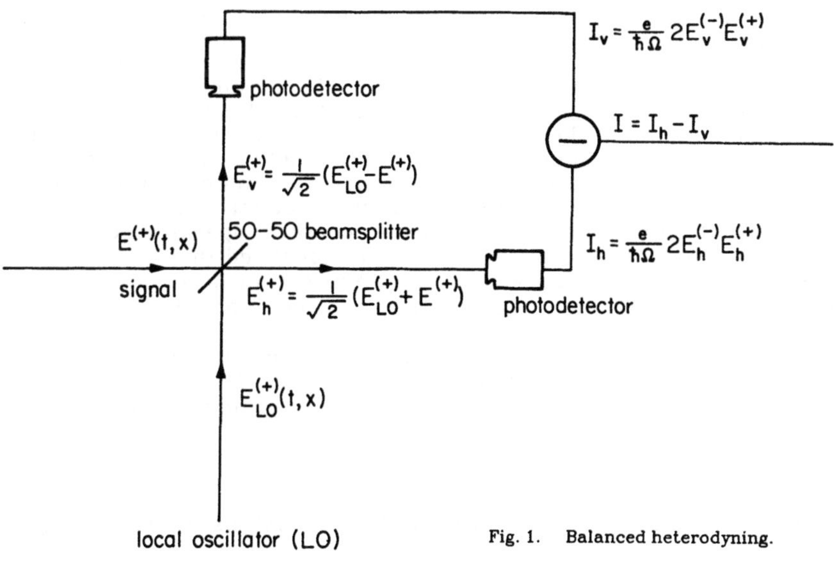

Fig. 1. Balanced heterodyning.

from the two photodetectors are differenced coherently. Balanced heterodyning is preferred because of its ability to suppress noise from the LO, which appears as a common-mode signal in the two photocurrents and thus disappears from the differenced photocurrent. Our broadband treatment of balanced heterodyning is based on YURKE's [8] analysis (for a complementary approach, see [16]).

The LO field within the bandwidth of interest has positive-frequency part

$$E_{LO}^{(+)}(t,x) = \int_B \frac{d\omega}{2\pi}(\hbar\,\omega/2)^{1/2} a_{LO}(\omega) e^{-i\omega(t-x/c)} . \tag{2.4}$$

Here and below x is used in a general sense to denote distance along a beam path. Assume that the LO produces a monochromatic excitation at frequency Ω:

$$\langle a_{LO}(\omega)\rangle = Ae^{i\phi}2\pi\delta(\omega-\Omega) , \tag{2.5}$$

$$\langle E_{LO}^{(+)}(t,x)\rangle = (\hbar\,\Omega/2)^{1/2}Ae^{i\phi}e^{-i\Omega(t-x/c)} . \tag{2.6}$$

In these expressions, ϕ is the phase of the LO wave, and A is the (real) amplitude; the power carried by the monochromatic excitation is $P_{LO} = \hbar\,\Omega A^2$. The expectation value of the LO electric field is

$$\langle E_{LO}(t,x)\rangle = \langle E_{LO}^{(+)}(t,x)\rangle + \langle E_{LO}^{(-)}(t,x)\rangle = (2\hbar\,\Omega)^{1/2}A\cos[\Omega(t-x/c)-\phi] . \tag{2.7}$$

It is useful to write $E_{LO}^{(+)}$ as the sum of its expectation value and a "noise operator" $\Delta E_{LO}^{(+)}$, i.e,

$$E_{LO}^{(+)}(t,x) = (\hbar\,\Omega/2)^{1/2}Ae^{i\phi}e^{-i\Omega(t-x/c)} + \Delta E_{LO}^{(+)}(t,x) . \tag{2.8}$$

The field operators for the fields emerging from the 50-50 beamsplitter have positive-frequency parts

$$\begin{aligned} E_{h}^{(+)}(t,x) &= 2^{-1/2}[E_{LO}^{(+)}(t,x) \pm E^{(+)}(t,x)] \\ &= \frac{1}{2}(\hbar\,\Omega)^{1/2}Ae^{i\phi}e^{-i\Omega(t-x/c)} + 2^{-1/2}[\Delta E_{LO}^{(+)}(t,x) \pm E^{(+)}(t,x)] , \end{aligned} \tag{2.9}$$

where $h(v)$ designates the field propagating horizontally (vertically) in Fig. 1.

In heterodyne detection, one is interested in beats between the LO frequency Ω and frequencies $\Omega \pm \varepsilon$. These beats appear in the Fourier component of the differenced photocurrent at frequency ε. There is a bandwidth B_{el}, the photodetectors' electronic bandwidth, over which one can see beats with frequency Ω in the photocurrents. For the present analysis we assume that B_{el} is somewhat larger than the bandwidth B of interest. The task now is to construct operators for the two photocurrents. To make the task easier, assume that the photodetectors are ideal. Surprisingly, there is a lively controversy over how to proceed — specifically, over the precise form of the photocurrents' Fourier components at nonzero frequencies ε. Often cited as an aid is the dictum that "an ideal photodetector emits one photoelectron for each incident photon," but this dictum provides little help, because the nonzero-frequency components of the photocurrents involve the arrival *rate* of photons, not just the total number of photons. One must somehow convert a quantity with units of intensity, derived from the field operators, to a quantity with units of photon intensity (photons per sec); multiplication by the electronic charge e then yields a photocurrent. Unfortunately, there is no accepted operator for photon intensity, for the good reason that a photon cannot be localized in spacetime. Any attempt to confine a photon wave packet within a finite time interval must fail, because the wave packet inevitably has negative-frequency components.

SHAPIRO and his collaborators [5,16,17] nevertheless construct an operator for photon intensity in the following way. They replace the electric-field operators introduced above with "photon-units" field operators; for example, the positive-frequency part (2.2) of the signal electric field is replaced by

$$\mathcal{E}^{(+)}(t,x) \equiv 2^{-1/2}\int_B \frac{d\omega}{2\pi}\, a(\omega)e^{-i\omega(t-x/c)} . \tag{2.10}$$

They then interpret the quantity $2\mathcal{E}^{(-)}\mathcal{E}^{(+)}$ as the "photon intensity" of the signal.

This procedure converts from intensity to photon intensity by doing a separate conversion for each Fourier mode — i.e., for each frequency ω. It assumes implicitly an average over an infinitely long time, so that each detected photon can be assigned a particular frequency within B. Thus, the relevance of the "photon intensity" $2\tilde{E}^{(-)}\tilde{E}^{(+)}$ to heterodyne detection is unclear.

In this lecture we adopt a different procedure. Since the photodetectors cannot distinguish frequencies within B, they should respond to all incident energy within B in the same way. Thus, to convert from intensity to photon intensity, we introduce a characteristic energy $\hbar\bar{\Omega}$, assumed constant over the bandwidth B. Ultimately, the frequency $\bar{\Omega}$ must be derived from the detailed properties of a particular photodetector. If the detector is ideal, however, we should choose $\bar{\Omega} = \Omega$. If we assume that the photodetectors are located at $x = 0$, the resulting photocurrent operators are

$$I_{\substack{h\\v}}(t) = (e/\hbar\Omega)2E_{\substack{h\\v}}^{(-)}(t,0)E_{\substack{h\\v}}^{(+)}(t,0)$$

$$= \frac{1}{2}eA^2 + e(2\hbar\Omega)^{-1/2}A\left[(\Delta E_{LO}^{(+)} \pm E^{(+)})e^{-i\phi}e^{i\Omega t} + (\Delta E_{LO}^{(-)} \pm E^{(-)})e^{i\phi}e^{-i\Omega t}\right]$$

$$+(e/\hbar\Omega)(\Delta E_{LO}^{(-)} \pm E^{(-)})(\Delta E_{LO}^{(+)} \pm E^{(+)}) . \tag{2.11}$$

In the latter inequality the second term on the right-hand side describes the beats between the LO and frequencies $\Omega \pm \varepsilon$. Throughout the following we assume A is sufficiently large that the last term in (2.11) can be neglected.

The difference between the two procedures centers on whether one retains square-root factors $(\omega/\Omega)^{1/2}$ in the beat terms of the photocurrents. The controversy is sometimes characterized as whether an ideal photodetector detects intensity $2E^{(-)}E^{(+)}$ or photon intensity $2\tilde{E}^{(-)}\tilde{E}^{(+)}$, but this characterization is somewhat misleading. No one contends that (2.11) is correct when a photodetector is exposed to two frequencies ω_1 and ω_2 so widely separated that there are no beats between them in the photocurrent — i.e., $|\omega_1 - \omega_2| \gg B_{el}$; in this case each frequency makes a separate contribution to the dc component of the photocurrent. The argument concerns the Fourier components of the photocurrent at nonzero frequencies — specifically, whether these components should contain the square-root factors [see (2.18) below]. BONDURANT [18] has recently argued in favor of using photon intensity, but YURKE's [9] analysis indicates that the square-root factors should be retained. Additional work is probably required to clarify the situation.

Consider now the operator for the differenced photocurrent, given by

$$I(t) = I_h(t) - I_v(t) = (2/\hbar\Omega)^{1/2}eAE_1(\phi;t,0) , \tag{2.12}$$

where

$$E_1(\phi;t,x) \equiv E^{(+)}(t,x)e^{-i\phi}e^{i\Omega(t-x/c)} + E^{(-)}(t,x)e^{i\phi}e^{-i\Omega(t-x/c)} \tag{2.13a}$$

is a quadrature phase of the signal field. As advertised, the LO noise operator ΔE_{LO} disappears from $I(t)$. It is often useful to have available a second quadrature phase

$$E_2(\phi;t,x) \equiv E_1(\phi+\pi/2;t,x)$$

$$= -iE^{(+)}(t,x)e^{-i\phi}e^{i\Omega(t-x/c)} + iE^{(-)}(t,x)e^{i\phi}e^{-i\Omega(t-x/c)} , \tag{2.13b}$$

which is orthogonal to $E_1(\phi)$.

A shortcut to understanding the form of $I(t)$ arises from writing the signal electric field in terms of the orthogonal quadrature phases. To do so, first notice that

$$E^{(\pm)}(t,x) = \frac{1}{2}[E_1(\phi;t,x) \pm iE_2(\phi;t,x)]e^{\pm i\phi}e^{\mp i\Omega(t-x/c)} ; \tag{2.14}$$

this expression shows that $[E_1(\phi) + iE_2(\phi)]e^{i\phi}$ can be regarded as the complex amplitude of the signal field, defined with respect to frequency Ω. Now write the electric field (2.1) of the signal as

$$E(t,x) = E_1(\phi;t,x)\cos[\Omega(t-x/c)-\phi] + E_2(\phi;t,x)\sin[\Omega(t-x/c)-\phi] . \tag{2.15}$$

Comparison with (2.7) shows that balanced heterodyning does indeed detect the quadrature phase of the signal which is in phase with the LO. It is interesting to write the intensity in terms of the quadrature phases:

$$2E^{(-)}E^{(+)} + \int_B \frac{d\omega}{2\pi}\,\frac{1}{2}\,\hbar\,\omega = \frac{1}{2}\left[E_1{}^2(\phi) + E_2{}^2(\phi)\right] . \qquad (2.16)$$

The second term on the left-hand side is the zero-point contribution.

We are often interested in the Fourier components of $I(t)$, so it is useful to introduce Fourier expansions for the quadrature phases:

$$E_m(\phi;t,x) = \int_R \frac{d\varepsilon}{2\pi}\,(\hbar\,\Omega)^{1/2}\left[\alpha_m(\phi;\varepsilon)e^{-i\varepsilon(t-x/c)} + \alpha_m{}^\dagger(\phi;\varepsilon)e^{i\varepsilon(t-x/c)}\right] . \qquad (2.17)$$

Here $m = 1,2$, and the integral runs over the frequency interval R defined by $0 \leqq \varepsilon \leqq \varepsilon_{max}$. The Fourier components $\alpha_1(\phi;\varepsilon)$ and $\alpha_2(\phi;\varepsilon)$, called quadrature-phase amplitudes [10,11], are related to the creation and annihilation operators by

$$\alpha_1(\phi;\varepsilon) = 2^{-1/2}\left[(1+\varepsilon/\Omega)^{1/2}\,a(\Omega+\varepsilon)e^{-i\phi} + (1-\varepsilon/\Omega)^{1/2}a^\dagger(\Omega-\varepsilon)e^{i\phi}\right] , \qquad (2.18a)$$

$$\begin{aligned}\alpha_2(\phi;\varepsilon) &= 2^{-1/2}\left[-i(1+\varepsilon/\Omega)^{1/2}a(\Omega+\varepsilon)e^{-i\phi} + i(1-\varepsilon/\Omega)^{1/2}a^\dagger(\Omega-\varepsilon)e^{i\phi}\right] \\ &= \alpha_1(\phi+\pi/2;\varepsilon) . \qquad (2.18b)\end{aligned}$$

The Fourier expansion of $I(t)$ takes the form

$$I(t) = 2^{1/2}e\,A \int_R \frac{d\varepsilon}{2\pi}\left[\alpha_1(\phi;\varepsilon)e^{-i\varepsilon(t-x/c)} + \alpha_1{}^\dagger(\phi;\varepsilon)e^{i\varepsilon(t-x/c)}\right] . \qquad (2.19)$$

Thus, the Fourier component of the differenced photocurrent at frequency ε, $2^{1/2}eA\,\alpha_1(\phi;\varepsilon)$, reflects the beating between the LO and signal modes at frequencies $\Omega \pm \varepsilon$. The square-root factors $(1 \pm \varepsilon/\Omega)^{1/2}$ in (2.18) would be absent if one used Shapiro's photon intensity to calculate the differenced photocurrent; the presence of these factors here is, of course, the crux of the aforementioned controversy.

The commutators for an orthogonal set of quadrature-phase amplitudes $\alpha_1(\phi;\varepsilon)$ and $\alpha_2(\phi;\varepsilon)$ are

$$[\alpha_1(\phi;\varepsilon),\alpha_1(\phi;\varepsilon')] = [\alpha_1(\phi;\varepsilon),\alpha_2(\phi;\varepsilon')] = [\alpha_2(\phi;\varepsilon),\alpha_2(\phi;\varepsilon')] = 0 , \qquad (2.20a)$$

$$[\alpha_1(\phi;\varepsilon),\alpha_1{}^\dagger(\phi;\varepsilon')] = [\alpha_2(\phi;\varepsilon),\alpha_2{}^\dagger(\phi;\varepsilon')] = (\varepsilon/\Omega)2\pi\delta(\varepsilon-\varepsilon') , \qquad (2.20b)$$

$$[\alpha_1(\phi;\varepsilon),\alpha_2{}^\dagger(\phi;\varepsilon')] = [\alpha_1{}^\dagger(\phi;\varepsilon),\alpha_2(\phi;\varepsilon')] = i2\pi\delta(\varepsilon-\varepsilon') . \qquad (2.20c)$$

These commutators impose constraints, discussed below, on the noise in the differenced photocurrent.

Broadband squeezing occurs when the noise in the signal electric field is not distributed randomly within the sinusoidal oscillation $e^{-i\Omega(t-x/c)}$. Equivalently, broadband squeezing means that the noise varies from one quadrature phase to another, or, operationally, that the noise in the differenced photocurrent varies with LO phase. In describing broadband squeezing, one does not care how the noise in each Fourier component of $I(t)$ is distributed in phase. Therefore, it is simplest to assume that each Fourier component of $I(t)$ has random-phase noise and that the noise in the different Fourier components is uncorrelated. This assumption is codified by the conditions

$$\left\langle \Delta\alpha_1(\phi;\varepsilon)\Delta\alpha_1(\phi;\varepsilon')\right\rangle = 0 , \qquad (2.21a)$$

$$\left\langle \Delta\alpha_1(\phi;\varepsilon)\Delta\alpha_1{}^\dagger(\phi;\varepsilon')\right\rangle_{sym} = \Sigma(\phi;\varepsilon)2\pi\delta(\varepsilon-\varepsilon') , \qquad (2.21b)$$

where "sym" denotes a symmetrized product, and $\Sigma(\phi;\varepsilon)$ is a dimensionless power spectral density for $E_1(\phi)$ (dimensions of number of photons per sec per Hz). Equations (2.21) are equivalent to saying that $I(t)$ has time-stationary (TS) noise; its two-point correlation function is given by

$$\begin{aligned}\left\langle \Delta I(t)\Delta I(t+\tau)\right\rangle_{sym} &= \frac{(2eA)^2}{\hbar\,\Omega}\,\frac{1}{2}\left\langle \Delta E_1(\phi;t,0)\Delta E_1(\phi;t+\tau,0)\right\rangle_{sym} \\ &= (2eA)^2\int_R \frac{d\varepsilon}{2\pi}\Sigma(\phi;\varepsilon)\cos\varepsilon\tau ; \qquad (2.22)\end{aligned}$$

thus, $(2eA)^2\Sigma(\phi;\varepsilon)$ is the spectral density of $I(t)$.

Equations (2.21) hold for all values of ϕ — i.e, $I(t)$ has TS noise for all ϕ — if the following conditions are satisfied for any particular ϕ:

$$\left\langle \Delta\alpha_m(\phi;\varepsilon)\Delta\alpha_n(\phi;\varepsilon')\right\rangle = 0 \,, \tag{2.23a}$$

$$\left\langle \Delta\alpha_m(\phi;\varepsilon)\Delta\alpha_n{}^\dagger(\phi;\varepsilon')\right\rangle_{\rm sym} = \Sigma_{mn}(\phi;\varepsilon)2\pi\delta(\varepsilon-\varepsilon') \,. \tag{2.23b}$$

Here $m,n = 1,2$ and $\Sigma_{mn}(\phi;\varepsilon)$ is a dimensionless power spectral-density matrix for the orthogonal quadrature phases $E_1(\phi)$ and $E_2(\phi)$ $[\Sigma_{11}(\phi;\varepsilon) = \Sigma(\phi;\varepsilon)$, $\Sigma_{22}(\phi;\varepsilon) = \Sigma(\phi+\pi/2;\varepsilon)]$. Since the conditions (2.23) ensure that any quadrature phase of the signal has TS noise, we say that the signal has time-stationary quadrature phase (TSQP) noise [10,11] whenever conditions (2.23) are satisfied. Actually, the conditions (2.23) are a bit stronger than is necessary to ensure that $I(t)$ has TS noise for all ϕ, but they are a neat set of conditions, satisfied by a set of Gaussian states that exhibit broadband squeezing [10,11,13], so we adopt them as the definition of TSQP noise.

The commutators (2.20) place the following constraints on the spectral densities [8,10,13]:

$$\Sigma(\phi;\varepsilon) \geqq \varepsilon/2\Omega \,, \tag{2.24a}$$

$$\Sigma_{11}(\phi;\varepsilon)\Sigma_{22}(\phi;\varepsilon) = \Sigma(\phi;\varepsilon)\Sigma(\phi+\pi/2;\varepsilon) \geqq 1/4 \,. \tag{2.24b}$$

The constraint (2.24a) follows from (2.20b); there would be no such minimum value for $\Sigma(\phi;\varepsilon)$ if one used photon intensity to construct the photocurrent operators. The uncertainty principle (2.24b) follows from (2.20c).

Broadband squeezing means that the spectral density $\Sigma(\phi;\varepsilon)$ varies with LO phase ϕ. Such variation is a consequence of a special correlation between signal modes at frequencies $\Omega \pm \varepsilon$. If the signal is in the vacuum state, then it has random-phase noise, and $\Sigma(\phi;\varepsilon)$ has the minimum value consistent with the uncertainty principle (2.24b):

$$\Sigma(\phi,\varepsilon) = \frac{1}{2} \quad \text{for all } \phi \quad \text{(vacuum)} \,. \tag{2.25}$$

The factor 1/2 in (2.25) corresponds to the zero-point (vacuum) noise in the following way: each frequency $\Omega \pm \varepsilon$ contributes a half-quantum per sec per Hz of zero-point noise to the noise at frequency ε, and each of two orthogonal quadratures carries half of the resulting total noise of one quantum per sec per Hz. Real quantum squeezing occurs when $\Sigma(\phi;\varepsilon)$ dips below the vacuum level of 1/2 for some range of values of ϕ; of course, the uncertainty principle (2.24b) then forces $\Sigma(\phi+\pi/2;\varepsilon)$ to rise above the vacuum level.

3. Degenerate and Nondegenerate Squeezing

Broadband squeezing involves pairs of Fourier modes, each pair consisting of modes with frequencies $\Omega \pm \varepsilon$. Squeezing is a consequence of a special correlation between modes in a pair; modes in different pairs are uncorrelated. Thus, to simplify the description, it is desirable to concentrate on one pair of modes at a time. Fourier modes are not really suited to this purpose, however, because they form a continuum. Strictly speaking, it is not possible to pick out two modes (a set of measure zero) from this continuum. A better set of modes for this purpose consists of discrete "wave-packet" modes, each of which occupies a finite bandwidth $\Delta\omega$ and has an effective duration $T \equiv 2\pi/\Delta\omega$.

To define the discrete modes, partition the bandwidth B into bins B_j of width $\Delta\omega \equiv 2\pi/T$. Let the center frequency of the jth bin be $\omega_j \equiv \Omega + j\Delta\omega \equiv \Omega + \varepsilon_j$; thus, B_j consists of frequencies ω such that $\omega_j - \Delta\omega/2 < \omega \leqq \omega_j + \Delta\omega/2$. Now introduce new annihilation operators

$$a_{jk} \equiv T^{1/2}\int_{B_j}\frac{d\omega}{2\pi}\, a(\omega)e^{-ik\omega T} \,. \tag{3.1}$$

Using the orthogonality relation

$$T \int_{B_j} \frac{d\omega}{2\pi}\, e^{i(k'-k)\omega T} = \delta_{kk'} , \tag{3.2}$$

one can demonstrate the canonical commutation relations

$$[a_{jk}, a_{j'k'}{}^{\dagger}] = \delta_{jj'}\delta_{kk'} ; \tag{3.3}$$

and using the completeness property

$$T \sum_{k=-\infty}^{+\infty} e^{ik(\omega-\omega')T} = 2\pi \sum_j \delta(\omega - \omega' - j\,\Delta\omega) , \tag{3.4}$$

one can invert (3.1) to obtain

$$a(\omega) = T^{1/2} \sum_{k=-\infty}^{\infty} a_{jk} e^{ik\omega T} \quad , \quad \omega \text{ lies in } B_j . \tag{3.5}$$

In terms of the new annihilation operators, $E^{(+)}$ takes the form

$$E^{(+)}(t,x) = \sum_{j,k} (\hbar\,\omega_j/2T)^{1/2} a_{jk} e^{-i\omega_j(t-x/c-kT)} f_j(t - x/c - kT) , \tag{3.6}$$

$$f_j(t) \equiv T \int_{B_j} \frac{d\omega}{2\pi} (\omega/\omega_j)^{1/2} e^{-i(\omega-\omega_j)t} . \tag{3.7}$$

Each wave-packet mode $e^{-i\omega_j(t-x/c-kT)} f_j(t - x/c - kT)$ is labeled by two discrete indices: the index j, which runs over a finite range, specifies the mode's frequency bin B_j, and the index k, which runs from $-\infty$ to ∞, specifies the position of the mode's "envelope function" $f_j(t - x/c - kT)$.

In the following we assume that T is so large that $\Delta\omega/\omega_j \simeq \Delta\omega/\Omega \ll 1$ can always be neglected relative to 1. With this assumption, $f_j(t)$ becomes

$$f_j(t) = \frac{\sin(\pi t/T)}{\pi t/T} \equiv f(t) . \tag{3.8}$$

The function $f(t)$ has a maximum value of 1 at $t = 0$, and it has effective duration T. Thus the use of wave-packet modes partitions the time-frequency "phase space" into cells of "volume" $\simeq T(\Delta\omega) = 2\pi$.

By confining our experiments to a region of spacetime near $t - x/c = 0$, we can dispense with all modes except those with $k = 0$, and we can approximate $f(t - x/c)$ by 1. The effect of other modes, all of which have a zero at $t - x/c = 0$, can be made as small as desired within the relevant spacetime region by choosing T sufficiently large. Hence, we are led to the following expansion of $E^{(+)}$ in terms of discrete modes with $k = 0$ ($a_j \equiv a_{j0}$):

$$E^{(+)}(t,x) = \sum_j (\hbar\,\omega_j/2T)^{1/2} a_j e^{-i\omega_j(t-x/c)} . \tag{3.9}$$

The corresponding discrete-mode expansion for the quadrature phase (2.13a) is

$$E_1(\phi;t,x) = (\hbar\,\Omega/T)^{1/2} \left[x_1(\phi) + \sum_{j>0} \left[\alpha_{1j}(\phi) e^{-i\varepsilon_j(t-x/c)} + \alpha_{1j}{}^{\dagger}(\phi) e^{+i\varepsilon_j(t-x/c)} \right] \right] , \tag{3.10a}$$

where $\varepsilon_j \equiv j\,\Delta\omega$, and

$$x_1(\phi) \equiv 2^{-1/2}(a_0 e^{-i\phi} + a_0{}^{\dagger} e^{i\phi}) = T^{1/2} \int_{R_0} \frac{d\varepsilon}{2\pi} [\alpha_1(\phi;\varepsilon) + \alpha_1{}^{\dagger}(\phi;\varepsilon)] , \tag{3.10b}$$

$$\begin{aligned} \alpha_{1j}(\phi) &\equiv 2^{-1/2}[(1 + \varepsilon_j/\Omega)^{1/2} a_{+j} e^{-i\phi} + (1 - \varepsilon_j/\Omega)^{1/2} a_{-j}{}^{\dagger} e^{i\phi}] \\ &= T^{1/2} \int_{R_j} \frac{d\varepsilon}{2\pi} \alpha_1(\phi;\varepsilon) , \quad j > 0 , \end{aligned} \tag{3.10c}$$

are discrete quadrature-phase amplitudes. In (3.10b), R_0 consists of frequencies ε such that $0 \leq \varepsilon \leq \Delta\omega/2$, and in (3.10c), R_j for $j > 0$ consists of frequencies ε such that $(j - \frac{1}{2})\Delta\omega < \varepsilon \leq (j + \frac{1}{2})\Delta\omega$. Just as before, we can introduce orthogonal quadrature-phase amplitudes

$$x_2(\phi) \equiv x_1(\phi + \pi/2) = 2^{-1/2}(-ia_0 e^{-i\phi} + ia_0{}^\dagger e^{i\phi}) \,, \tag{3.10d}$$

$$\begin{aligned}\alpha_{2j}(\phi) &\equiv \alpha_{1j}(\phi + \pi/2) \\ &= 2^{-1/2}[\, -i(\, 1 + \varepsilon_j/\Omega)^{1/2} a_{+j} e^{-i\phi} + i(\, 1 - \varepsilon_j/\Omega)^{1/2}\, a_{-j}{}^\dagger\, e^{i\phi}\,]\,,\ j > 0\,.\end{aligned} \tag{3.10e}$$

Since the differenced photocurrent (2.12) is proportional to $E_1(\phi;t,x)$, the content of (3.10) should be clear: within a narrow bandwidth $\Delta\omega/2$ at zero frequency, balanced heterodyning (actually homodyning) detects the Hermitian quantity $x_1(\phi)$ — Hermitian because the dc component of the differenced photocurrent is specified by a single number; within a narrow bandwidth $\Delta\omega$ about frequency ε_j, balanced heterodyning detects the complex quantity $\alpha_{1j}(\phi)$ — complex because the differenced photocurrent at frequency ε_j is characterized by an amplitude and a phase.

The nonzero commutators for an orthogonal set of discrete quadrature-phase amplitudes are

$$[x_1(\phi), x_2(\phi)] = i \,, \tag{3.11a}$$

$$[\alpha_{1j}(\phi), \alpha_{1j}{}^\dagger(\phi)] = [\alpha_{2j}(\phi), \alpha_{2j}{}^\dagger(\phi)] = \varepsilon/\Omega \,, \tag{3.11b}$$

$$[\alpha_{1j}(\phi), \alpha_{2j}{}^\dagger(\phi)] = [\alpha_{1j}{}^\dagger(\phi), \alpha_{2j}(\phi)] = i \,. \tag{3.11c}$$

These commutators imply a set of constraints [10,13]:

$$\left\langle [\Delta x_1(\phi)]^2 \right\rangle \left\langle [\Delta x_2(\phi)]^2 \right\rangle \geqq \frac{1}{4} \,, \tag{3.12a}$$

$$\left\langle |\Delta\alpha_{1j}(\phi)|^2 \right\rangle \geqq \varepsilon/2\Omega \quad , \quad \left\langle |\Delta\alpha_{2j}(\phi)|^2 \right\rangle \geqq \varepsilon/2\Omega \,, \tag{3.12b}$$

$$\left\langle |\Delta\alpha_{1j}(\phi)|^2 \right\rangle \left\langle |\Delta\alpha_{2j}(\phi)|^2 \right\rangle \geqq \frac{1}{4} \,. \tag{3.12c}$$

In (3.12) the quantity

$$\left\langle |\Delta\alpha|^2 \right\rangle \equiv \left\langle \Delta\alpha\Delta\alpha^\dagger \right\rangle_{\rm sym} = \frac{1}{2}\left\langle \Delta\alpha\Delta\alpha^\dagger + \Delta\alpha^\dagger\Delta\alpha \right\rangle$$

is the mean-square uncertainty of the complex operator α. If the signal is in the vacuum state, then

$$\left\langle [\Delta x_1(\phi)]^2 \right\rangle = \left\langle |\Delta\alpha_{1j}(\phi)|^2 \right\rangle = 1/2 \quad \text{for all } \phi \qquad \text{(vacuum)}\,; \tag{3.13}$$

the 1/2 in each case represents the zero-point (vacuum) noise.

If we assume again that the signal has TSQP noise, then the two-point correlation function of $I(t)$ is given in terms of discrete modes by

$$\left\langle \Delta I(t)\Delta I(t+\tau) \right\rangle_{\rm sym} = (2eA)^2 \left[\frac{1}{2T} \left\langle [\Delta x_1(\phi)]^2 \right\rangle + \frac{1}{T} \sum_{j>0} \left\langle |\Delta\alpha_{1j}(\phi)|^2 \right\rangle \cos\varepsilon_j\tau \right] \tag{3.14}$$

Notice that for TSQP noise, the natural "variance" for the discrete quadrature-phase amplitude $\alpha_{1j}(\phi)$ is the mean-square uncertainty $\langle |\Delta\alpha_{1j}(\phi)|^2 \rangle$. To go from the continuum-mode two-point correlation function (2.22) to the discrete-mode result (3.14), one makes the replacements

$$\int_{R_0} \frac{d\varepsilon}{2\pi}\, \Sigma(\phi;\varepsilon) \;\rightarrow\; T^{-1}\frac{1}{2} \left\langle [\Delta x_1(\phi)]^2 \right\rangle , \tag{3.15a}$$

$$\int_{R_j} \frac{d\varepsilon}{2\pi}\, \Sigma(\phi;\varepsilon) \;\rightarrow\; T^{-1}\left\langle |\Delta\alpha_{1j}(\phi)|^2 \right\rangle ,\ j > 0 \,. \tag{3.15b}$$

The factor of 1/2 for $j = 0$ has a ready explanation: the noise for $j = 0$ comes from the bin B_0 with bandwidth $\Delta\omega$; the noise for $j > 0$ comes from two bins B_j and B_{-j}, with total bandwidth $2\Delta\omega$.

Degenerate and nondegenerate squeezing are distinguished only by where one looks in the Fourier spectrum of $I(t)$. Within a narrow bandwidth $\Delta\omega/2$ at dc, balanced heterodyning (homodyning) detects $x_1(\phi)$; in this case squeezing means that $\langle[\Delta x_1(\phi)]^2\rangle$ varies with ϕ — a kind of squeezing called degenerate, or single-mode, squeezing because it involves only the single mode $j = 0$. Within a narrow bandwidth $\Delta\omega$ about frequency ε_j, balanced heterodyning detects $\alpha_{1j}(\phi)$; in this case squeezing means that $\langle|\Delta\alpha_{1j}(\phi|^2\rangle$ varies with ϕ — a kind of squeezing called nondegenerate, or two-mode, squeezing because it is a consequence of correlation between modes j and $-j$. In either case, real quantum squeezing occurs when the relevant variance drops below the vacuum level of 1/2.

This is a good place to clear up a potential puzzle. In the preceding analysis the modes with $k = 0$ seem to account for all the noise. What about the noise in all the modes with $k \neq 0$? This puzzle arises from approximating $f(t - x/c)$ by 1 in (3.9); the results obtained using this approximation are valid only in the spacetime region near $t - x/c = 0$. The modes with $k \neq 0$ are required to account for the noise away from $t - x/c = 0$.

4. Quadrature Squeezing [12,13]

Restrict attention now to the single discrete mode at degeneracy ($j = 0$) so that the electric-field operator assumes the form [see (2.1) and (3.9)]

$$E(t,x) = (\hbar\Omega/2T)^{1/2}[a_0 e^{-i\Omega(t-x/c)} + a_0^\dagger e^{i\Omega(t-x/c)}] ,$$
$$= (\hbar\Omega/T)^{1/2}\Big[x_1(\phi)\cos[\Omega(t - x/c) - \phi] + x_2(\phi)\sin[\Omega(t - x/c) - \phi]\Big] , \quad (4.1)$$

where $x_1(\phi)$ and $x_2(\phi)$ are the Hermitian real and imaginary parts of $2^{1/2}a_0e^{-i\phi}$ [see (3.10b) and (3.10d)]:

$$2^{1/2}a_0e^{-i\phi} = x_1(\phi) + ix_2(\phi) . \quad (4.2)$$

If the noise in the electric field is distributed randomly in phase (TS noise in the signal field), then

$$\langle[\Delta x_1(\phi)]^2\rangle = \langle|\Delta a_0|^2\rangle + \frac{1}{2}\langle(\Delta a_0)^2\rangle e^{-2i\phi} + \frac{1}{2}\langle(\Delta a_0^\dagger)^2\rangle e^{2i\phi} \quad (4.3a)$$

is independent of ϕ, a condition equivalent to

$$\langle(\Delta a_0)^2\rangle = 0 . \quad (4.3b)$$

Degenerate squeezing occurs when $\langle[\Delta x_1(\phi)]^2\rangle$ varies with LO phase ϕ — a variation that shows up in balanced homodyne detection.

Consider now an analogous situation. Restrict attention to a single pair of discrete modes j and $-j$, and for simplicity drop the index j. The quadrature phases simplify to the form [see (3.10a), (3.10c), and (3.10e)]

$$E_m(\phi;t,x) = (\hbar\Omega/T)^{1/2}[\alpha_m(\phi)e^{-i\varepsilon(t-x/c)} + \alpha_m^\dagger(\phi)e^{i\varepsilon(t-x/c)}]$$
$$= (2\hbar\Omega/T)^{1/2}\Big[\alpha_{m1}(\phi,\theta)\cos[\varepsilon(t - x/c) - \theta]$$
$$+ \alpha_{m2}(\phi,\theta)\sin[\varepsilon(t - x/c) - \theta]\Big] , \qquad m = 1,2 , \quad (4.4)$$

where the quantities

$$\alpha_{m1}(\phi,\theta) \equiv 2^{-1/2}[\alpha_m(\phi)e^{-i\theta} + \alpha_m^\dagger(\phi)e^{i\theta}] , \quad (4.5a)$$

$$\alpha_{m2}(\phi,\theta) \equiv 2^{-1/2}[-i\alpha_m(\phi)e^{-i\theta} + i\alpha_m^\dagger(\phi)e^{i\theta}] = \alpha_{m1}(\phi,\theta + \pi/2) \quad (4.5b)$$

are the Hermitian real and imaginary parts of $2^{1/2}\alpha_m(\phi)e^{-i\theta}$:

$$2^{1/2}\alpha_m(\phi)e^{-i\theta} = \alpha_{m1}(\phi,\theta) + i\alpha_{m2}(\phi,\theta) . \quad (4.5c)$$

The quantities $\alpha_{m1}(\phi,\theta)$ and $\alpha_{m2}(\phi,\theta)$ are the quadrature components of the quadrature phase $E_m(\phi)$; notice that the first index specifies which quadrature phase and the second index specifies which quadrature component of the quadrature phase.

Balanced heterodyning yields a differenced photocurrent $I(t)$ proportional to $E_1(\phi;t,x)$. Imagine now that one homodynes $I(t)$ down to dc by mixing it with a second local oscillator that has time dependence $\cos(\varepsilon t - \theta)$. The angle θ is the phase of the second LO. As a result of this homodyning of $I(t)$, one detects a quantity proportional to $\alpha_{11}(\phi,\theta)$. If the noise in the quadrature phases (4.4) is distributed randomly in phase (TS noise in the quadrature phases; TSQP noise in the signal field), then the variance of $\alpha_{11}(\phi,\theta)$,

$$\langle[\Delta\alpha_{11}(\phi,\theta)]^2\rangle = \langle|\Delta\alpha_1(\phi)|^2\rangle + \frac{1}{2}\langle[\Delta\alpha_1(\phi)]^2\rangle e^{-2i\theta} + \frac{1}{2}\langle[\Delta\alpha_1{}^\dagger(\phi)]^2\rangle e^{2i\theta}, \tag{4.6}$$

is independent of θ — i.e., $\langle[\Delta\alpha_1(\phi)]^2\rangle = 0$; hence, in this case, the noise in the homodyne photocurrent is characterized completely by the mean-square uncertainty $\langle|\Delta\alpha_1(\phi)|^2\rangle$. Quadrature squeezing means that $\langle[\Delta\alpha_{11}(\phi,\theta)]^2\rangle$ varies with θ — i.e., that the noise after homodyning $I(t)$ to dc varies with the phase of the second LO.

To investigate the trade-offs involved in quadrature squeezing, one requires the commutator algebra for the four Hermitian operators $\alpha_{11}(\phi,\theta)$, $\alpha_{12}(\phi,\theta)$, $\alpha_{21}(\phi,\theta)$, and $\alpha_{22}(\phi,\theta)$. To simplify the notation, we drop the reference to ϕ and θ. The nonzero commutators are given by

$$[\alpha_{11},\alpha_{12}] = [\alpha_{21},\alpha_{22}] = i\varepsilon/\Omega\,, \tag{4.7a}$$

$$[\alpha_{11},\alpha_{21}] = [\alpha_{12},\alpha_{22}] = i\,. \tag{4.7b}$$

These commutators enforce a set of uncertainty principles

$$\langle(\Delta\alpha_{11})^2\rangle\langle(\Delta\alpha_{12})^2\rangle \geqq \varepsilon^2/4\Omega^2\,, \quad \langle(\Delta\alpha_{21})^2\rangle\langle(\Delta\alpha_{22})^2\rangle \geqq \varepsilon^2/4\Omega^2\,, \tag{4.8a}$$

$$\langle(\Delta\alpha_{11})^2\rangle\langle(\Delta\alpha_{21})^2\rangle \geqq \frac{1}{4}\,, \quad \langle(\Delta\alpha_{12})^2\rangle\langle(\Delta\alpha_{22})^2\rangle \geqq \frac{1}{4}\,. \tag{4.8b}$$

For nondegenerate squeezing one is primarily interested in the case where the signal field has TSQP noise, for which

$$\langle|\Delta\alpha_m|^2\rangle = \langle(\Delta\alpha_{m1})^2\rangle = \langle(\Delta\alpha_{m2})^2\rangle\,, \quad m = 1,2 \qquad \text{(TSQP noise)}\,. \tag{4.9}$$

In that case, if one reduces $\langle|\Delta\alpha_1|^2\rangle$ below the vacuum level of 1/2, then $\langle|\Delta\alpha_2|^2\rangle$ must be above the vacuum level [see (3.12c)]; the reduction in $\langle|\Delta\alpha_1|^2\rangle$ is ultimately limited by (3.12b) [or (4.8a)], i.e., $\langle|\Delta\alpha_1|^2\rangle \geqq \varepsilon/2\Omega$. For quadrature squeezing there is more freedom: one can reduce $\langle(\Delta\alpha_{11})^2\rangle$ below 1/2, the price again being that $\langle(\Delta\alpha_{21})^2\rangle$ must be above 1/2 [see (4.8b)]; the difference is that $\langle(\Delta\alpha_{11})^2\rangle$ can be made smaller than $\varepsilon/2\Omega$ — in fact, as small as desired — at the price of increasing $\langle(\Delta\alpha_{12})^2\rangle$ above $\varepsilon/2\Omega$ [see (4.8a)].

For a thorough discussion of Gaussian states that exhibit degenerate, nondegenerate, and quadrature squeezing, the reader is referred to [13].

References

[1] H. P. Yuen and J. H Shapiro: IEEE Trans. Inform. Theory **IT-24**, 657 (1978).

[2] J. H. Shapiro, H. P. Yuen, and J. A. Machado Mata: IEEE Trans. Inform. Theory **IT-25**, 179 (1979).

[3] H. P. Yuen and J. H Shapiro: IEEE Trans. Inform. Theory **IT-26**, 78 (1980).

[4] C. M. Caves: Phys. Rev. D **23**, 1693 (1981).

[5] R. S. Bondurant and J. H. Shapiro: Phys. Rev. D **30**, 2548 (1984).

[6] J. R. Klauder, S. L. McCall, and B. Yurke: Phys. Rev. D, in press.

[7] R. E. Slusher, L. W. Hollberg, B. Yurke, J. C. Mertz, and J F. Valley: Phys. Rev. Lett **55**, 2409 (1985).

[8] B. Yurke: Phys. Rev. A **32**, 300 (1985).

[9] B. Yurke: Phys. Rev. A **32**, 311 (1985).

[10] C. M. Caves and B. L. Schumaker: Phys. Rev. A **31**, 3068 (1985).

[11] B. L. Schumaker and C. M. Caves: Phys. Rev. A **31**, 3093 (1985).

[12] B. L. Schumaker: Digest of Technical Papers, Optical Society of America Annual Meeting, 14-18 October 1985, Washington, D. C., p. 102.

[13] B. L. Schumaker: Phys. Reports, in press.

[14] H. P. Yuen and V. W. S. Chan: Opt. Lett. **8**, 177 (1983).

[15] B. L. Schumaker: Opt. Lett. **9**, 189 (1984).

[16] J. H. Shapiro: IEEE J. Quantum Electron. **QE-21**, 237 (1985).

[17] J. H. Shapiro and S. S. Wagner: IEEE J. Quantum Electron. **QE-20**, 803 (1984).

[18] R. S. Bondurant: Phys. Rev. A **32**, 2797 (1985).

Theory of Squeezed Light Generation

M.D. Reid, A. Lane, and D.F. Walls
Physics Department, University of Waikato, Hamilton, New Zealand

#1 Introduction

There is presently a considerable experimental effort being made to generate squeezed states of light (1). Current experiments include four wave mixing in optical fibers (2), four wave mixing in atomic sodium (3,4), and intracavity second harmonic generation (5). The results of three of these experimental groups are described in the articles by Levenson and Shelby, Slusher and Yurke, and Kimble and Hall in these proceedings. In this article we shall discuss the theory relevant to two of the aforementioned experiments.

We shall discuss in some details the quantum theory of non-degenerate four wave mixing in a medium of two level atoms. This theory is directly applicable to the experiments performed by Slusher and his co-workers in atomic sodium at Bell Laboratories (4). We shall also describe the theory pertinent to the experiments on squeezing in intracavity second harmonic generation being conducted by Kimble and Hall at the Joint Institute for Laboratory Astrophysics in Boulder (5).

#2 Non-Degenerate Four Wave Mixing

We begin with a general description of non-degenerate four wave mixing in an optical cavity. This theory is a generalization of the theory of degenerate four wave mixing developed by Reid and Walls (6). The medium is modelled as N two level atoms and is interacting with the radiation field at three frequencies. We write the model Hamiltonian in the electric dipole and rotating wave approximations as follows:

$$H = \sum_{\mu=0}^{4} H_\mu$$

$$H_0 = \sum_{j=1}^{3} \hbar\omega_j a_j^\dagger a_j + \sum_{i=1}^{N} \frac{\hbar\omega_0}{2} \sigma_{z i}$$

$$H_1 = i\hbar g \sum_{i=1}^{N} \Big[\sigma_i \left(a_1^\dagger e^{-ik_1 r_i} + a_2^\dagger e^{-ik_2 r_i} + a_3^\dagger e^{-ik_3 r_i}\right)$$

$$-\sigma_i^\dagger \left(a_1 e^{ik_1 r_i} + a_2 e^{ik_2 r_i} + a_3 e^{ik_3 r_i}\right)\Big] \qquad (1)$$

$$H_2 = i\hbar\left(a_1^\dagger \epsilon e^{-i\omega_p t} - a_1 \epsilon^* e^{i\omega_p t}\right)$$

$$H_3 = \sum_{i=1}^{N} \left\{\Gamma\sigma_i^\dagger + \Gamma^\dagger\sigma_i + \Gamma_p\sigma_{zi}\right\}$$

$$H_4 = \sum_{i=1}^{3} \left(\Gamma_c^\dagger a_j + \Gamma_c a_j^\dagger\right)$$

where a_1 is a boson annihilation operation of the pump mode at frequency ω_1, a_2 and a_3 are annihilation operators describing cavity modes at frequencies ω_2 and ω_3, such that $2\omega_1 = \omega_2 + \omega_3$. We denote $\epsilon = \omega_2 - \omega_1$. σ_i, $\sigma_i^\dagger$, σ_{zi} are pseudo spin operators for the ith two level atom with atomic resonance frequency ω_0. The atomic reservoirs Γ and Γ_p describe energy loss from the atoms via spontaneous emission and phase damping or collisional processes respectively.
$\gamma_\perp$ and $\gamma_\parallel$ are the atomic transverse and longitudinal decay rates respectively. The dissipation of the cavity field modes is described by the field reservoir Γ_c. The cavity decay rates of all modes are assumed equal to κ. The pump mode a_1 is a resonant cavity mode driven by an external coherent input field ϵ of frequency ω_R. In the case of a single cavity for both pump and sideband modes, we have assumed that the cavity detuning $\omega_1 - \omega_p$ is much smaller than the separation in frequency $\epsilon = \omega_2 - \omega_1$ between adjacent cavity modes. A Fokker Planck equation derived for the distribution function is converted to C number Langevin equations by Ito calculus.

In the non-degenerate situation the atomic polarization ν oscillates at an infinite number of frequencies. We derive equations describing the gain of the weak sideband modes in the presence of the strong pump mode. We therefore consider the limit where the strong pump mode α_1 is treated correctly to all orders describing completely the saturation of the medium while the expressions for the weak fields α_2, α_3 are kept to first order only. Then the polarization ν will oscillate at three dominant frequencies ω_p, $\omega_p + \epsilon$ and $\omega_p - \epsilon$.

In the limit of a high Q cavity where the atomic decay rates exceed the cavity decay rate, one is justified in adiabatically eliminating the atomic variables. This gives the following stochastic differential equations for the field modes.

$$\dot{\alpha}_1 = \epsilon - \kappa(1+i\phi)\alpha_1 - \frac{2c\kappa\alpha_1}{(1+i\Delta)\pi(0)} + F_1(t)$$

$$\dot{\alpha}_2 = -\kappa\gamma(\delta)\alpha_2 + \kappa\chi(\delta)\alpha_3^\dagger e^{2i\theta_0} + F_2(t) \qquad (2)$$

$$\dot{\alpha}_3 = -\kappa\gamma(-\delta)\alpha_3 + \kappa\chi(-\delta)\alpha_2^\dagger e^{2i\theta_0} + F_3(t)$$

where $\pi(0) = 1 + I/(1 + \Delta^2)$, $\Delta = (\omega_0 - \omega_p)/\gamma_\perp$, $\delta = -\epsilon/\gamma_\parallel$

the loss coefficient $\gamma(\delta) = \left(1 + i\phi + \gamma_R(\delta) + i\gamma_I(\delta)\right)$, $\phi = (\omega_1 - \omega_p)/\kappa$

and the nonlinear gain coefficient $\chi(\delta) = \left(\chi_R(\delta) + i\chi_I(\delta)\right)$

$F_i(t)$ are fluctuating forces such that $\langle F_i(t)\rangle = 0$ and the non-zero noise correlations for the sidebands are

$$\langle F_2(t)F_3(t')\rangle = \kappa R e^{2i\theta_0}\delta(t-t'),\quad R = R_R + iR_I$$

$$\langle F_2(t)F_2^\dagger(t')\rangle = \langle F_3(t)F_3^\dagger(t')\rangle \tag{3}$$

$$= \kappa\Lambda\ \delta(t-t')$$

The parameters γ, χ, R and Λ are functions of: the scaled detuning δ of the sidebands from the pump; and Δ the detuning of the pump from the medium; the cooperativity parameter $C = \dfrac{g^2N}{2\gamma_\perp\kappa}$ of the cavity, the scaled steady-state pump intensity $I = \dfrac{4g^2|\alpha_1^{ss}|^2}{\gamma_\parallel\gamma_\perp}$ (where $\alpha_1^{ss} = |\alpha_1^{ss}|e^{i\theta_0}$ is the steady-state solution of eq(2)); and the measure $f = \dfrac{\gamma_\parallel}{2\gamma_\perp}$ of the collisional damping present (which we take to be one assuming no collisions). The explicit solutions are presented in reference (7) and (15).

The expressions for $\gamma(\delta)$ and $\chi(\delta)$ are identical to those derived previously by Fu and Sargent (8) and Boyd et al (9) in a semiclassical study of non-degenerate four wave mixing in a two level medium. The quantum aspects of the field are included via the quantum noise terms $F_j(t)$ and their correlations R and Λ.

The steady-state cavity pump intensity I satisfies

$$I_p = I\left[\left(1 + \frac{2C}{(1+\Delta^2)\pi(0)}\right)^2 + \left(\phi - \frac{2C\Delta}{(1+\Delta^2)\pi(0)}\right)^2\right] \tag{4}$$

where $I_p = 4g^2\left|\frac{\epsilon}{\kappa}\right|^2/\gamma_\parallel\gamma_\perp$ is the scaled external driving field. The phase θ_0 of the steady-state cavity pump is shifted from the phase θ_p of the external driving field $\left(\epsilon = |\epsilon|e^{i\theta_p}\right)$ by $\tan(\theta_p-\theta_0) = \left[\phi - \frac{\Delta 2c}{(1+\Delta^2)\pi(0)}\right]\Big/\left[1+ \frac{2c}{(1+\Delta^2)\pi(0)}\right]$

The behaviour of a two level atom, resonance frequency ω_0, when irradiated by a laser field at frequency ω_p has been well studied. At low field intensities inelastic scattering of laser photons occurs at the frequencies $2\omega_p - \omega_0$. At higher intensities the atom saturates and Stark shifting of the atomic energy levels occurs. A splitting of each energy level by the generalised Rabi frequency Ω $\left(\frac{\Omega}{\gamma_\perp} = \sqrt{2I+\Delta^2}\right)$ occurs. Photons are now scattered inelastically at the frequencies ω_p, $\omega_p \pm \Omega$.

Plots of the absorptive $\gamma_R(\delta)$, dispersive $\gamma_I(\delta)$ and nonlinear coupling $\chi(\delta)$ profiles are given by Boyd et al (9).

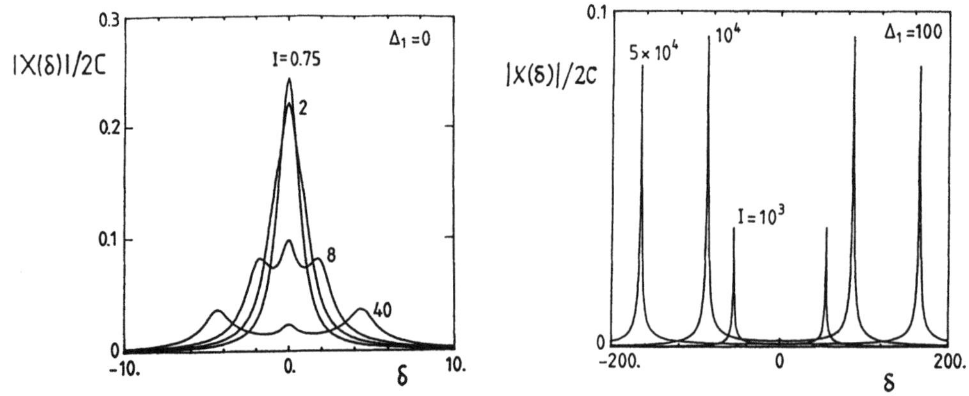

Fig. 1(a) Fig. 1(b)

$|X(\delta)|/2C$ versus δ (a) $\Delta = 0$ (b) $\Delta = 100$

In Fig.(1) we plot $|\chi(\delta)|$. The coupling response $|\chi(\delta)|$ may be thought of as the spectrum for four wave mixing between sidebands. On resonance with the atom ($\Delta = 0$), and below threshold maximum coupling occurs near $\delta = 0$ (degenerate four wave mixing). The spectrum splits into three peaks as saturation occurs. With a detuned atom, below threshold the maximum coupling $|\chi(\delta)|$ is at the doublet $\omega_p \pm \Omega$ (ω_0, $2\omega_p - \omega_0$), with a small central peak occurring at higher intensities. The large enhancement at the sidepeaks is due to the scattering process described above. Important to note is the relatively flat spectral range of $|\chi(\delta)|$ for $|\delta| < \Omega$. $|\chi(\delta)|$ is still quite large for the region between the sidepeaks. This is particularly noticeable on the plot for $\Delta = 100$ where the central peak is very small.

The noise coefficient Λ may be interpreted as the fluorescent spectrum of a pumped two level atom, with no coupling between sidebands. If thermal noise $\bar{n}$ from the cavity reservoir is present, the thermal noise coefficient adds to the term Λ. The Λ is thus a noisy term detracting from squeezing. The noise coefficient R on the other hand represents phase-sensitive quantum noise which may give rise to squeezing. If one considers a prototype Hamiltonian to produce ideal squeezed states

$$H = \hbar\chi(a_2^\dagger a_3^\dagger + a_2 a_3) \tag{5}$$

the stochastic differential equations are

$$\begin{aligned} \dot{\alpha}_2 &= -i\chi\alpha_3^\dagger + F_2(t) \\ \dot{\alpha}_3 &= -i\chi\alpha_2^\dagger + F_3(t) \end{aligned} \tag{6}$$

where $\langle F_2(t)F_3(t')\rangle = -i\chi\delta(t-t')$.

Thus the perfect squeezing situation corresponds to $\Lambda = 0$ and $R = \chi$ and no loss $\gamma(\delta)$.

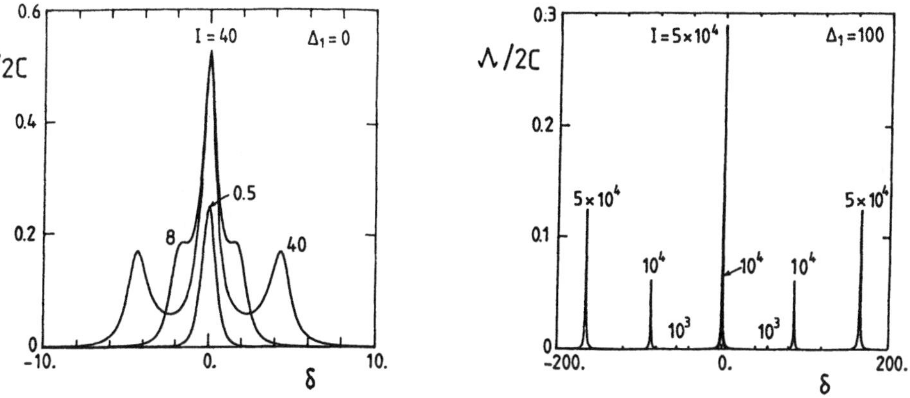

Fig. 2(a) Fig. 2(b)
Λ/2C versus δ (a) Δ = 0 (b) Δ = 100

Plots of the function Λ are shown in Fig.(2). Fig.(2a) is the absorptive case $\Delta = 0$, for various pump intensities I. At high I, we have the resonance fluorescence spectrum with the sidebands corresponding to the Rabi frequencies $|\delta| \sim \frac{\Omega}{2\gamma_\perp} = \frac{1}{2}\sqrt{2I}$. Figure (2b) shows the behaviour for a detuned atom ($\Delta = 100$). Again in the high intensity I regime we have the usual fluorescence spectrum, showing three peaks at $\delta = 0$ and $\delta = \pm\frac{\Omega}{2\gamma_\perp} = \pm\frac{1}{2}\sqrt{2I + \Delta^2}$. The effect of the detuning Δ has been to reduce the heights of the three peaks, but to increase the relative height of the sidebands to the central peak. This has been predicted for the fluorescence of a detuned two level atom. Interesting to note is the spectrum plotted for $\Delta = 100$ and its comparison with that for $|\chi|$ (Fig.1). The spectrum for Λ at high intensity I shows three sharp peaks at $\delta = 0, \pm\frac{\Omega}{2\gamma_\perp}$ with very low fluorescence between peaks. The result for $|\chi|$ on the other hand has a relatively flat spectrum between the Rabi sidebands, being greater than Λ in this region. This result is important to note, bearing in mind that $|\chi|$ generates squeezing while Λ detracts from squeezing. Thus regimes of δ where $|\chi| \gg \Lambda$ produce better squeezing. The behaviour of the phase-sensitive noise term R is displayed in Fig.(3).

Plot 3(a) is for the atom pumped on resonance ($\Delta = 0$). For this case R_I is zero. R the "sideband" spectrum should be compared with Λ for $\Delta = 0$. R has a negative central peak and positive Rabi sidebands. This is in agreement with results derived by Holm et al.(10). As one increases the detuning Δ the imaginary component R_I becomes nonzero. At $\Delta = 100$, R_I becomes the same as χ_I and dominates over R_R for regions away from the Rabi and central peaks.

The results of this section are in agreement with previous work. The parameters γ and χ agree with the semiclassical calculation of Fu and Sargent (8) and Boyd et al (9). Comparisons with the work of Sargent et al.(10) for $\Delta = 0$ show agreement for the coefficients Λ and R.

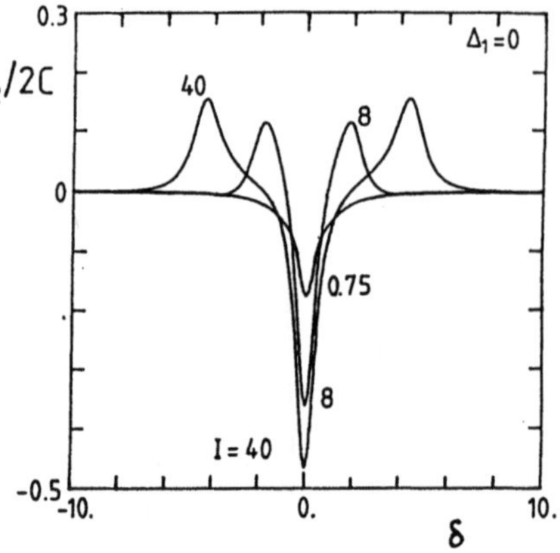

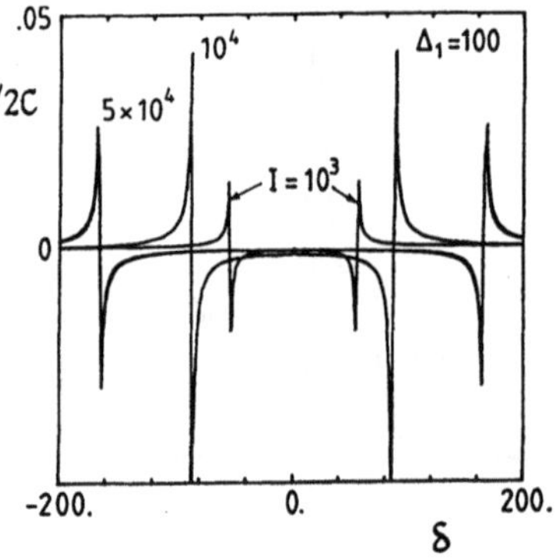

Fig. 3(a) Fig. 3(b)
R/2C versus δ (a) Δ = 0 (b) R_I;Δ = 100

Squeezing Spectrum

We shall now consider four wave mixing in a cavity and calculate the squeezing in the output field. Squeezing is observed in a homodyne detection scheme where the output (a_{2out}, a_{3out}) of the two sidebands at frequency $\omega_p \pm \epsilon$ beat with a local oscillator $E_{Lo} = \epsilon\, e^{i\theta_{Lo}}$ provided by, and phase shifted θ_{Lo} with respect to the external pump at frequency ω_p. The sidebands and local oscillator impinge on the surface of a photodetector forming a photocurrent $i(\epsilon)$. A spectrum analyser allows measurement of the fluctuations in the current at frequency ϵ. One finds the fluctuations to be proportional to $V(X_\theta,\delta)$, the variance in the quadrature phase amplitude defined by (11,12)

$$X_\theta = a_{2OUT}e^{-i(\theta_{Lo}+\theta_p)} + a^+_{OUT}e^{i(\theta_{Lo}+\theta_p)}, V(X_0,\delta) = \tfrac{1}{2}\langle X_0X_0^+ + X_0X_0^+\rangle - \langle X_0\rangle\langle X_0^+\rangle \qquad (7)$$

Of interest to us then is the spectral fluctuations of the output field.

The Langevin equation describing the cavity modes α_2 and α_3 may be rewritten in the following matrix form

$$\frac{d}{dt}\underset{\sim}{\alpha} = -\underline{A}\,\underset{\sim}{\alpha} + \underline{D}^{\frac{1}{2}}\,\zeta(t) \qquad (8)$$

where $\underset{\sim}{\alpha} = (\alpha_2, \alpha_2^\dagger, \alpha_3, \alpha_3^\dagger)$. $\zeta(t)$ is a delta-correlated noise force. $\underline{A}$ is the drift matrix derivable from eq(2), and D is the diffusion matrix whose elements determine the noise correlations as $\langle\Gamma_i(t)\Gamma_j(t')\rangle = D_{ij}\delta(t-t')$. The spectrum of the correlation matrix $\langle\alpha_i(t)\alpha_j(0)\rangle$, obtained by linearising about a stable state $\underset{\sim}{\alpha}^{ss}$, is given by (13)

$$S_{ij}(\omega,\delta) = \int_{\infty}^{\infty} e^{-i\omega t}\langle\alpha_i(t)\alpha_j(0)\rangle dt + \alpha^{ss}_i\alpha^{ss}_j\delta(\omega)$$

$$= \left[(\underline{A} + i\omega\underline{I})^{-1}\,\underline{D}(\underline{A}^T - i\omega\underline{I})^{-1}\right]_{ij} + \alpha^{ss}_i\alpha^{ss}_j\delta(\omega) \qquad (9)$$

where I is the identity matrix. This spectrum $S(\omega,\delta)$ is about a fixed sideband frequency ϵ or δ, and $\omega = 0$ corresponds precisely to this sideband frequency. In the high Q cavity considered here, the spectrum $S(\omega,\delta)$ has a Lorentzian profile, width κ. We are interested in the case $\omega = 0$, corresponding to the sideband frequencies. One finds for the fluctuations

$$S_{21}(0,\delta) = \frac{\left\{\Lambda\left(|\gamma(-\delta)|^2 + |\chi(\delta)|^2\right) + R\gamma^*(-\delta)\chi^*(\delta) + R^*\gamma(-\delta)\chi(\delta)\right\}}{\kappa|\gamma(\delta)\gamma^*(-\delta) - \chi(\delta)\chi(-\delta)^*|^2} = S_{12}(0,\delta)$$

$$S_{43}(0,\delta) = S_{12}(0,-\delta) = S_{34}(0,\delta)$$

$$S_{13}(0,\delta) = e^{2i\theta_0}\bar{S}_{13}(0,\delta) = S_{31}(0,\delta)\ , \ S_{24}(0,\delta) = S_{13}^*(0,\delta) = S_{42}(0,\delta) \tag{10}$$

$$\bar{S}_{13}(0,\delta) = \frac{\left\{\Lambda\left(\gamma^*(-\delta)\chi(-\delta) + \gamma^*(\delta)\chi(\delta)\right) + R\gamma^*(\delta)\gamma^*(-\delta) + R^*\chi(\delta)\chi(-\delta)\right\}}{\kappa|\gamma(\delta)\gamma^*(-\delta) - \chi^*(\delta)\chi(-\delta)|^2}$$

The statistics of the output sideband modes can then be deduced from the boundary conditions at the cavity mirrors. The squeezing in the combined output sideband modes is then determined by the variance (12)

$$V(X_\theta, \delta) = 1 + 2\kappa\left[S_{21}(0,\delta) + S_{43}(0,\delta) + e^{-2i\theta}\bar{S}_{13}(0,\delta) + e^{2i\theta}\bar{S}_{13}^*(0,\delta)\right] \tag{11}$$

where $\theta = \theta_{L0} + \theta_p - \theta_0$ and where we have taken the optimal situation of a single port cavity. Squeezing is realized when $V(X_\theta,\delta) < 1$ and perfect squeezing corresponds to $V(X_\theta,\delta) = 0$.

The best squeezing is thus, for optimal θ

$$V(X_\theta,\delta) = 1 + 2\kappa\left[S_{12}(0,\delta) + S_{34}(0,\delta) - 2|\bar{S}_{13}(0,\delta)|\right] \tag{12}$$

which is obtained for θ given by

$$\cos 2\theta = -\frac{\mathrm{Re}(\bar{S}_{13}(0,\delta))}{|\bar{S}_{13}(0,\delta)|} \text{ and } \sin 2\theta = \frac{-\mathrm{Im}(\bar{S}_{13}(0,\delta)|}{|\bar{S}_{13}(0,\delta)|} \tag{13}$$

Some of the results for the squeezing spectrum have been discussed by Reid and Walls (15).

It is possible in the first instance to calculate the intensity fluctuations $S_{21}(0,\delta)$ as a function of δ (we will now call this the "intensity spectrum" and abbreviate to $S_{21}(\delta)$), if we have no coupling between the weak fields (i.e. we put $\chi = R = 0$). The result is $S_{21}(\delta) = \Lambda/\kappa|\gamma(\delta)|^2$. Thus in a very low C limit, the transmitted intensity (apart from the central delta function component) spectrum $\kappa S_{21}(0,\delta)$ is simply Λ, Figure 2. Calculation of the variance $V(X_\theta,\delta)$ as a function of δ (we will call this the squeezing spectrum) in this limit shows (since $S_{13}(0,\delta) = 0$) $V(X_\theta,\delta) = 1 + 2\Lambda$, and is phase insensitive. Thus Λ represents the desqueezing noise due to fluorescence.

As discussed with reference to equation (5), it is the mixing terms χ and R which produces the squeezing. The advantage of nondegenerate four wave mixing over degenerate ($\delta = 0$) four wave mixing is that $\chi(\delta)$ and R can be enhanced relative to Λ as one increases δ to the Rabi frequency $\delta = \Omega/2\gamma_{\perp}$. This is due to the scattering effect discussed above. The scattering process couples photons at $+\delta$ and $-\delta$ via two pump photons, and is enhanced at the Rabi frequencies. Thus one can increase the pump intensity and saturate the atoms and avoid the desqueezing effects associated with increasing fluorescence by increasing δ appropriately. One can thus attain good squeezing at saturation intensities, and also at much lower atomic detunings (e.g. $\Delta \sim 4$), lower C values, and for a broader range of the parameters Δ, C, I and ϕ than was possible in the degenerate situation $\delta = 0$.

Figures 4 and 5 plots the squeezing spectrum for two values of C with an atomic detuning $\Delta = 100$. The first value of C is such that one has bistability of the central pump mode. The intensity $I \sim 5 \times 10^4$ corresponds to atoms well saturated. The intensity spectrum $S_{21}(0,\delta)$ at such high intensities shows three peaks at $\delta = 0$ and $\delta = \pm\frac{1}{2}\sqrt{2I+\Delta^2}$ corresponding to the generalised Rabi frequencies. We see that no squeezing is observed for the central peak because of the large central peak of the fluorescence spectrum Λ. But on increasing δ to the sidepeaks, squeezing is attained due to the enhanced coupling between sidebands.

As the intensity I is decreased to approach the unstable region from above, the sidepeaks move in towards the central peak, a collective feature of the

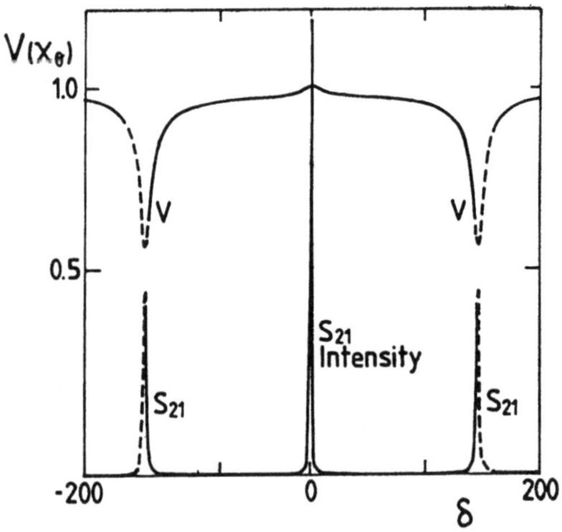

Fig. 4
Squeezing spectra $V(X_\theta,\delta)$.
$C = 1100$, $I = 5 \times 10^4$, $\Delta = 100$.
----- unstable regions. Intensity spectrum $S_{21}(\delta)$ shown also.

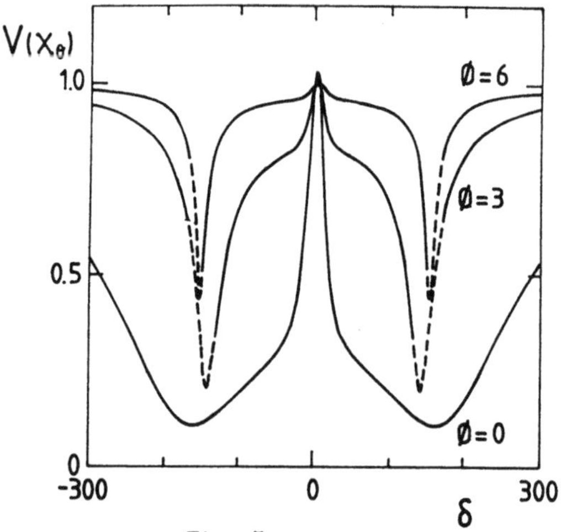

Fig. 5
Squeezing spectra $V(X_\theta,\delta)$.
$C = 250$, $I = 5 \times 10^4$, $\Delta = 100$
----- unstable regions.

transmitted spectrum noted in previous works. We attain considerable enhancement of squeezing as we increase δ to coincide with these sidepeaks of the intensity spectrum.

At lower intensities, on the lower branch, where the nonlinearity of the cavity is reduced ($|\chi(\delta)| \sim 2CI/\Delta^3 \ll 1$), the squeezing drops considerably. Little advantage is obtained by increasing δ in this situation, the central fluorescence peak Λ vanishing at low intensities.

The squeezing spectrum for the lower C case, where one has not "clear" bistability in the pump mode, shows similar effects (Figure 5). Good squeezing is attainable at sufficiently high intensities, and for a range of ϕ values. The effect of ϕ is to alter the width of the side-peaks and hence of the squeezing spectrum. ϕ~5 corresponds approximately to the pump resonant with the linear dressed cavity.

If we were to increase the C values such that $2C \to \Delta^2$, a reduction in the amount of squeezing possible is noticed. This is a result of increased atomic absorption $\gamma_R(\delta)$.

The experimental situation of Slusher et al. (4) corresponds to a standing wave cavity rather than the ring cavity described in the theory here. The approximation, valid for highly reflecting mirrors, has been made that one can neglect spatial fluctuations. A modification of the result (2) has been necessary to better describe the standing wave cavity mode situation. The simplest mode function (~ sin kr) has been taken and averages calculated over the standing wave phase. Another approximation has been that of a high Q cavity,

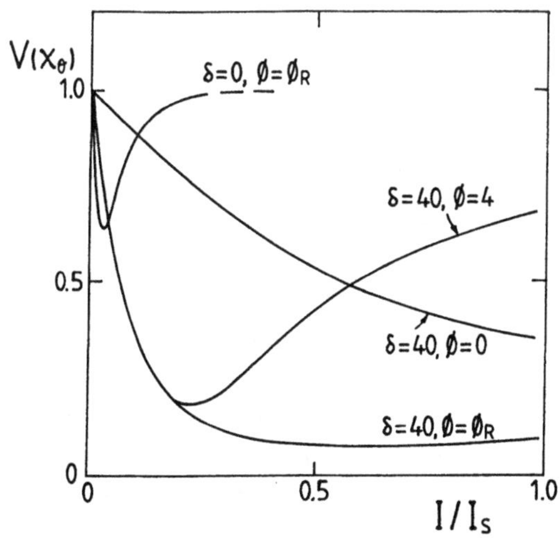

Fig.6 Comparison with experiment: Squeezing V versus the intensity in units of the saturation intensity I_s, for various cavity detuning ϕ. $\phi_R = -\gamma_I(0)$ and corresponds to resonance with the dressed cavity, at all intensities. $\Delta = -300$, $C = 750$.

$\kappa << \gamma_\perp$. This is not strictly true in the experiment, but further calculations made (see Section 3) indicate the result for the squeezing at $\omega = 0$ precisely to be insensitive to this assumption. We have also assumed fluctuations to be small so that a linear treatment of fluctuations about a stable steady-state cavity solution is valid. This approximation will break down near a point of instability. Also assumed is the absence of collisions and Doppler broadening.

With these approximations, Figure 6 plots the result for optimal squeezing obtained, versus intracavity intensity I, for parameters C and Δ believed to correspond to the experiment. Also plotted is the result predicted for a degenerate experiment $\delta = 0$. The experiment is presently operating in a regime of low intensity ($I/I_s = 0.02$, well below saturation), where there is little difference between degenerate and non-degenerate predictions. A theory, based on the degenerate $\delta = 0$ noise terms, has been presented by Klauder et al. (16) for the experiment, and includes effects such as phase jitter. It shows good agreement with the present experiment.

#3 Squeezing in a low Q Cavity

The results discussed above hold in the high Q cavity limit where the atoms have been adiabatically eliminated. This is possible when the atomic decay times are much shorter than the cavity decay time ($\kappa << \gamma_\perp$). In order to discuss situations where this is not the case it is necessary to consider the full equations for the coupled field and atomic variables. We consider a single-mode cavity driven by an external driving field, such as considered in optical bistability. The system is thus the same as considered in (1) but with no intracavity sidemodes present. The linearized equations obtained without adiabatic elimination have been solved numerically.

We present results (Figures 7, 8, 9) for the linearised squeezing spectrum ($V(X_\theta,\omega)$ optimised for best squeezing at each frequency ω and centered at the pump frequency ω_p. We select $\Delta = 100$ and $C = 1100$, and values of intracavity pump intensity I are chosen corresponding to: below, but quite near threshold ($I = 450$); above threshold ($I = 3000$); and well above threshold ($I = 5 \times 10^4$).

An analytical treatment (17,18) of the limiting high Q ($\kappa/\gamma_\perp \rightarrow 0$) cavity indicates best squeezing to be attained near the turning points at the central frequency ω_p. As one moves above threshold, a considerable reduction in squeezing at this central frequency is noticed. The reason is the increased fluorescence Λ at the pump frequency $\delta = 0$, discussed in the previous section. These effects are apparent in our results for the high Q cavity where $\kappa = \gamma_\perp/50$.

The plot of the squeezing spectrum in a low Q ($\kappa = \gamma_\perp/0.05$) cavity shows two important features. The first is that the result for squeezing at the central frequency ω_p is unchanged from the high Q case. Thus one sees a degrading of squeezing at the central pump frequency above threshold.

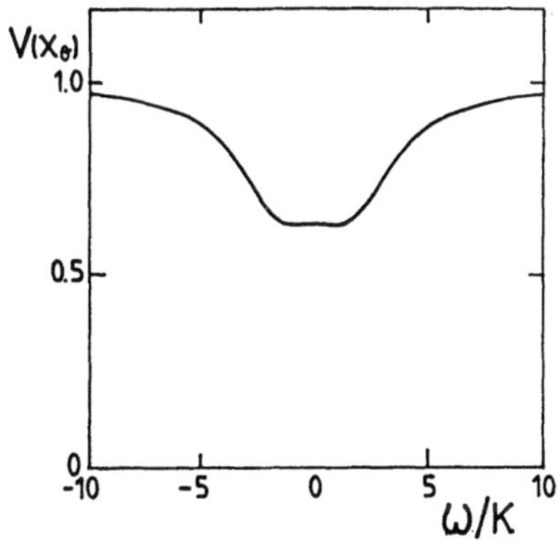

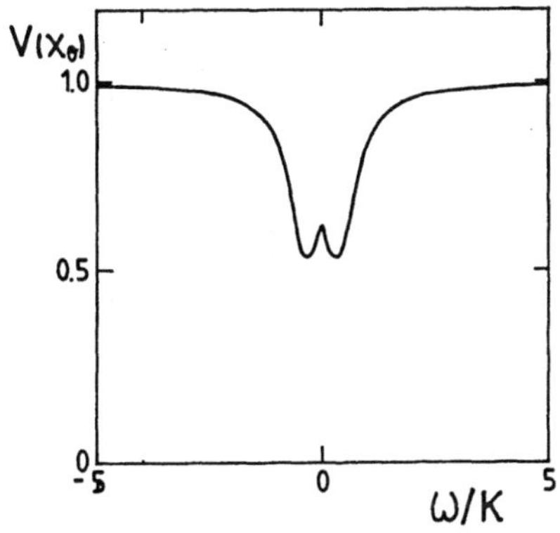

Fig. 7(a) Fig. 7(b)

Squeezing spectrum $V(X_\theta,\omega)$: Lower Branch $I = 450$ $\Delta = 100$, $C = 1100$, $\phi = 18$.
(a) high Q cavity ($\kappa = \gamma_\perp/50$) (b) low Q cavity ($\kappa = 20\gamma_\perp$)

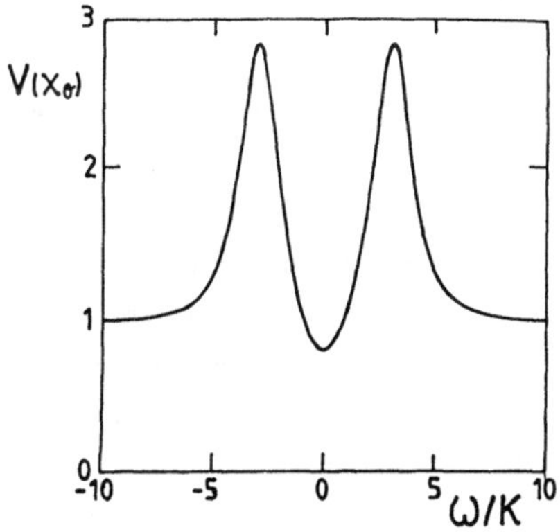

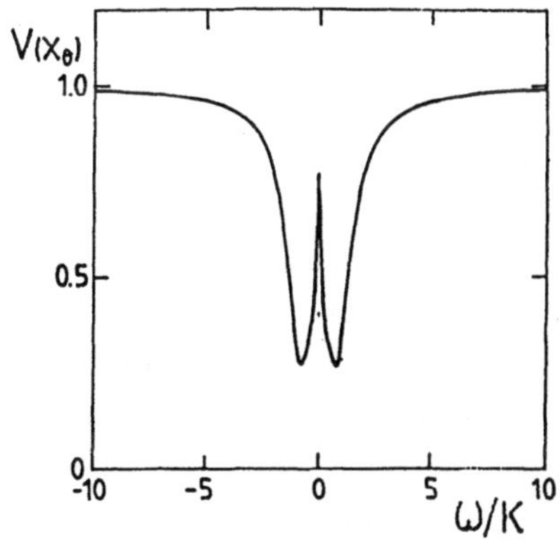

Fig. 8(a) Fig. 8(b)

Squeezing spectrum $V(X_\theta,\omega)$: Upper Branch $I = 3000$. $\Delta = 100$, $C = 1100$, $\phi = 18$
(a) high Q cavity ($\kappa = \gamma_\perp/50$) (b) low Q cavity ($\kappa = 20\gamma_\perp$)

An analytical treatment of dispersive optical bistability in a perfect low Q cavity ($\kappa/\gamma_\perp \to \infty$) has been given by Carmichael et al.(19). The theory calculates the transmitted intensity spectrum $S_{21}(\omega)$ In a low Q cavity the atomic resonance ω_0 is a lot closer to the pump frequency ω_p than the cavity resonance ω_1. The intensity spectrum $S_{21}(\omega)$ thus exhibits features of atomic fluorescence with additional collective effects, such as linewidth narrowing near points of instability and cooperative broadening along the lower branch, apparent.

Sufficiently far above threshold the spectrum tends to that of usual detuned atomic fluorescence, with three peaks, width ~ $\gamma_\perp$, at $\omega = 0$ and at the Rabi frequencies $\omega/\kappa = \pm\frac{\gamma_\perp}{\kappa}\Omega = \pm\frac{\gamma_\perp}{\kappa}\sqrt{2I+\Delta^2}$. The squeezing spectrum corresponding to this

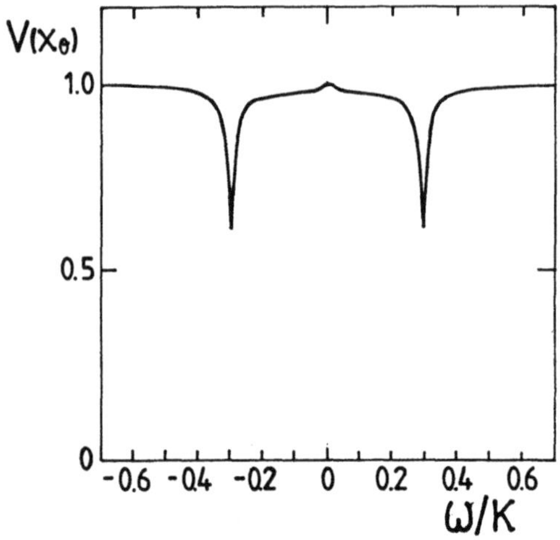

Fig. 9 Squeezing spectrum $V(X_\theta,\omega)$: Upper Branch (with atoms well saturated). Δ =100, C = 1100, ϕ = 19. I = 5×10^4 Low Q cavity ($\kappa = 10^3\ \gamma_\perp$).

regime is illustrated in Figure 9. The central peak at ω = 0 (the pump frequency) shows no squeezing.

However, on increasing ω out to the Rabi sidepeaks, there is considerable reduction in noise and squeezing is possible. This is due to a coupling between photons on either side $\pm\omega$ of the spectrum brought about by the inelastic scattering process described in the preceding section. This coupling $\chi(\omega)$ is analogous to the coupling $\chi(\delta)$ discussed in the preceding section. There is an enhancement in the coupling as one approaches the Rabi frequency.

Approaching the region of instability from above (I = 3000), the low Q cavity exhibits an intensity spectrum with sidepeaks moving in towards the central peak. Good squeezing (Figure 8) is attained upon increasing ω to these sidepeaks. On the lower branch one sees, in the low Q cavity intensity spectrum, a broadened doublet. The squeezing spectrum in Figure 7(b) shows enhancement at this doublet. The reduction in squeezing at the central frequency ω = 0 is not so significant as observed above threshold since the fluorescence does not show a large central peak at low intensities I below saturation. Thus there is little difference in high Q and low Q cavity results in terms of optimal squeezing attainable. The order of squeezing decreases below threshold as one decreases I since the nonlinearity of the cavity is reduced.

The enhancement of three wave coupling and reduction of noisy fluorescence in saturating atoms, as one moves away from the pump frequency and out towards the δ frequency of the Rabi sidepeaks, has been discussed in Section II. The results of this section indicate these effects to be also observed in the output squeezing spectrum of a single-mode low Q cavity. In fact comparison of Figures 8 and 4 indicate the infinitely low Q squeezing spectrum $V(X_\theta,\omega)$ to be identical with $V(X_\theta,\delta)$. Thus one can obtain significantly better squeezing in a low Q cavity than in a high Q cavity by increasing the intensity towards atomic saturation.

It is possible that the low Q cavity is more advantageous experimentally than the three-mode cavity discussed in #2, since in the latter one needs to tune the cavity spacing δ correctly. However, we stress our theory still assumes a 'high Q' cavity in the sense that the mirrors are highly reflecting and the single cavity mode approximation still holds. Good squeezing is attainable in a low Q cavity for much smaller atomic detuning ($\Delta \sim 4$) and thus greatly reduced C values ($C \sim 5$). Also the sensitivity of squeezing to system parameters such as ϕ is reduced in the low Q as compared to the high Q cavity. Importantly, the advantages of the low Q cavity are present even for moderate values of $\kappa/\gamma_\perp (\sim 5)$.

#4 Second Harmonic Generation

We shall consider briefly the generation of squeezed states in second harmonic generation, since this system is being investigated experimentally by Kimble and Hall. The results of their experiments are reported by Kimble in this volume. Suggestions to consider second harmonic generation as a candidate for squeezing were made by Mandel (20) and Lugiato, Strini and de Martini (21). A calculation of the squeezing spectra of the output light emerging from intracavity second harmonic generation was given by Collett and Walls (14). The Hamiltonian describing intracavity second harmonic generation is

$$\begin{aligned} H = {} & h\omega a_1^\dagger a_1 + 2h\omega a_2^\dagger a_2 \\ & + h\frac{\kappa}{2}\left(a_1^{\dagger 2} a_2 - a_1^2 a_2^\dagger\right) \\ & + ih\left(E_1 a_1^\dagger e^{-i\omega t} - E_1^* a_1 e^{i\omega t}\right) \\ & + a_1\Gamma^\dagger + a_1^\dagger\Gamma_1 + a_2\Gamma^\dagger + a_2^\dagger\Gamma_2 \end{aligned} \tag{14}$$

where a_1 and a_2 are the boson operators for modes of frequency ω and 2ω respectively. κ is the coupling constant proportional to the second order nonlinear susceptibility of the medium Γ_1, Γ_2 are heat bath operators which represent cavity losses for the two modes and E_1 is the amplitude of the coherent arriving field at frequency ω. The details of the calculation are given in reference (14).

In Fig. (10a) we plot the maximum squeezing in the output field at the fundamental frequency $V(X_\theta,\omega_m)$ (ω_m is the frequency at which the maximum squeezing occurs) against the driving field amplitude E. The graphs shown are for different ratios of the cavity damping rates $\frac{\gamma_1}{\gamma_2}$ of the fundamental and second harmonic modes. At $E = E_c$ the steady state becomes unstable to a self oscillating behaviour. However, before this threshold is reached one sees that a plateau of good squeezing may be reached for $\gamma_1 \gg \gamma_2$.

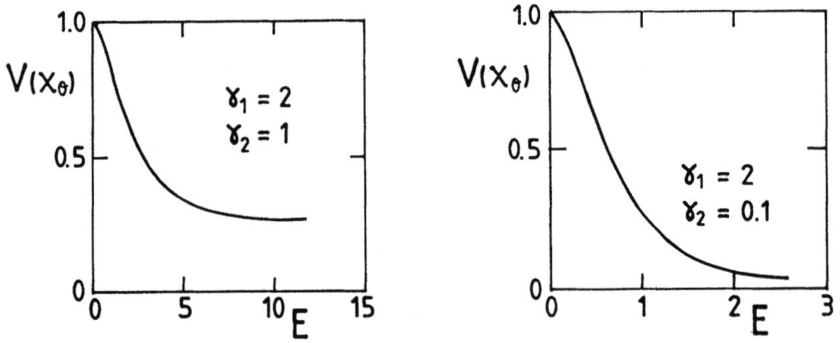

Fig. 10(a) Optimal squeezing in output field at the fundamental frequency $V(X_\theta, \omega_m)$ versus driving field amplitude E.

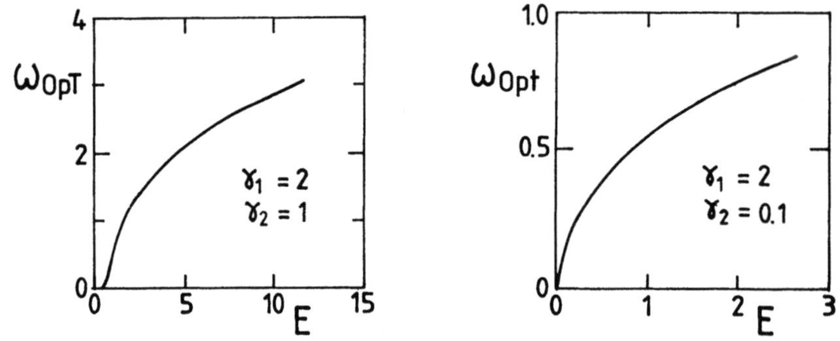

Fig.10(b) Plot of ω_m versus driving field E.

In Fig.(10b) we plot ω_m against the driving field E. This shows that the squeezing spectrum is split into two peaks symmetric about $\omega = 0$ as the driving field is increased.

References

(1) D.F. Walls, Nature 306, 141 (1983).
(2) M.D. Levenson, R.M. Shelby, A. Aspect, M.D. Reid and D.F. Walls, Phys. Rev. A32, 1550 (1985).
(3) R.S. Bondurant, P. Kumar, J.M. Shapiro and M. Maeda, Phys. Rev. 30A, 343 (1984).
(4) R.E. Slusher, L. Hollberg, B. Yurke, J.C. Mertz and J.F. Valley, Phys. Rev. Letters 55, 2409 (1985).
(5) H.J. Kimble and J. Hall, "Proceedings of the M.I.T. Conference on Squeezed States of Light" (1985), P. Kumar and J. Shapiro Editors.
(6) M.D. Reid and D.F. Walls, Phys. Rev. A32, 396 (1985).
(7) M.D. Reid and D.F. Walls (to be published).
(8) T. Fu and M. Sargent, Opt.Lett 4, 366 (1979).
(9) R.W. Boyd, M.G. Raymer, P. Narum and D.J. Harten, Phys. Rev. 24A, 411 (1981).
(10) D. Holm, M. Sargent III and B. Capron, presented in Proceedings of the MIT Endicott Hour Workshop on Squeezed States of Light, Shapiro and Kumar (1985). M. Sargent III, D. Holm and M. Zubairy, Phys. Rev. A31, 3112 (1985), M. Sargent III, M. Zubairy and F. de Martini, Opt. Lett. 8, 76, (1983).

(11) C.M. Caves and B.L. Schumaker, Phys. Rev. A31, 3068 (1985), B.L. Schumaker and C.M. Caves, Phys. Rev. A31, 3093 (1985).
(12) B. Yurke, Phys. Rev. A32, 300 (1985).
(13) C.W. Gardiner, 'Handbook of Stochastic Methods' (Springer, Berlin, 1983).
(14) M.J. Collett and D.F. Walls, Phys. Rev. A32, 2887 (1985), M.J. Collett and C.W. Gardiner, Phys. Rev. A 30, 1386 (1984).
(15) M.D. Reid and D.F. Walls, Phys. Rev. A (March) (1986).
(16) J.R. Klauder, S. McCall and B. Yurke (to be published)
(17) P.D. Drummond and D.F. Walls, Phys. Rev. A 23, 2563 (1981).
(18) M.D. Reid and D.F. Walls, Phys. Rev. A32, 396, (1985).
(19) H.J. Carmichael, D.F. Walls, P.D. Drummond and S.S. Hassan, Phys. Rev. 27A, 3112 (1983).
(20) L. Mandel, Optics Comm. 42, 437 (1982).
(21) L.. Lugiato, G. Strini and F. de Martini, Optics Letter 8, 256, (1983).

Experimental Studies of Squeezed-State Generation by Four-Wave Mixing in an Optical Cavity

B. Yurke and R.E. Slusher

AT & T Bell Laboratories, Murray Hill, NJ 07974, USA

Wideband squeezing below the vacuum noise level has been observed in light generated by four-wave mixing due to Na atoms in an optical cavity. The noise in the quadrature components was measured using a balanced homodyne detector. Here our observations are compared with the predictions of a wideband phenomenological model of the four-wave mixer and the detection system. Various noise sources which limit the degree of squeezing observed have been identified. Work is in progress to reduce these noise sources.

I. Introduction

There has been a considerable advance in theoretical understanding and experimental technique since the early days [1-5] when theorists began to suggest that it may be possible to reduce the noise in one quadrature component of the electromagnetic field via a parametric device such as a parametric amplifier or four-wave mixer and that it may be possible to observe this noise reduction via a homodyne detector [6].

Various experimental programs [7-9] are under way directed toward the generation of squeezed-coherent radiation. Key ingredients that led to our [7] successful generation and detection of squeezed-coherent radiation were the realization that: cavities could be used to enhance squeezing [10-12], balanced homodyne detection could be used to overcome local oscillator noise [13-15], and squeezing effects could be observed under nondegenerate conditions [16-21] where the contamination of the signal by scattered pump light can be reduced. We were also assured by theorists that squeezing was possible even when quantum fluctuations of the pump light [22-25] and nonidealities of the nonlinear medium [26-28] were taken into account.

Here a brief account of wideband squeezing is given. A wideband analysis of a ring-cavity-forward-four-wave mixer is presented which phenomenologically includes four-wave-mixing gain, cavity and medium losses, and spontaneous emission noise. A calculation of the response of a homodyne detector to the light generated by this four-wave mixer is carried out which includes detector inefficiencies and local-oscillator phase jitter. Our experimental results are compared with this phenomenological model in order to identify various noise sources which limit the degree of squeezing observed. Although the experiments were carried out for a backward-four-wave mixer in a Fabry-Perot cavity we choose here to compare our results with those of a ring cavity with forward-four-wave mixing because the analysis of this configuration is easier and leads to more compact analytic expressions. Further, the quantitative difference between the behavior of the two four-wave-mixing configurations is small enough to be ignored over the parameter ranges and precision appropriate for the present experiments.

The comparison of our experimental results with this phenomenological model suggests that local-oscillator phase jitter, detector losses, and cavity losses not intrinsic to the four-wave-mixing medium degrade the amount of squeezing observed to the extent that the 7% (in power) squeezing observed would amount to 23% squeezing if these nonidealities are eliminated. Work is currently in progress to install more efficient photodetectors, to stabilize the local oscillator phase, and to eliminate other noise sources. Further, theoretical calculations by REID and WALLS [28] and KLAUDER, McCALL and YURKE [29] indicate that much greater squeezing can be obtained from an atomic Na beam over the parameter range accessible with our present apparatus. The details of our apparatus have been published elsewhere [7,30]. Here only those aspects of the experiment pertinent to the present discussion will be described.

II. The Apparatus

A cavity four-wave mixer employing a beam of atomic Na for the nonlinear medium was used to generate the squeezed light. A schematic diagram of the four-wave mixer geometry is depicted in Fig. 1. The atomic Na beam is pumped in the backward-four-wave-mixing configuration via two counter propagating pump beams that are perpendicular to the atomic beam. The standing wave generated by these opposing pump beams produces a time-varying index of refraction at $2\nu_L$ in the four-wave-mixing medium, where ν_L is the pump frequency. This leads to the generation of highly correlated photon pairs at frequencies symmetrically located about the pump frequency. Those pairs having frequency components at

$$\nu_{1,2} = \nu_L \pm n\nu_{sc} \tag{2.1}$$

where n is an integer and ν_{sc} is the cavity mode-spacing frequency are coherently built up by the multipass effect [10-12]. This light exhibits wideband squeezing upon exiting the output mirror. The experimental cavity is confocal with a mirror separation near 107 cm. The squeezing cavity mode-spacing frequency ν_{sc} is 140.5 MHz. We have observed squeezing for n = 3. The observation of squeezed light in Fabry-Perot mode pairs closer to the pump frequency have been hampered

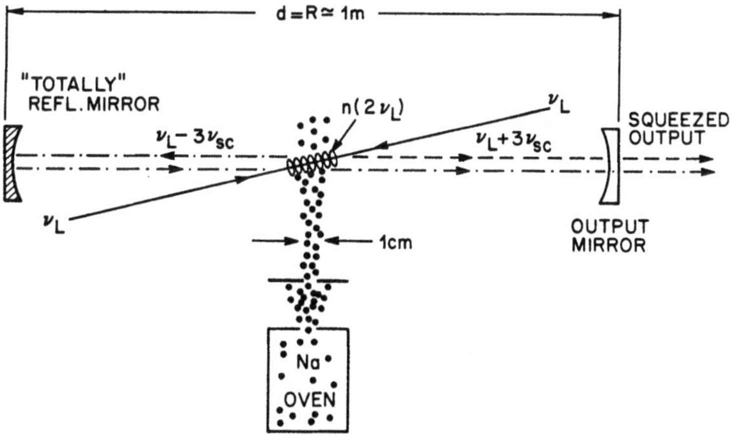

Fig. 1. A schematic diagram of the four-wave-mixing geometry employed to generate squeezed light. A beam of atomic sodium is pumped in a backward-four-wave-mixing configuration via two counter propagating pump beams at frequency ν_L. Pairs of counter propagating photons are generated in the four-wave-mixing process. Pairs having frequency components at $\nu_L + 3\nu_{sc}$ and $\nu_L - 3\nu_{sc}$, where ν_{sc} is the frequency spacing for the cavity modes, are coherently built up and exit the output mirror as two-mode or wideband squeezed states.

by spontaneous emission noise [31-33] from Doppler-shifted Na atoms due to the 5 degree divergence of the Na beam. The "totally" reflecting mirror had a reflectivity of 0.995 while the output mirror had a transmission of 2%. Vacuum fluctuations entering through the totally reflecting mirror degrade the amount of squeezing produced. Hence, in order not to be limited by this noise source, it is important that the reflectivity of this mirror be much higher than that of the output mirror.

The sodium beam density is typically a few times 10^{11} cm^{-3} where the cavity and pump mode overlap. The cavity and pump beams are at an angle of 0.86 degrees with respect to each other and overlap in the sodium beam over a distance of 1 cm.

The frequencies characterizing our experiment are depicted in Fig. 2. The continuous curve represents the light transmitted through the sodium beam. There is strong absorption at the D_2 hyperfine pair near 589 nm. The pump light frequency ν_L is tuned approximatly 1.5 GHz to the high-frequency side of this hyperfine pair. The position of the cavity modes are also depicted. A probe beam consisting of light shifted 281 MHz above the pump frequency is injected into the cavity mode at $\nu_L + 2\nu_{sc}$ and used to lock the cavity modes to the pump frequency. The error signal for the feedback to stabilize the cavity modes is obtained by the method discussed by DREVER, et al. [34] in which FM sidebands (at $\pm$ 11 MHz in our experiment) are placed on the probe beam before it is injected into the Fabry-Perot Cavity. This technique avoids having to dither the cavity length which would reduce the build-up of squeezed light in the cavity.

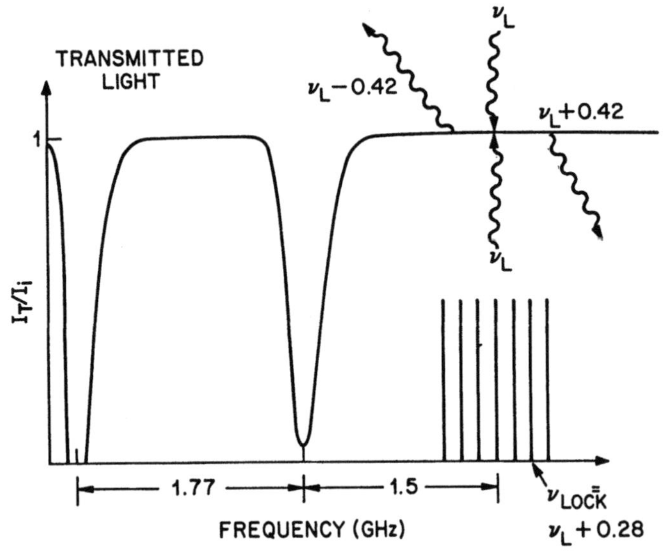

Fig. 2. Frequencies of the Na absorption, cavity modes, pump frequency, and conjugate photon pairs associated with four-wave mixing are shown for the present experimental parameters. The transmitted to incident light intensity ratio I_T/I_i is shown for small intensities near the hyperfine pair. The dye laser pump frequency ν_L is located approximately 1.5 GHz to the high side of the weaker Na resonance. Four-wave mixing generates conjugate photon pairs at frequencies near $\nu_L \pm 0.42$ coincident with a pair of cavity resonances shown as the vertical lines at the right of the figure.

Wideband squeezing is observed using a balanced homodyne detector whose local oscillator light is split off from the pump. Squeezing below the vacuum noise level manifests itself when the noise power of the homodyne detector at a given frequency drops below that when no light enters the signal port of the homodyne detector. The homodyne detector signal at 421.5 MHz is generated from light in the cavity modes at $\nu_L \pm 3\nu_{sc}$. The four-wave mixer output beam and the local oscillator light are mode matched. We typically achieve a mode-matching efficiency of 0.75. The efficiency of the photodetectors (RCA Type 30971 Si pin photodiodes) employed in the homodyne detector is 0.7 for the light near 589 nm. These homodyne detector nonidealities introduce noise which degrades the degree of squeezing that can be observed. Further, phase jitter of the local oscillator light relative to the pump also degrades the amount of squeezing that can be observed. These phase shifts arise as the pump frequency jitters relative to the four-wave mixer cavity modes. For our pump, a Coherent Radiation Model 699-21 dye laser, this frequency jitter is due to dye stream fluctuations. The frequency jitters over a 1 MHz bandwidth at frequencies up to 50 kHz. At present our stabilization loop compensates for phase jitter at rates up to 10 kHz. Further frequency stabilization of the pump laser is planned using an acousto-optic and electro-optic modulator as described by HALL and HANSCH [35].

III. Two-Mode Analysis and Quadrature Components

In this section it will be convenient to use angular frequencies $\omega = 2\pi\nu$. A nondegenerate four-wave-mixing process couples a signal mode at frequency $\omega_p + \omega$ with an idler mode at $\omega_p - \omega$, where ω_p is the pump frequency. Light at frequency $\omega_p + \omega$ or $\omega_p - \omega$ when directed to a homodyne detector whose local oscillator has the frequency ω_p will give rise to photocurrents at frequency ω. As emphasized by CAVES and SCHUMAKER [17-18] these considerations lead one to conclude that most phenomena involving wideband squeezing can be understood by considering only two modes of the electromagnetic field, a mode at frequency $\omega_p + \omega$ and its conjugate at frequency $\omega_p - \omega$. The electric field associated with these two modes will have the form

$$E(t) = E[a(+\omega)e^{-i(\omega_p+\omega)t} + a(-\omega)e^{-i(\omega_p-\omega)t} + h.c.]/\sqrt{2} \tag{3.1}$$

where $a(+\omega)$ and $a(-\omega)$ are the annihilation operators for respectively the mode at frequency $\omega_p + \omega$ and the mode at frequency $\omega_p - \omega$ and satisfy the usual boson commutation relations

$$[a(+\omega),a^\dagger(+\omega)] = [a(-\omega),a^\dagger(-\omega)] = 1 , \tag{3.2}$$

where the other commutation relations are zero.

In writing (3.1) it has been assumed that $\omega_p >> \omega$ so that factors of the form $(\omega_p \pm \omega)^{1/2}$ could be approximated by $\omega_p^{1/2}$ and absorbed in the constant E. The effects of including these factors have been explored by CAVES, SCHUMAKER, and YURKE [16-20,36] and are small enough to be inconsequential for the present generation of experiments.

With straightforward algebra (3.1) can be recast into the component form

$$E(t) = E[X_1(t,\theta)\cos(\omega_p t+\theta) + X_2(t,\theta)\sin(\omega_p t+\theta)] \tag{3.3}$$

where the quadrature field component operators $X_1(t,\theta)$ and $X_2(t,\theta)$ are given by

$$\begin{aligned} X_1(t,\theta) &= [(a(+\omega)e^{-i\omega t} + a(-\omega)e^{i\omega t})e^{i\theta} + h.c.]/\sqrt{2} \\ X_2(t,\theta) &= [-i(a(+\omega)e^{-i\omega t} + a(-\omega)e^{i\omega t})e^{i\theta} + h.c.]/\sqrt{2} \quad . \end{aligned} \tag{3.4}$$

As will be discussed later, these operators correspond to the physical quantity measured by a homodyne detector where θ is the local oscillator phase and hence is under the control of the experimenter. Since $\omega << \omega_p$ the time variation of X_1 or X_2 is slow compared to an optical cycle and can be detected electronically. It is also worth noting that $X_2(t,\theta) = X_1(t,\theta-\pi/2)$, that is, either quadrature of the electromagnetic field can be viewed via a homodyne detector by properly adjusting the local oscillator phase.

The quadrature component operators satisfy the commutation relation

$$[X_1,X_2] = 2i \tag{3.5}$$

and hence satisfy the uncertainty relationship

$$\Delta X_1 \Delta X_2 \geq 1 \ . \tag{3.6}$$

For coherent light with frequency components at $\omega_p + \omega$ and $\omega_p - \omega$ the equality holds, that is, coherent light is in a minimum uncertainty state for the uncertainty relation (3.6). For such light the variance of X_1 and X_2 is one. As will be discussed in the next section, four-wave mixers can be constructed which generate a class of states for which the variance of X_1 or X_2 is reduced below the coherent state or vacuum state variance of 1. Such states are called wideband or two-mode squeezed states and are of considerable interest, since they can be used to enhance the sensitivity of interferometers [37-39].

IV. The Ring-Cavity-Four-Wave Mixer

In this section the output field of a four-wave mixer consisting of a forward-four-wave-mixing medium [26] imbedded in a ring cavity will be calculated. The behavior of this system should be quite similar to the behavior of a backward-four-wave-mixing medium in a Fabry-Perot cavity provided the round-trip distance of the ring cavity is twice the length of the Fabry-Perot cavity. Further, since in the Fabry-Perot configuration the light passes through the nonlinear medium twice during each round trip, the gain and loss of the four-wave mixing medium in the ring cavity should be taken to be twice that for the medium in the Fabry-Perot cavity when simulating one device by the other.

The ring-cavity-four-wave mixer to be analyzed here is depicted in Fig. 3. The beam path labels a_{in}, b, c, c_n, d, d_n, e, f and a_{out} also denote the annihilation operators for the light at various points along the

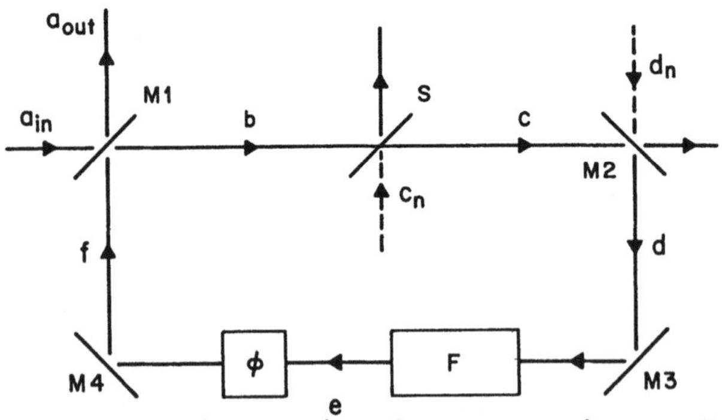

Fig 3. A ring-cavity-four-wave mixer. F is an ideal forward-four-wave mixer. Medium losses are simulated phenomenologically by the beam splitter S. Signals enter the ring cavity through mirrors M1 and M2. The phase accumulated by the light as it propagates around the ring has been lumped into the phase shifter ϕ.

ring cavity. The mirror M1 is the input mirror and has the scattering matrix

$$b(\omega) = K^{1/2}a_{in}(\omega) + (1-K)^{1/2}f(\omega)$$
$$a_{out}(\omega) = -(1-K)^{1/2}a_{in}(\omega) + K^{1/2}f(\omega) \qquad (4.1)$$

where K is the mirror's transmission coefficient. A second mirror M2 through which a probe beam d_n can be injected into the cavity has also been included. The scattering matrix for this mirror has the same form as (4.1) and in particular

$$d(\omega) = (1-T)^{1/2}c(\omega) + T^{1/2}d_n(\omega) \qquad (4.2)$$

where T is the transmission coefficient. Four-wave-mixing-medium losses have been included phenomenologically by a beam splitter S which extracts light from the ring cavity. The beam splitter provides a port c_n through which vacuum fluctuations can enter the cavity. This beam splitter also has a scattering matrix of the form (4.1) and in particular

$$c(\omega) = (1-L)^{1/2}b(\omega) + L^{1/2}c_n(\omega) \qquad (4.3)$$

where L is the loss presented to the cavity by the beam splitter. The forward-four-wave mixer F is taken to be ideal with a scattering matrix of the form [26]

$$e(\omega) = \cosh(\beta)d(\omega) + \sinh(\beta)d^{\dagger}(-\omega) \qquad (4.4)$$

where β is the nonlinear phase shift for the four-wave-mixing medium. For calculational convenience the phase shifts suffered by the light as it propagates around the ring cavity have been lumped into a single phase shifter whose scattering matrix is given by

$$f(\omega) = e^{i\phi(\omega)}e(\omega) \quad . \qquad (4.5)$$

The linear equations (4.1)-(4.5) can be solved to express the output field a_{out} in terms of the input fields a_{in}, c_n, and d_n. One finds

$$a_{out}(\omega) = A_1(\omega)a_{in}(\omega) + B_1(\omega)a_{in}(-\omega) + A_2(\omega)c_n(\omega) + B_2(\omega)c_n(-\omega)$$
$$+ A_3(\omega)d_n(\omega) + B_3(\omega)d_n(-\omega) \quad \text{where} \qquad (4.6)$$

$$A_1(\omega) = [\, -(1-k)^{1/2} - e^{-i\phi(-\omega)}K(1-T)^{1/2}(1-L)^{1/2}\cosh(\beta)$$
$$+ (e^{i\phi(\omega)} + e^{-i\phi(-\omega)})(1-T)^{1/2}(1-L)^{1/2}\cosh(\beta)$$
$$- e^{i[\phi(\omega)-\phi(-\omega)]}(1-K)^{1/2}(1-T)(1-L)\ ]/D(\omega) \qquad (4.7)$$

$$B_1(\omega) = e^{i\phi(\omega)}K(1-T)^{1/2}(1-L)^{1/2}\sinh(\beta)/D(\omega) \qquad (4.8)$$

$$A_2(\omega) = [\ e^{i\phi(\omega)}K^{1/2}L^{1/2}(1-T)^{1/2}\cosh(\beta) \qquad (4.9)$$
$$- e^{i[\phi(\omega)-\phi(-\omega)]}K^{1/2}(1-K)^{1/2}L^{1/2}(1-L)^{1/2}(1-T)\ ]/D(\omega)$$

$$B_2(\omega) = e^{i\phi(\omega)}K^{1/2}L^{1/2}(1-T)^{1/2}\sinh(\beta)/D(\omega) \qquad (4.10)$$

$$A_3(\omega) = [\ e^{i\phi(\omega)}K^{1/2}T^{1/2}\cosh(\beta) \qquad (4.11)$$
$$-e^{i[\phi(\omega)-\phi(-\omega)]}K^{1/2}(1-K)^{1/2}T^{1/2}(1-T)^{1/2}(1-L)^{1/2}\ ]/D(\omega)$$

$$B_3(\omega) = e^{i\phi(\omega)}K^{1/2}T^{1/2}\sinh(\beta)/D(\omega) \qquad (4.12)$$

and the resonant denominator $D(\omega)$ is given by

$$D(\omega) = \{ 1 - [e^{i\phi(\omega)}+e^{-i\phi(-\omega)}](1-K)^{1/2}(1-T)^{1/2}(1-L)^{1/2}\cosh(\beta) + e^{i[\phi(\omega)-\phi(-\omega)]}(1-K)(1-T)(1-L) \} \quad . \tag{4.13}$$

Equations (4.7)-(4.13) constitute an explicit expression for the output field annihilation operator a_{out} in terms of the input field annihilation operators a_{in}, c_n, d_n and their Hermitian conjugates. The moments of X_1 for the output field can now be evaluated for various incoming signals.

V. The Homodyne Detector

The frequency component at frequency ω of an ideal homodyne detector's output [13-20] is a measure of $X_1(t,\theta)$ defined by (3.4). When the homodyne detector's output is fed into a spectrum analyzer the spectrum analyzer reports the power spectrum

$$S(\omega,\theta) = \langle X^2(t,\theta)\rangle \quad . \tag{5.1}$$

Note that from the definition (3.4) $X_1(t,\theta)$ is a function of ω .

Here $S(\omega,\theta)$ will be evaluated for the state $|\psi\rangle$ defined by

$$a_{in}(\omega)|\psi\rangle = 0 \qquad a_{in}(-\omega)|\psi\rangle = 0 \tag{5.2}$$

$$c_n(\omega)c_n(\omega)|\ \rangle = N|\psi\rangle \qquad c_n(-\omega)c_n(-\omega)|\psi\rangle = M|\psi\rangle \tag{5.3}$$

and

$$d_n(\omega)|\psi\rangle = \alpha|\psi\rangle \qquad d_n(-\omega)|\psi\rangle = 0 \quad . \tag{5.4}$$

From (5.2) one sees that this state is the vacuum state for the input field a_{in}, that is, no light enters the input port of the cavity. The state $|\psi\rangle$, as indicated by (5.3), is a number operator eigenstate for the fields entering the loss port c_n. The spontaneous emission light injected into the cavity by Doppler-shifted Na atoms can be phenomenologically modeled by thermally averaging over the number of quanta N and M entering c_n. The state $|\psi\rangle$, according to (5.4), is a coherent state for the mode entering d_n at frequency $\omega_p + \omega$ and a vacuum state for the mode entering d_n at $\omega_p - \omega$. The response of a homodyne detector to the four-wave mixer when a probe beam is injected into the cavity can be modeled in this way. This will facilitate comparison with experiment, since we measure the classical gain of our four-wave mixer by injecting a probe beam at frequency $\nu_L + 3\nu_{sc}$ into the cavity through the "totally" reflecting mirror [7,30] of Fig. 1.

Upon evaluating $\langle X^2(t,\theta)\rangle$ for the state $|\psi\rangle$ and then thermally and time averaging, one finds that $S(\omega,\theta)$ can be expressed in the form

$$S(\omega,\theta) = A(\omega) + \mathrm{Re}\{e^{2i\theta}B(\omega)\} \tag{5.5}$$

where

$$\begin{aligned} A(\omega) = \{ & |A_1(\omega)|^2 + |B_1(-\omega)|^2 + |A_1(-\omega)|^2 + |B_1(\omega)|^2 \\ & + (1+f_{se})[\ |A_2(\omega)|_2 + |B_2(-\omega)|^2 + |A_2(-\omega)|^2 + |B_2(\omega)|^2\ ] \\ & + (2|\alpha|^2+1)[\ |A_3(\omega)|^2 + |B_3(-\omega)|^2\ ] \\ & + |A_3(-\omega)|^2 + |B_3(\omega)|^2 \}/2 \end{aligned} \tag{5.6}$$

and

$$B(\omega) = A_1(\omega)B_1(-\omega) + A_1(-\omega)B_1(\omega) + (1+f_{se})[\, A_2(\omega)B_2(-\omega) + A_2(-\omega)B_2(\omega)\,] + (2|\alpha|^2+1)A_3(\omega)B_3(-\omega) + A_3(-\omega)B_3(\omega) \quad (5.7)$$

where f_{se} is a parameter proportional to the amount of spontaneous emission light injected into the cavity.

The effects of phase jitter on the power spectrum $S(\omega)$ can be treated as follows. Let θ_o denote the phase which minimizes $S(\omega,\theta)$, i.e.

$$e^{2i\theta_0}B(\omega) = -|B(\omega)| \quad . \quad (5.8)$$

Then writing $\theta = \theta_0 + \delta$, eq. (5.7) can be put into the form

$$S(\omega,\theta) = A(\omega) - \cos(2\delta)|B(\omega)| \quad . \quad (5.9)$$

If δ jitters rapidly about zero on the time scale during which the spectrum analyzer constructs the power spectrum, then the power spectrum reported for the quadrature with the least amount of power (the in phase quadrature) will have the form

$$S_-(\omega) = A(\omega) - f|B(\omega)| \quad (5.10)$$

where f is the average of $\cos(2\delta)$ over the probability distribution for δ. Similarly the power spectrum for the quadrature with the maximum power (the out of phase quadrature) has the form

$$S_+(\omega) = A(\omega) + f|B(\omega)| \quad . \quad (5.11)$$

If homodyne detector losses are included the power spectra of the in-phase and out of phase quadratures are given by [19,20]

$$S_{\pm}(\omega) = \eta[A(\omega) \pm f|B(\omega)|] + 1 - \eta \quad (5.12)$$

where η is the overall efficiency of the homodyne detector, i.e. if η_m denotes the mode-matching efficiency and η_p the photodetector efficiency then $\eta = \eta_m\eta_p$. Equation (5.12) together with (5.6) and (5.7) constitute the basic model we have used to determine the effects of losses and phase jitter on our homodyne detector.

VI. Comparison With Experiment

In sections IV and V a phenomenological model of a four-wave mixer and a homodyne detector was presented which we have used to understand various features of our experimental observations. A number of the parameters entering the model can be determined directly. The transmission coefficients for the output mirror and the "totally" reflecting mirror have been measured directly. Hence, the transmission coefficient for the mirrors M1 and M2 are respectively K = 0.02 and T = 0.005. The mode-matching efficiency and the photodetector efficiencies have also been measured and have the values 0.75 and 0.7 respectively. This implies an overall homodyne detector efficiency η of 0.5. The Fabry-Perot cavity has a length of 1 m. Consequently the length d of the ring cavity should be chosen to be 2 meters. The phase shift $\phi(\omega)$ as a function of frequency is

$$\phi(\omega) = \phi_o + \tau\omega \quad (6.1)$$

where $\tau = d/c$ and c is the speed of light. For maximum gain at line center $\phi_o = 2\pi n$ where n is an integer. The amount of spontaneous emission noise f_{se} is also measured directly [31] and amounts to 1/12 photon per unit bandwidth compared to the 1/2 photon per unit bandwidth for vacuum fluctuation noise.

This leaves three parameters β, L, and f to be extracted from the homodyne detector power spectra. It is worth noting that $S_+(\omega)$ is relatively insensitive to the value of the phase jitter factor f (f = 1 when there is no phase jitter), whereas, $S_-(\omega)$ being the difference of two numbers of comparable size is quite sensitive to f. As can be seen from (5.6), (5.7), and (5.12) the value of $S_+(\omega)$ when $|\alpha| >> 1$ is primarily determined by $A_3(\omega)$ and $B_3(-\omega)$ and consequently has a different functional dependence on β and L than when $|\alpha| = 0$. Hence, β and L can be determined by measuring $S_+(\omega)$ at line center when no light is injected into the cavity through mirror M2 ($|\alpha| = 0$) and when probe light is injected into the cavity ($|\alpha| >> 1$).

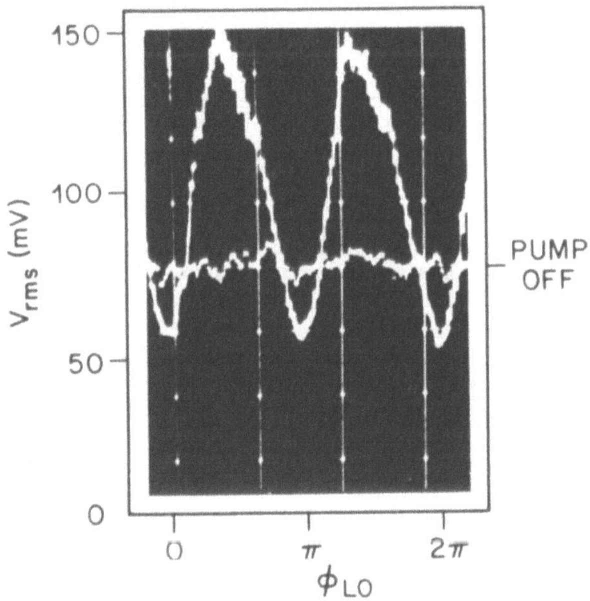

Fig. 4. A measurement of the classical four-wave-mixing gain at line center as a function of the phase of the local oscillator beam ϕ_{LO}. With the pump off, the cavity output (due to an injected laser beam at frequency $\nu_L+3\nu_{sc}$ where $3\nu_{sc}$ = 420.6 MHz) is constant as a function of ϕ_{LO}. With the laser pump on, the light in the squeezing cavity is alternately amplified and deamplified with a period of π in ϕ_{LO}. The averaging time for these data is 1ms.

Data for the case when light is injected into the cavity at line center for the mode $\nu_L + 3\nu_{sc}$ is shown in Fig. 4. Here $|\alpha| >> 1$. The steady trace is obtained when the pump light to the four-wave-mixing medium is blocked, i.e. $\beta = 0$. The oscillatory curve is obtained with the pump light on (four-wave mixing present). The ratio of the peak voltage to the voltage when the pump is off is nearly 2. Hence, acceptable values of β and L are those which give at line center

$$[S_+(\omega)/S_+(\omega,\beta=0)]^{1/2} = 2 \quad \text{when } |\alpha| >> 1 \quad . \tag{6.2}$$

Fig. 5 shows the homodyne detector noise spectra when the probe beam has been turned off, $|\alpha| = 0$. As frequency is swept, the homodyne detector phase ϕ_{LO} is also varied. When the signal port of the homodyne detector is blocked, the steady curve is obtained which is due primarily to vacuum fluctuations entering the signal port of the homodyne detector plus a 10% in power contribution from the amplifier noise. The oscillatory curve is obtained when the cavity light is directed to the four-wave mixer and exhibits squeezing at frequency shifts from the cavity center frequency of 0.3 MHz. The ratio of the pump intensity I

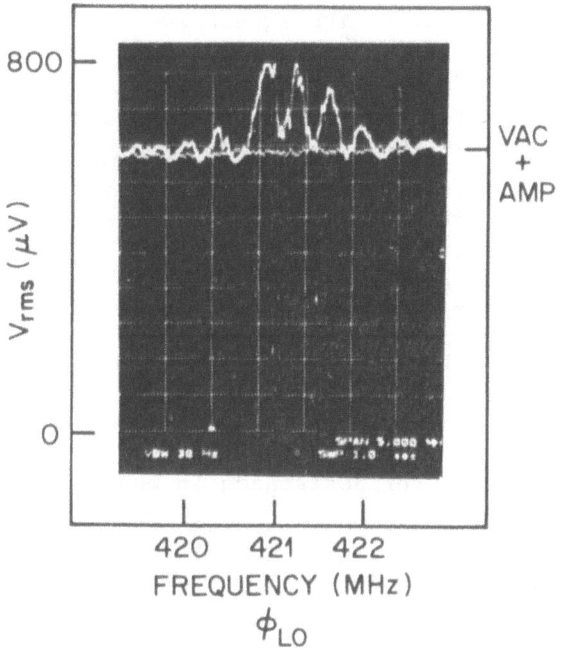

Fig. 5 Homodyne detector noise spectra due to a pair of cavity resonances spaced at $\pm 3\nu_{sc}$ from the pump frequency. Pump light is blocked for the nearly flat trace (dim trace labeled VAC+AMP) and the pump light is on for the brighter modulated trace. The amplifier noise level (AMP) is obtained by blocking the LO beam. Both spectrum analyzer frequency and LO phase are linearly swept. The spectrum analyzer rf bandwidth was 300 kHz and the video bandwidth was 30 Hz.

to the saturation intensity I_s was $I/I_s = 0.01$ when this data was taken. A more precise measure of the noise squeezing ($\sim$7% in power) at a fixed frequency shift of 0.3 MHz is shown in Fig. 12 of reference 30.

As indicated by Fig. 5 the ratio of the peak voltage excursion at the center of the four-wave mixers passband to the vacuum voltage level is typically 3/2. Hence, in the phenomenological model β and L must have values such that at line center

$$[S_+(\omega)/S_+(\omega,\beta=0)]^{1/2} = 3/2 \quad \text{when } |\alpha| = 0. \tag{6.3}$$

Equations (6.2) and (6.3) determine β and L. Good fits with the data were obtained when $\beta = 0.0135$ and $L = 0.03$. This value for β is consistent with the values of the reflectivity of the four-wave-mixing medium that have been measured directly experimentally. The loss L is greater than expected for the atomic beam. Additional sources of loss are now being identified.

Having determined β and L, f can now be determined from the line center value of $S_-(\omega)$ when $|\alpha| = 0$. At mode center no squeezing is observed but the noise does drop to nearly the vacuum noise level for suitable values of the local oscillator phase. Hence, f is determined by the requirement that $S_-(\omega)$ be close to 1 at line center for $|\alpha| = 0$. Fig. 6 shows a plot of $S_+(\omega)$ and $S_-(\omega)$ for a phase jitter factor $f = 0.85$, the other model parameters being those already given in the text. The model predicts 6% squeezing (in power) at a frequency 470 kHz away from the cavity resonance center frequency. This compares favorably with the 7% squeezing we have observed at 300 kHz from the center frequency [7,30]. An f parameter of 0.85 implies a phase jitter of the order ± 27 degrees. This is consistent with other measurements of phase jitter that we have made [30].

To conclude, the model presented here provides a self-consistent picture by which the degree of squeezing can be predicted from other measurements and these predictions compare favorably with observation.

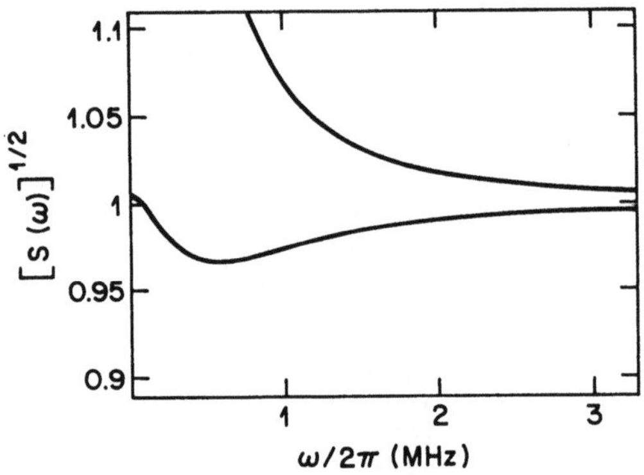

Fig. 6 The square root of $S_+(\omega)$ (upper trace) and $S_-(\omega)$ (lower trace) as a function of $\omega/2\pi$ (in MHz) for the four-wave mixer and detector model with the parameter values described in the text. Squeezing in excess of 3% (6% in power) is indicated at a frequency 470 kHz away from the center frequency ($\omega = 0$). This compares favorably with the 7% squeezing observed at 300 kHz from the center frequency. Note that there is no squeezing at the center frequency due to phase jitter.

We thank J. R. Klauder and S. L. McCall for assistance in modeling the experiment.

1 H. Takahasi: Adv. Commun. Syst. 1, 227 (1965).
2 E.Y.C. Lu: Nuovo Cimento Lett. 2, 1241 (1971); 3, 585 (1972).
3 D. Stoler: Phys. Rev. Lett. 23, 1397 (1974).
4 H.P. Yuen: Phys. Rev. A 13, 2226 (1976).
5 H.P. Yuen and J.H. Shapiro: Opt. Lett. 4, 334 (1979).
6 H.P. Yuen and J.H. Shapiro: IEEE Trans. Inform. Theory IT-26, 78 (1980).
7 R.E.Slusher, L. Hollberg, B. Yurke, J.C. Mertz and J.F. Valley: Phys. Rev. Lett., 55, 2409 (1985).
8 M.D. Levenson, R.M. Shelby and S.M. Perlmutter: Opt. Lett. 10, 514 (1985).
9 M.D. Maeda, P. Kumar and J.H. Shapiro: in Proceedings of the Topical Meeting on Instabilities and Dynamics of Lasers and Nonlinear Optical Systems, Rochester, New York, June 1985.
10 B. Yurke: Phys. Rev. A 29, 408 (1984).
11 S. Reynaud and A. Heidmann: Opt. Comm. 50, 271 (1984).
12 M.J. Collett and C.W. Gardiner: Phys. Rev. A, 30, 1386 (1984).
13 H.P. Yuen and V.W.S. Chan: Opt. Lett. 8, 177 (1983).
14 L. Mandel: Phys. Rev. Lett. 49, 136 (1982).
15 B.L. Schumaker: Opt. Lett. 9, 189 (1982).
16 C.M. Caves: Phys. Rev. D 26, 1817 (1982).
17 C.M. Caves and B.L. Schumaker: Phys. Rev. A, 31, 3068 (1985).
18 B.L. Schumaker and C.M. Caves: Phys. Rev. A, 31, 3093 (1985).
19 B. Yurke: Phys. Rev. A 32, 300 (1985).
20 B. Yurke: Phys. Rev. A 32, 311 (1985).
21 M.J. Collett and D.F. Walls: Phys. Rev. A, 32, 2887 (1985).
22 G. Milburn and D.F. Walls: Opt. Comm. 39, 401 (1981).
23 K. Wodkiewicz and M.S. Zubairy: Phys. Rev. A, 27, 2003 (1983).
24 G.J. Milburn, D.F. Walls and M.D. Levenson: J. Opt. Soc. Am. B 1, 390 (1984).
25 G. Scharf and D.F. Walls: Opt. Comm. 50, 245 (1984).
26 P. Kumar and J.H.Shapiro: Phys. Rev. A 30, 1568 (1984).
27 M.D. Reid and D.F. Walls: Opt. Comm. 50, 406 (1984).
28 M.D. Reid and D.F. Walls: Phys. Rev. A 31, 1622 (1985).
29 J.R. Klauder, S.L. McCall and B. Yurke: Submitted to Phys. Rev. A
30 R.E. Slusher and B. Yurke: Frontiers in Quantum Optics, ed. S. Sarkar and E.R. Pike, Adam Hilger Ltd, 1986.
31 R.E. Slusher, L.W. Hollberg, B. Yurke, J.C. Mertz and J.F. Valley: Phys. Rev. A 31, 3512 (1985).

32 R.E. Slusher, L.W. Hollberg, B. Yurke, and J.C. Mertz: Proceedings of Topical Meeting on Instabilities and Dynamics of Lasers and Nonlinear Optical Systems, Rochester, N.Y. (1985).
33 R.E. Slusher, B. Yurke, L.W. Hollberg, J.C. Mertz: Conference Proc. SEICOLS'85, Maui, Hawaii (1985).
34 R.W.P. Drever, J.L. Hall, F.W. Kowalski, J. Hough, G.M. Ford, A.J. Munley and H. Ward: Applied Phys. B 31, 97 (1983).
35 J.L. Hall and T.W. Hänsch: Opt. Lett. 9, 502 (1984).
36 B. Yurke: Phys. Rev. A 29, 1419 (1984).
37 C.M. Caves: Phys. Rev. D 23, 1693 (1981).
38 R.S. Bondurant, J.H. Shapiro: Phys. Rev. D 30, 2548 (1984).
39 B. Yurke, S.L. McCall and J.R. Klauder: submitted to Phys. Rev. A.

Intracavity Frequency Doubling for the Generation of Squeezed States of Light

H.J. Kimble[1] *and J.L. Hall*[2]

[1]Department of Physics, University of Texas, Austin, TX 78712, USA
[2]Joint Institute for Laboratory Astrophysics, University of Colorado, Boulder, CO 80309, USA

I. Introduction

While squeezed states of light represent a manifestation of the nonclassical character of the electromagnetic field and hence are of great intrinsic interest in quantum optics, there are as well a number of applications in measurement science associated with the possibility of sensitivity beyond the standard quantum limit. Given the landmark initial observation of this phenomenon by Slusher et al. [1], a number of experimental groups are concentrating on making their own "first" observations and on subsequently exploring limitations on the degree of achievable and detectable squeezing [2,3]. Most of the experimental research thus far has concentrated on squeezing produced by four-wave mixing, either in atomic vapors or in condensed media. The tack that we have taken is to investigate a completely different technique with a new set of potential advantages (and disadvantages). Our experiments attempt to produce squeezed states of light by second harmonic generation within an optical cavity resonant at both fundamental and harmonic frequencies. Of course the Hamiltonian describing the two-mode coupled system in the limit of a lossless medium is related to a broad class analyzed by Yuen[4], and hence it is not surprising that squeezing is predicted in this system. However, an issue that is of prime importance in our investigation is the extent to which these parametric models of quantized nonlinear processes are realistic. This issue is naturally entangled with a number of other technological and scientific questions that we are only beginning to unravel. We begin in Section II with an overview of the relevant theoretical predictions before turning in Section III to the actual experiment. Other contributions to this volume as well as review articles [5] should be consulted for a more comprehensive view of the field.

II. Theoretical Overview

The configuration that we wish to consider is shown in Fig. 1. A frequency doubling crystal contained inside a cavity resonant at both fundamental and second harmonic frequencies is driven by a coherent field of amplitude ε_1 and frequency ω_L. The classical equations of motion for this system have been obtained in Ref. [6] and references therein and are given by

$$\dot{\alpha}_1 = -i\Delta_1\alpha_1 - \Gamma_1\alpha_1 + K\alpha_1^*\alpha_2 + \varepsilon_1 \ , \qquad (1)$$

$$\dot{\alpha}_2 = -i\Delta_2\alpha_2 - \Gamma_2\alpha_2 - \frac{1}{2}K\alpha_1^2 \ . \qquad (2)$$

The intracavity amplitudes α_i at the fundamental (i=1) and harmonic (i=2) frequencies are normalized such that $|\alpha_i|^2$ expresses the mean photon number in the cavity mode i. The detunings Δ_i from the cavity resonances at ω_i are given by $\Delta_1 = \omega_1 - \omega_L$ and $\Delta_2 = \omega_2 - 2\omega_L$, while the mode amplitudes are damped at rates Γ_1 and Γ_2. Of course the terms responsible for the nonlinear physics are those involving the coupling coefficient K, which is proportional to the second order susceptibility of the intracavity crystal.

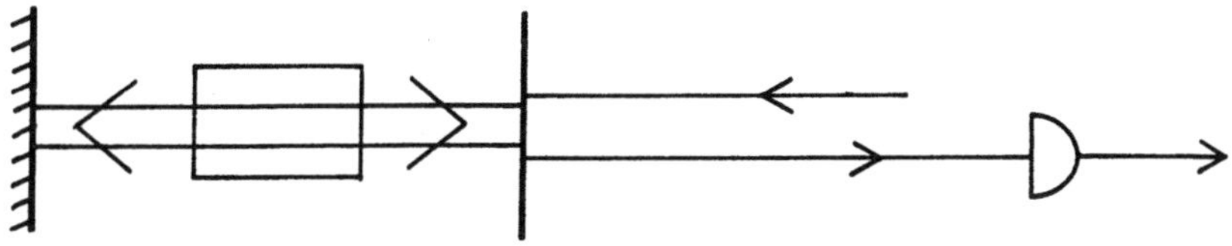

Figure 1 - Nonlinear crystal inside an optical cavity resonant at fundamental and harmonic frequencies and pumped by an external coherent field. The reflected field is directed onto a photodiode for spectral analysis.

The rich variety of nonequilibrium phenomena associated with Eqns. (1-2) has been explored in a series of theoretical papers [6-8]. Included are predictions of dispersive bistability, of hysteresis in limit cycle oscillation, and of period-doubling bifurcations to chaos. We will focus our attention on excitation below the threshold for limit-cycle oscillation, which occurs at a driving field

$$K|\varepsilon_1^c| = (2\Gamma_1 + \Gamma_2)[2\Gamma_2(\Gamma_1 + \Gamma_2)]^{1/2} \tag{3}$$

and frequency

$$\Omega c = [\Gamma_2(2\Gamma_1 + \Gamma_2)]^{1/2} \tag{4}$$

for the case $\Delta_1=0=\Delta_2$[6]. This instability is associated with a Hopf bifurcation to a self-pulsing state. The equations of motion obtained from Eqns. (1-2) for the moduli and phases of the complex fields α_i reveal that this bifurcation results from an instability in the phase variables for $\Delta_1=0=\Delta_2$. As the threshold for limit cycle oscillation is approached, the damping coefficient for phase deviations approaches zero. Amplitude deviations are on the other hand still strongly damped. One might therefore surmise that squeezing in this system could arise from the combination of the limiting action of the sub/second harmonic generation on the amplitude of the field and of the associated dynamic detuning of the cavity by the nonlinear phase deviations.

This qualitative picture of the nature of the quantum fluctuations in this system has been made quantitative by Milburn and Walls [9] and by Lugiato, Strini, and De Martini [10], who calculated the variances in the quadrature components of the intracavity fields. For the fundamental, these are given by

$$\hat{x}_+ = \frac{1}{2}[\hat{\alpha}_1 e^{-i\theta} + \hat{\alpha}_1^+ e^{i\theta}],$$

$$\hat{x}_- = \frac{1}{2i}[\hat{\alpha}_1 e^{-i\theta} - \hat{\alpha}_1^+ e^{i\theta}] , \tag{5}$$

with the caret (^) denoting a Hilbert space operator. The treatment by Lugiato et al. [10] for the case $\Delta_1=0=\Delta_2$ predicted that appreciable squeezing could be obtained in the amplitude quadrature ($\theta=0$, $\hat{x}_+$) at modest levels of the incident driving field. It is this work that largely stimulated the present investigation.

The analyses of Ref. [9-10] dealt with the intracavity field. Yurke[11] made the extremely important contribution of noting the distinction between the statistics of the field inside and outside the cavity. An input-output theory of damped quantum systems has been developed by Collett and Gardiner[12] and applied to the problem of intracavity harmonic generation by Collett and Walls[13]. Central to the analysis is the concept of the spectrum of the squeezing representing a Fourier decomposition of the field fluctuations. We begin by introducing the amplitude correlation functions of the intracavity fields defined in (5),

$$A_{\pm}(\tau) \equiv \langle T{:}\hat{x}_{\pm}(t),\hat{x}_{\pm}(t+\tau){:}\rangle \, , \tag{6}$$

where the colons denote normal ordering, T stands for time ordering, the covariance $\langle a,b\rangle \equiv \langle ab\rangle - \langle a\rangle\langle b\rangle$, and the time t is sufficiently long to ensure a steady state. The spectrum of squeezing $S_{\pm}(\Omega)$ of the output field from an ideal single-ended cavity of damping rate Γ_1 is then given in terms of the spectral density $A_{\pm}(\Omega)$,

$$A_{\pm}(\Omega) = \frac{1}{2\pi}\int_{-\infty}^{\infty} d\tau e^{-i\Omega\tau} A_{\pm}(\tau) \, , \tag{7}$$

$$S_{\pm}(\Omega) = 2\Gamma_1 \, A_{\pm}(\Omega) \, . \tag{8}$$

The critical steps in this derivation are found in the work of Ref. [12-13] and relate the moments of the output field to those of the intracavity field. Note that excitation by a field in a coherent state is assumed. We shall demonstrate shortly the relevance of the quantity $S_{\pm}(\Omega)$ to the actual detection process. For the moment simply note that $S_{\pm}(\Omega) = 0$ for a coherent state, while $S_{\pm}(\Omega) = -0.25$ for perfect squeezing.

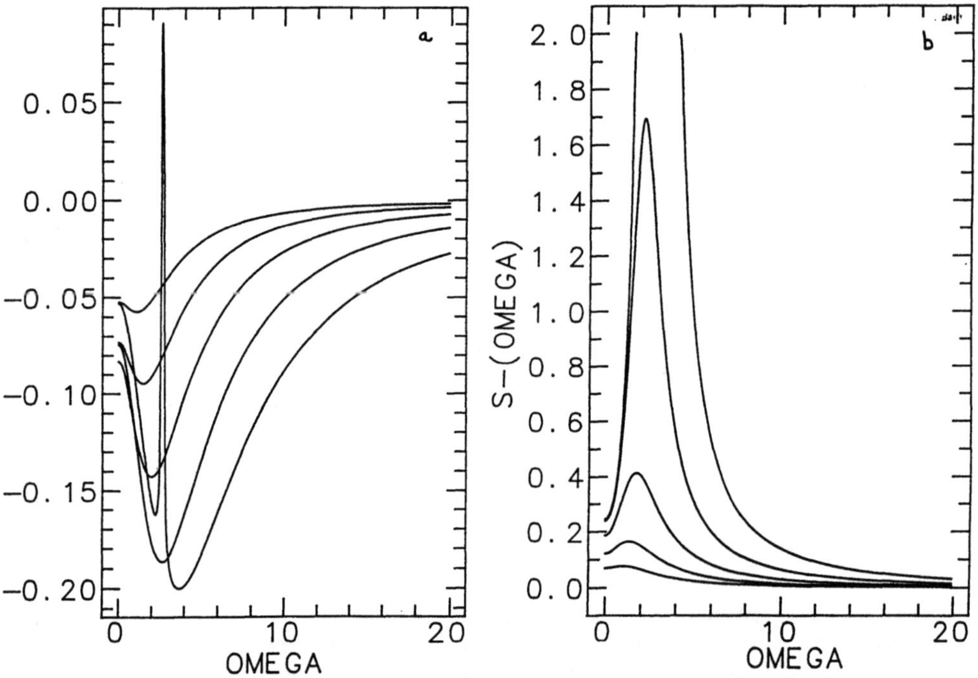

Figure 2 - Spectrum of squeezing $S_{\pm}(\Omega)$ vs. Ω with Ω normalized by Γ_2. The curves are drawn for successive doubling of the intracavity intensity and terminate at an intensity 2.3% below the point of bifurcation to oscillation. f=0.33, θ=0. The sharp spike in $S_{+}(\Omega)$ is associated with nonzero detunings δ_1=0.02, $\delta_2=\delta_1/f$; for zero detunings it is absent. All other curves are for $\delta_1=0=\delta_2$.

Figure [2] displays the quantities $S_{\pm}(\Omega)$ versus Ω for a set of parameters relevant to our experimental investigation. The curves are drawn from a generalization of Eqn. (4.4) of Ref. [13] to include nonzero detunings and are symmetric for $\Omega \rightarrow -\Omega$. The abscissa gives the frequency in units of Γ_2, which in our work is approximately $2\pi \times 10^7$/sec. Since we are interested in squeezing in the fundamental field, the ratio $f \equiv \Gamma_2/\Gamma_1$ should be small to minimize losses at the second harmonic. For example, as f increases from 0.33 to 1.0 the minimum value of S_+ increases from -0.20 to -0.15. The position of the minimum in the spectrum approaches the frequency Ω_c given by Eqn. (4) as ε_1 tends to ε_1^c. Note that whether or not one is able to achieve good squeezing in the quadrature $S_+(\Omega)$, the fluctuations in $S_-(\Omega)$ diverge as the point of instability is approached.

The amplitude fluctuations of the output field from the cavity can be analyzed in one of two ways. Slusher et al. [1] have successfully implemented a balanced homodyne detector [15], which yields directly the quantity $S_{\pm}(\Omega)$. In our case we would require a local oscillator of power much greater than the total reflected power from the cavity. Unless an auxiliary cavity is used to reduce this reflected power, this scheme implies powers too large for the photodiodes. Alternatively one might imagine employing the large coherent reflection from the cavity as the local oscillator. This is essentially the course that we have taken as indicated in Fig. [1]. In this case the reflected field E_r is of the form

$$E_r = rE_i + tE \;, \tag{9}$$

where E_i and E are the incident and intracavity fields and r and t are the complex reflection and transmission coefficients of the input coupling mirror, respectively. Unfortunately, it is not the case that $|rE_i| >> |tE|$ in our work, so that the usual analysis involving a strong local oscillator rE_i and a weak signal field tE is not applicable. Instead, the computation of the photocurrent fluctuations must include all orders in the local oscillator and cavity leakage fields.

We have carried out such an analysis for the photocurrent fluctuations for an arrangement as in Fig. [1]. One finds that the spectral density $\Phi(\Omega)$ of the photocurrent fluctuations ($\Delta i \equiv i - \langle i \rangle$) is related to the field fluctuations in the following manner.

$$\begin{aligned}\Phi(\Omega) &\equiv \frac{1}{2\pi} \int \langle \Delta i(t) \Delta i(t+\tau) \rangle \, e^{-i\Omega\tau} d\tau \\ &= e\langle i \rangle \left[1 + \alpha \frac{C(\Omega)}{\langle \hat{E}_r^+ \hat{E}_r \rangle}\right] \\ &\equiv e\langle i \rangle \left[1 + G(\Omega)\right] . \end{aligned} \tag{10}$$

Here e is the charge carried by each photopulse, $\langle i \rangle$ the average current, and α the detection efficiency for propagation from the output of the cavity to the photodetector, including the detector's quantum efficiency ($0<\alpha<1$). The function $C(\Omega)$ is given following Glauber's work [16] on photoelectric detection by

$$\begin{aligned} C(\tau) &= \langle T\!:\! \Delta\hat{I}_r(t+\tau) \Delta\hat{I}_r(t) \!:\rangle , \\ C(\Omega) &= \frac{1}{2\pi} \int d\tau e^{-i\Omega\tau} C(\tau) \;, \end{aligned} \tag{11}$$

with T denoting time ordering, $\Delta\hat{I} = \hat{I} - \langle I \rangle$, $\hat{I} = \hat{E}^+ \hat{E}$ and the subscript "r" referring to the total field reflected from the cavity, Eqn. (9). For an ideal cavity tuned to resonance, the phase relation between E_i and E simplifies leading to

$$\Phi(\Omega) = e\langle i\rangle \; [1 + \frac{4\alpha}{1+g} S_+(\Omega)] \; , \tag{12}$$

where S_+ is evaluated for $\theta=0$. g expresses the ratio of loss rate γ_1' by all avenues other than through the input coupler to the loss rate through the input coupler γ_1, $g = \gamma_1'/\gamma_1 (\gamma_1 + \gamma_1' = \Gamma_1$; g=0 for a single-ended ideal cavity). Note that for perfect squeezing in the amplitude quadrature, $S_+(\Omega) = -0.25$, and for unit detection efficiency, $\alpha=1$, so that

$$\Phi(\Omega) = 0 \; , \tag{13}$$

as expected.

The input-output analysis of Collett and Gardiner [12] together with a knowledge of the function $C(\Omega)$ allows us to make a quantitative assessment of the photocurrent fluctuations that might be expected in an intracavity frequency doubling experiment. While we will not quote its explicit form, the function $C(\Omega)$ is given in terms of the usual components of the spectral matrix in the linearized approximation. As will be discussed in the next section, we require both $S_\pm(\Omega)$ and $C(\Omega)$ for nonzero detunings $\Delta_1 \neq 0$, $\Delta_2 \neq 0$. We have therefore extended the existing calculations to include detunings other than zero [17]. This work is based upon the formalism of Ref. [12-14].

Figure [3] illustrates the dependence of the functions $S_+(\Omega)$ and $\phi(\Omega) \equiv \Phi(\Omega)/e\langle i\rangle$ on cavity detuning. The normalized detunings $\delta_i \equiv \Delta_i/\Gamma_i$ have been introduced and δ_2 eliminated in favor of an offset $\delta_o = \Delta_o/\Gamma_2$ with

$$\Delta_2 = 2\Delta_1 + \Delta_o, \; \delta_2 = 2\delta_1/f + \delta_o, \tag{14}$$

which specifies the detuning of the harmonic given δ_1 and δ_o. For example, if the cavity length were scanned for a fixed frequency of the driving field, δ_1 and δ_2 would both be swept with a constant offset δ_o during the scan. In Fig. [3] $\phi(\Omega)$ has been plotted both for an ideal cavity and detection scheme ($g=0$, $\alpha=1.0$) and for a set of parameters that approximates our experimental configuration ($g=0.23$, $\alpha=0.7$). Note that if $K=0$ (no nonlinear coupling), the ratio of reflected intensity at $\delta_1=0$ to that at $\delta_1 \gg 1$ is given by

$$D_o = \left(\frac{1-g}{1+g} \right)^2 . \tag{15}$$

Since a variation of total power on the diode during a scan of δ_1 would lead to a corresponding change in the baseline level of $\Phi(\Omega)$ even for a coherent state input, we in practice stabilize the total reflected power to better than 1% by servo control of the incident power during the scan. The calculation leading to Fig. [3] attempts to mimic this experimental procedure by likewise adjusting the incident and hence intracavity intensity during a scan in δ_1 to hold the reflected intensity fixed. A deficiency in the current calculation is that the adjustment is made relative to the properties of the empty cavity and not to those of the full nonlinear problem. The consequence of this procedure is that, at each point except $\delta_1=0$ in Fig. [3], the intracavity field is underestimated and hence the nonclassical deviations tend to be somewhat larger than those shown in the figure. However, the shapes of $S_+(\Omega)$ and $\phi(\Omega)$ for δ_1 small are not significantly affected.

Figure [3a] is drawn for a weak incident field corresponding to operation at only 5% of the incident threshold power. In this case the qualitative features of the plot are relatively insensitive to the scan offset δ_o over a range of ±1.0. However, as the degree of squeezing increases, the requirements on the definition of the scan offset become more stringent, as illustrated in

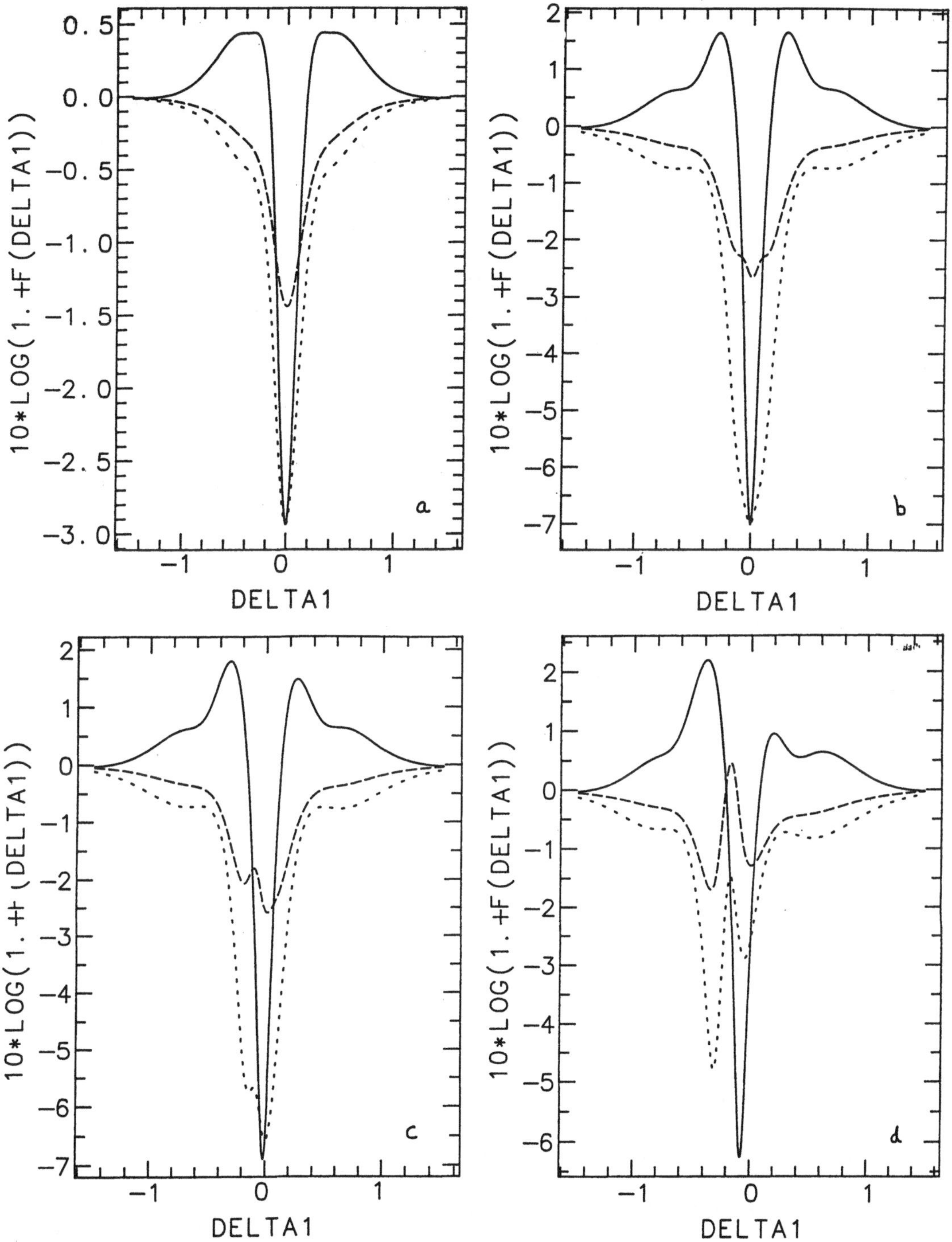

Figure 3 - 10 log $(1+F(\delta_1))$ vs. detuning δ_1, with $F(\delta_1)=S_+/0.25$ (solid curve); $F(\delta_1)=G$ (dotted curve), g=0, α=1.0; $F(\delta_1)=G$ (dashed curve), g=0.22, α=0.7. In all cases f=0.33, $\delta_2=\delta_1/f + \delta_0$. (a) Incident excitation =0.05 $|\varepsilon_1^c|^2$, $\delta_0=0$, $\Omega/\Gamma_2=2$. (b)-(d) 0.55$|\varepsilon_1^c|^2$, $\Omega/\Gamma_2=3$, (b) $\delta_0=0$, (c) $\delta_0=0.25$, (d) $\delta_0=1.0$.

Fig. [3b-d]. The incident excitation level corresponds to 0.55 that of the critical value at $\delta_1=0=\delta_2$. Offsets δ_o as small as ±0.25 now significantly affect the spectral features. As threshold is approached, not only is the squeezing improving in one quadrature, but as Fig. [2b] illustrates the fluctuations in the complementary quadrature are becoming very large. The phase θ (Eqn. (5)) necessary for optimal squeezing must be precisely defined to avoid contamination of one quadrature component by the other. Only over a very narrow range in δ_1 is the phase between the cavity and incident fields suitable for observing squeezing in a given quadrature of the reflected field. (In Fig. [2] the sharp spike is absent for $\delta_1=0=\delta_2$, but present for $\delta_1=0.02$, $\delta_2=0.06$.)

Before turning to a description of the experiment, we return to Eqn. (1-2) to evaluate the coupling coefficient K somewhat more carefully. It is of course this coefficient together with the cavity losses that sets the scale for the power level at threshold and hence for the feasibility of the experiment. In general K is expressed in terms of an overlap integral of the cavity mode functions $\Psi_i(\underline{r})(i=1,2)$ as

$$K = \frac{2d}{n^3}\left(\frac{\hbar\omega_1^3}{\varepsilon_o^3}\right)^{1/2} \int_\Gamma \Psi_2^*(\underline{r})[\Psi_1(\underline{r})]^2 d^3x, \qquad (16)$$

with Γ as the interaction volume and d the nonlinear coefficient [18]. By substituting the TEM_{oo} mode functions suitable for a standing-wave interferometer filled with medium of index n, we find by neglecting double refraction (90° phase matching) and linear absorption that

$$|K|^2 = \frac{\pi c^4 \hbar}{L^3 \lambda_2 n^3} E . \qquad (17)$$

Here L = cavity length and λ_i is the free-space wavelength at frequency ω_i. The nonlinear efficiency E is given by

$$E = \frac{16\pi^2 \ell d^2}{\varepsilon_o^3 c n^2 \lambda_1^3} g(\sigma, \xi) , \qquad (18)$$

$$g(\sigma,\xi) = \frac{\pi^2}{\xi} |\mathrm{Im}[H(\sigma, \xi)e^{-i(\phi_2 - 2\phi_1)}]|^2 , \qquad (19)$$

with ℓ = crystal length, $\xi = \ell/2Z_o$, Z_o = cavity confocal parameter, $\sigma = Z_o(2k_1 - k_2)$, $k_i = 2\pi n/\lambda_i$. The phases ϕ_i are those of the cavity mode functions ψ_i within the crystal; the factor $(\phi_2 - 2\phi_1)$ expresses the relationship of the standing-wave patterns of the two modes. $H(\sigma,\xi)$ is defined as in Eqn. (2.16) of Ref. [19],

$$H(\sigma,\xi) \equiv \frac{1}{2\pi} \int_{-\xi}^{\xi} dx \frac{e^{i\sigma x}}{1+ix} . \qquad (20)$$

In the plane-wave limit, $\xi \to 0$ and

$$g(\sigma,\xi) \to \xi \left(\frac{\sin\sigma\xi}{\sigma\xi}\right)^2 \sin^2(\phi_2 - 2\phi_1) . \qquad (21)$$

The first factor expresses the usual condition for phase matching in a single pass. The second results in a modulation of the coupling efficiency with relative phase of the standing waves within the crystal and arises from the 90° phase shift between the polarization wave and field at ω_2[20]. In the general case we require for optimal coupling

$$\phi_2 - 2\phi_1 = \pm\frac{\pi}{2} + \beta , \qquad (22)$$

with β the phase of the complex function H. When this condition is satisfied, the function $g(\sigma,\xi)$ reduces to $h(\sigma,\xi)$, with $h(\sigma,\xi)$ as defined and analyzed in Ref. [19], Eqn. (2.22) and Fig. (2).

Combining (3), (17) and (22) together with Eqn. (A7) of Ref. [6], we arrive at the critical incident power for self-oscillation in an optimized standing-wave cavity,

$$P_c = \frac{1}{4E}\,\frac{(1-\tau_1\sqrt{R_1R_1'})^3(1-\tau_2\sqrt{R_2R_2'})}{T_1}\,(2+f)^2(1+f), \qquad (23)$$

with R_i the reflectivity of the input coupler at frequency ω_i, R'_i the reflectivity of the rear reflector, and τ_i the single-pass transmission of the intracavity medium at frequency ω_i. T_1 is the transmissivity of the input coupler at ω_1.

III. Preliminary Experimental Results

Figure [4] is a simplified diagram of the experiment. The intracavity medium is a potassium dihydrogen phosphate crystal (KDP) of 30 mm length contained between two mirrors M and M′. These mirrors are made from plano-convex substrates (R=20mm, R =30mm) with the convex surface highly reflecting at 515 and 257 nm. Typical transmission coefficients are $T_1=7\times10^{-2}$, $T_1'=7\times10^{-3}$, $T_2=2.9\times10^{-3}$, and $T_2'=1.2\times10^{-3}$. Index matching between the KDP crystal and the plano surfaces of the mirrors is provided by filling the intracavity space with hexane ($n\sim1.38$). The measured single-pass absorption loss of the crystal at 515 nm is quite small, $\tau_1<0.005$, while $\tau_2\sim2\times10^{-2}$. The measured nonlinear conversion efficiency for extracavity single-pass generation is approximately 7×10^{-4}/Watt. For the assembled system operating at -12°C, the observed finesse F_1 is in the range 50-70 depending upon the particular mirror set employed, while $0.25<D_0<.45$. The finesse F_2 at ω_2 can be inferred from the dependence of second harmonic power on differential detuning as the two cavity transmission profiles are tuned relative to one another and is greater than 100.

These values of losses and conversion efficiency allow us to estimate the threshold power for instability and with it to set the scale for the degree of squeezing expected at a given incident power. From Eqn. (23) we find that P_c should lie in the range 100-200 mW. In fact we have observed limit cycle oscillation at an incident power level of approximately 100 mW. The predicted frequency of oscillation for $\Delta_1=0=\Delta_2$ is found from (4) to be roughly 25 MHz. The observed frequencies of oscillation for a range of detunings fall from 5 to

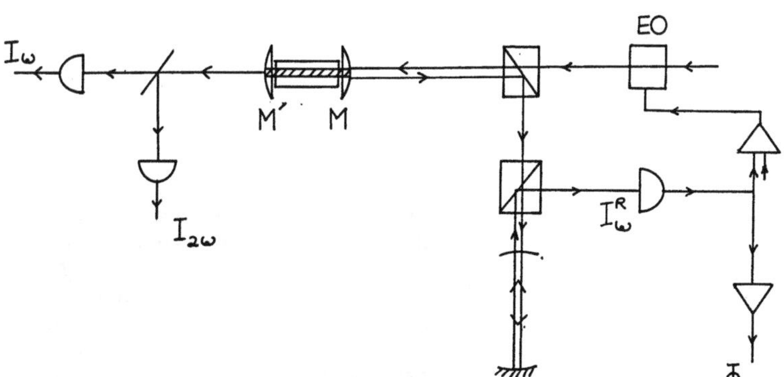

Figure 4 - Block diagram of the experiment. Not shown is the servo arrangement for frequency stabilization of the laser.

30 MHz. There is thus reasonable qualitative agreement between the predictions of the classical theory and our observations, leading us to believe that we are operating in a regime in which significant squeezing could be observed for only modest levels of excitation ($P \sim 50$ mW). Note that the analysis of Ref. [10,13] leads to an initial growth of squeezing

$$S_{+}(o) = -\frac{P}{P_c}[(2+f)^2(1+f)], \quad \frac{P}{P_c} << 1. \tag{24}$$

For $P_c = 150$ mW and $f = 0.5$ we thus expect an initial growth in squeezing at a rate of 6×10^{-2}/mW.

The excitation source in the experiments is the light from a frequency-stabilized argon ion laser whose amplitude fluctuations reach the shot noise level for frequencies beyond about 100 kHz. As Fig. [4] indicates, a fast servo holds the average value of the photocurrent at the detection diode constant to better than 1% by control of the incident power. An auxiliary cavity is often employed to phase shift the coherent carrier of the reflected cavity field relative to the broad spectrum of the amplitude fluctuations. The overall efficiency α from the input/output coupler M to the photodiode is approximately 0.7, including detector quantum efficiency.

The search for squeezing is conducted by scanning through the cavity resonance and spectrally analyzing the fluctuations in photocurrent produced by the total reflected field from the cavity. This scanning procedure is required because of problems associated with intracavity thermal processes. Heating of the intracavity medium and the concomitant changes in refractive index lead to hysteresis in the cavity transmission as the cavity length is swept. The culprit seems to be the index-matching fluid and not the KDP crystal.

Figure [5] illustrates the noise characteristics of the reflected cavity field in the absence of harmonic generation. The top trace records the RF noise density at fixed analysis frequency and the bottom the cavity transmission at ω_1 as the cavity length is swept. A coherent state would produce a flat noise spectrum, as indeed our laser does if directed onto a cavity without intracavity fluid or crystal. The rise in the noise spectrum on either side of the cavity resonance is associated with classical phase fluctuations produced by the

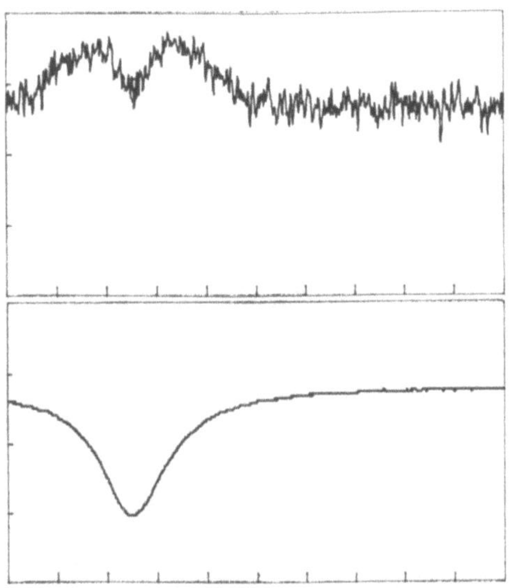

Figure 5 - Bottom trace: Cavity transmission at frequency ω_L vs. Δ_1 (increasing intensity downward). Top trace: $\Phi(\Omega)$ vs. Δ_1 at RF analysis frequency of 27.8 MHz with no UV (257 nm) generation, vertical scale 2.5 dB/div.

intracavity medium (again, probably the fluid). A simple model that treats the intracavity phase as a Gaussian random variable reproduces the gross qualitative features of Fig. [5]. In addition to this noise effect, we have also observed bright lines in the RF spectrum (0-15MHz) for cw excitation that we tentatively associate with high Q acoustic modes of crystal vibration. We attempt to avoid appreciable excitation of these modes by scanning at sufficiently high rates.

We conclude this section with some tantalizing but inconclusive observations of the spectrum of photocurrent fluctuations for a case of approximately 500 mW of circulating power at ω_1 and 5-10 mW at ω_2. Dominant features of Figure [6a] are the broad range over which the increased noise extends and the sharpness of the central feature as compared to the range of the cavity transmission function [6b]. These features presumably result from an interplay of the nonlinear processes and the classical intracavity phase noise described above. The curves in Fig. [6] were recorded with a digital storage oscilloscope. If the resolution is improved by expanding the time base to examine the central feature of Fig. [6a], we see in Fig. [6c] an apparent impressive reduction below the shot noise level, which is defined in the limit of large detuning in Fig. [6a]. We have confirmed that a given DC photocurrent corresponds to the same RF noise level (over the RF range of interest in these experiments) for three different sources of illumination, namely an incandescent lamp, a He-Ne laser, and the argon ion laser used in the experiments. In particular the noise level in Fig. [6a] at large detuning is just that obtained from these three sources in the absence of the nonlinear cavity but for the same DC photocurrent.

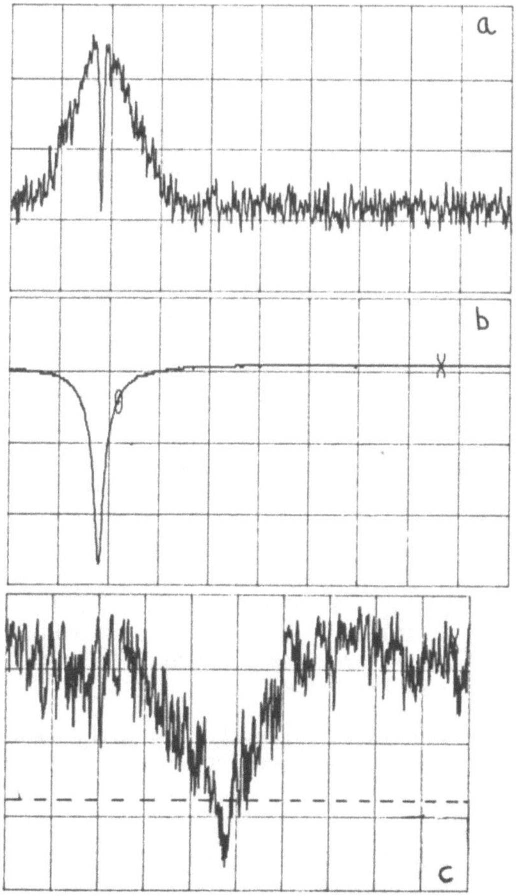

Figure 6 - Middle trace: Cavity transmission at frequency ω_L vs. Δ_1 (increasing intensity downward). Top trace: $\Phi(\Omega)$ vs. Δ_1 at RF analysis frequency of 31.4 MHz, optimized UV (257 nm) generation, vertical scale 1.25 dB/div. (16 trace average). Bottom trace: As in top trace, but expanded view of central feature near cavity resonance. The dashed line indicates the "shot-noise" or "vacuum" level inferred from the limit of large detuning in the top trace (4 trace average).

The qualitative character of this central "dip" is in accord with the theoretical discussion of the previous section and represents a reduction in the noise power of approximately 15%. The feature is persistent in that we have recorded many traces similar to Fig. [6] with appreciable deviations below the vacuum level. However it is not reproducible in that we are unable to systematically control its appearance. The dilemma that we currently face is that on the one hand the statistical significance of these features is uncertain due to the limited signal averaging employed, while on the other hand increased averaging tends to destroy the features altogether. It is not clear whether this is because of a lack of precision in the control of the system over the accumulation interval or because these features are in fact nothing more than statistical deviations about the vacuum level that tend to zero as the number of samples averaged is increased. We do find that the dependence of the noise spectrum on RF analysis frequency does tend to follow the general trend indicated by Fig. [2]. This observation is again inconclusive since the frequency scale for any intracavity noise process would largely be set by the cavity linewidths (Γ_1, Γ_2).

The phase dependence of the fluctuations in the reflected field from the cavity has been investigated with the help of the auxiliary cavity shown in Fig. [4], which is composed of an input coupler of reflectivity 0.97 at ω_1 and a highly reflective rear mirror (cavity free spectral range = 250 MHz). By scanning the narrow resonance feature of this cavity relative to the fixed incident frequency, we are able to phase shift the coherent carrier relative to the broad distribution of the spectrum of amplitude fluctuations[21]. Our observations of the dependence of the noise spectrum on local oscillator phase as defined by the detuning of the phase cavity are in qualitative agreement with the theoretical treatment outlined in Section II. In particular the field quadrature complementary to that shown in Fig. [6] shows enhanced fluctuations near $\delta_1=0$, i.e. at the position of the central feature in Fig. [6a,c].

A number of identifiable problems remain to be solved in this experiment, including more precise control of the spatial overlap of the cavity mode functions at ω_1 and ω_2 and better understanding of the intracavity thermal processes. Perhaps a time domain strategy of pulsing the intensity would allow tuning of the cavity over a narrower range and thus stretch out the time scale for RF spectral analysis. Nevertheless our preliminary results are highly encouraging, especially as regards the observation of dynamical instability. The approximate agreement between the calculated and measured values of critical power and frequency lends some credibility to the adoption of the theory of Section II to describe the quantum fluctuations in the experiment. The preliminary observations of the character of the RF noise spectrum can only be described as tantalizing.

The contributions of L. A. Wu to this work are gratefully acknowledged. The research was supported in part by the National Science Foundation (H.J.K.) and by the Office of Naval Research (J.L.H.).

References

1. R. E. Slusher, L. W. Hollberg, B. Yurke, J. C. Mertz, J. F. Valley, Phys. Rev. Lett. 55, 2409 (1985); R. E. Slusher and B. Yurke in Frontiers in Quantum Optics, S. Sarkar and E. R. Pike, eds. (Adam Hilger Ltd., 1986).
2. M. D. Levenson, R. M. Shelby, and S. H. Perlmutter, Opt. Lett. 10, 514 (1985).
3. M. W. Maeda, P. Kumar, J. H. Shapiro, in Optical Instabilities, R. W. Boyd, M. G. Raymer, and L. M. Narducci, eds. (Cambridge University Press, Cambridge, 1986), p. 370.
4. H. P. Yuen, Phys. Rev. A13, 2226 (1976).
5. D. F. Walls, Nature 306, 141 (1983).
6. P. D. Drummond, K. J. McNeil, and D. F. Walls, Optica Acta 27, 321 (1980).
7. P. Mandel and T. Erneux, Optica Acta 29, 7 (1982).

8. C. M. Savage and D. F. Walls, Optica Acta 30, 557 (1983).
9. G. J. Milburn and D. F. Walls, Phys. Rev. A27, 392 (1983).
10. L. A. Lugiato, G. Strini, and F. De Martini, Opt. Lett. 8, 256 (1983).
11. B. Yurke, Phys. Rev. A29, 408 (1984).
12. (a) M. J. Collett and C. W. Gardiner, Phys. Rev. A30, 1386 (1984); (b) C. W. Gardiner and M. J. Collett, Phys. Rev. A31, 3761 (1985).
13. M. J. Collett and D. F. Walls, Phys. Rev. A32, 2887 (1985).
14. P. D. Drummond, K. J. McNeil, and D. F. Walls, Optica Acta 28, 211 (1981).
15. H. P. Yuen and V. W. S. Chan, Opt. Lett. 8, 177; B. L. Schumaker, Opt. Lett. 9, 189 (1984).
16. R. J. Glauber, in Quantum Optics and Electronics, C. de Witt, A. Blandim and C. Cohen - Tannoudji, eds. (Gordon Breach, New York, 1965).
17. C. M. Savage and D. F. Walls have also considered the case of nonzero detuning (private communication).
18. F. Zernike and J. E. Midwinter, Applied Nonlinear Optics (John Wiley and Sons, New York, 1973).
19. G. D. Boyd and D. A. Kleinman, J. Appl. Phys. 39, 3597 (1968).
20. L. A. Wu and H. J. Kimble, J. Opt. Soc. Am. B2, 697 (1985).
21. M. D. Levenson, private communication.

Quantum Noise in Homodyne Detection

R. Loudon
Physics Department, Essex University, Colchester CO4 3SQ, England

1. Introduction

Optical homodyne detection theory is of considerable current interest in connection with the interpretation of experiments [1-3] on squeezed light. The theory has been developed in great detail and generality [4-6]. The aim of the present work is a clarification of the noise properties of the detection process. Pictorial representations are used to show the sources of noise and their contributions to the output noise in ordinary and balanced four-port homodyne detection, where only one signal quadrature is measured. A more general scheme is described for eight-port homodyne detection in which both quadratures are measured with arbitrary relative weights. The noise properties of this more complicated detector are also interpreted pictorially.

Homodyne detection corresponds to measurements with respect to a pair of Hermitian operators

$$\hat{x} = \tfrac{1}{2}\{\exp(i\arg\alpha_L)\hat{a}_S^\dagger + \exp(-i\arg\alpha_L)\hat{a}_S\}$$

$$\hat{y} = \tfrac{1}{2}i\{\exp(i\arg\alpha_L)\hat{a}_S^\dagger - \exp(-i\arg\alpha_L)\hat{a}_S\}, \tag{1}$$

where $\hat{a}_S^\dagger$ and $\hat{a}_S$ are the creation and destruction operators for the signal field mode, and the single local oscillator mode is excited in a coherent state $|\alpha_L\rangle$. The various operators have commutation relations

$$[\hat{a}_S,\hat{a}_S^\dagger] = 1 \quad \text{and} \quad [\hat{x},\hat{y}] = \tfrac{1}{2}i. \tag{2}$$

The two homodyne quadratures thus have an uncertainty relation

$$(\Delta x)_S^2(\Delta y)_S^2 \geq 1/16, \tag{3}$$

where the S subscripts signify that these are the basic input signal uncertainties.

Minimum-uncertainty measurements are realized for squeezed-state input signals [7-9]. The ideal squeezed state $|\alpha_S;s,\theta\rangle$ satisfies the eigenvalue equation

$$(\hat{a}_S\exp(-\tfrac{1}{2}i\theta)\cosh s + \hat{a}_S^\dagger\exp(\tfrac{1}{2}i\theta)\sinh s)|\alpha_S;s,\theta\rangle$$

$$= (\alpha_S\exp(-\tfrac{1}{2}i\theta)\cosh s + \alpha_S^*\exp(\tfrac{1}{2}i\theta)\sinh s)|\alpha_S;s,\theta\rangle. \tag{4}$$

It has a complex coherent amplitude α_S, with quadrature field expectation values

$$x_S \equiv \langle \hat{x} \rangle = |\alpha_S|\cos(\arg\alpha_S - \arg\alpha_L)$$

$$y_S \equiv \langle \hat{y} \rangle = |\alpha_S|\sin(\arg\alpha_S - \arg\alpha_L). \tag{5}$$

The corresponding variances are

$$(\Delta x)_S^2 = \tfrac{1}{4}\{\exp(-2s)\cos^2(\tfrac{1}{2}\theta-\arg\alpha_L) + \exp(2s)\sin^2(\tfrac{1}{2}\theta-\arg\alpha_L)\}$$

$$(\Delta y)_S^2 = \tfrac{1}{4}\{\exp(-2s)\sin^2(\tfrac{1}{2}\theta-\arg\alpha_L) + \exp(2s)\cos^2(\tfrac{1}{2}\theta-\arg\alpha_L)\}, \tag{6}$$

and these can be represented in terms of an uncertainty ellipse with semi-axes $\tfrac{1}{4}\exp(-s)$ and $\tfrac{1}{4}\exp(s)$ inclined at angles $\tfrac{1}{2}\theta - \arg\alpha_L$ to the homodyne axes. The simplest cases are

$$\left.\begin{aligned}(\Delta x)_S^2 &= \tfrac{1}{4}\exp(2s)\\ (\Delta y)_S^2 &= \tfrac{1}{4}\exp(-2s)\end{aligned}\right\}\quad \text{for } \tfrac{1}{2}\theta = \arg\alpha_L + \tfrac{1}{2}\pi$$

$$\left.\begin{aligned}(\Delta x)_S^2 &= \tfrac{1}{4}\exp(-2s)\\ (\Delta y)_S^2 &= \tfrac{1}{4}\exp(2s)\end{aligned}\right\}\quad \text{for } \tfrac{1}{2}\theta = \arg\alpha_L. \tag{7}$$

These variances have the minimum product allowed by (3).

2. Ordinary Homodyne Detection

Consider the configuration shown on the left of Fig. 1, where a lossless symmetric beam splitter couples the signal and local oscillator to two output modes. The relation between input and output mode destruction operators is

$$\begin{bmatrix}\hat{d}_1\\ \hat{d}_2\end{bmatrix} = \begin{bmatrix} r & t\\ t & r\end{bmatrix}\begin{bmatrix}\hat{a}_L\\ \hat{a}_S\end{bmatrix}, \tag{8}$$

where unitarity of the coupling matrix is ensured by the choice

$$r = i|r| \quad \text{and} \quad t = |t| \quad \text{with} \quad |r|^2 + |t|^2 = 1. \tag{9}$$

Only one of the output mode intensities is measured in ordinary homodyne detection, and this is here chosen to be mode $\hat{d}_1$. The intensity measurement, represented on the right of Fig. 1, is treated in the next section.

The input signal is assumed to be an ideal squeezed state, and maximum measured signal with minimum noise is achieved by the choice of phase angles

$$\arg\alpha_S = \tfrac{1}{2}\theta = \arg\alpha_L + \tfrac{1}{2}\pi. \tag{10}$$

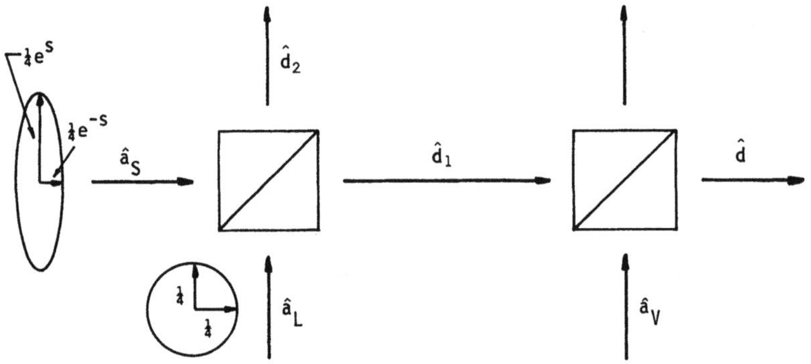

Fig. 1 Homodyne detector with squeezed signal and coherent local oscillator inputs

The uncertainty ellipse of the signal and the uncertainty circle of the coherent local oscillator are included in Fig. 1 with their semi-axis lengths shown. The contributions of the two inputs to the excitation of output mode $\hat{d}_1$ can be represented pictorially [10] as shown in Fig. 2. In this, and in subsequent diagrams of the same kind, the parallel mode amplitude contributions are always drawn as horizontal arrows, whatever their absolute phase angles, with uncertainty ellipses or circles attached to their tips. The two contributions are separated in the vertical direction for clarity. The semi-axis fluctuation lengths are shown in the figure, and it is important in some cases to keep track of the relative directions of the fluctuation arrows in different parts of the detector.

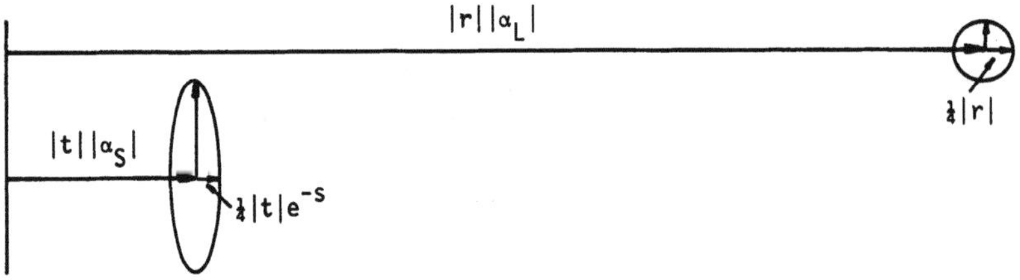

Fig. 2 Local oscillator and signal amplitude and noise diagrams for ordinary homodyne detection

Figure 2 is drawn for the conditions normally used in ordinary homodyne detection where

$$|t| \gg |r| \quad \text{and} \quad |r||\alpha_L| \gg |t||\alpha_S|, \tag{11}$$

so that almost all of the signal reaches the output arm but the local oscillator is so strong that even the small fraction reaching the output still dominates the signal contribution. The output intensity or photon number is obtained by squaring the sum of the two field contributions shown in Fig. 2. The mean photon number contains a cross term $2|r||t||\alpha_L||\alpha_S|$ that provides a measure of the signal amplitude, since the larger $|r|^2|\alpha_L|^2$ term is assumed to have a known magnitude. For the relative magnitudes assumed in (11), the uncertainty or noise on the mean photon number is mainly provided by the cross terms between the local oscillator amplitude $|r||\alpha_L|$ and the parallel components of the local oscillator and signal field fluctuations whose magnitudes are given in Fig. 2. These contributions to the uncertainty are therefore

$$\pm\tfrac{1}{2}|r|^2|\alpha_L| \quad \text{and} \quad \pm\tfrac{1}{2}|r||t||\alpha_L|\exp(-s). \tag{12}$$

The corresponding photon-number noises, defined in terms of the total uncertainty spreads, are denoted

$$N_L^{\frac{1}{2}} = |r|^2|\alpha_L| \quad \text{and} \quad N_S^{\frac{1}{2}} = |r||t||\alpha_L|\exp(-s), \tag{13}$$

and, since the two contributions are statistically independent, the total photon-number noise variance in mode $\hat{d}_1$ is

$$N_1 = N_L + N_S = |r|^2|\alpha_L|^2\{|r|^2 + |t|^2\exp(-2s)\}. \tag{14}$$

This result clearly demonstrates the possibility of reducing the homodyne noise to an arbitrarily low level [5] for sufficiently small $|r|$ and sufficiently large squeezing parameter s, while maintaining the local oscillator amplitude at a high enough level to satisfy (11).

The above results are valid only for the choice of phase angles given in (10). The same methods can however be used to derive the more complicated noise expressions that apply for more general angles.

3. Photodetection Noise

The measurement of output photon number in homodyne detection is modelled as a simple photon-counting process via a phototube with quantum efficiency η. Such complicating features as finite integration times and frequency-dependent detector responses are here ignored. The effects of a quantum efficiency of less than unity can be represented [4] by the second beam splitter shown on the right of Fig. 1. It couples mode $\hat{d}_1$ to a vacuum mode, with destruction operator $\hat{a}_V$, to form a final output operator $\hat{d}$ given by

$$\hat{d} = \eta^{\frac{1}{2}}\hat{d}_1 + (1-\eta)^{\frac{1}{2}}\hat{a}_V. \tag{15}$$

A measurement of mode $\hat{d}_1$ with quantum efficiency η is now translated into a measurement of mode $\hat{d}$ with unit quantum efficiency.

The noise properties of the detection process can be represented pictorially in the manner of Fig. 3. The contribution to the output mode $\hat{d}$ from the input $\hat{d}_1$ has an amplitude close to $\eta^{\frac{1}{2}}|r||\alpha_L|$, and a fluctuation obtained by scaling the uncertainty spread $N_1^{\frac{1}{2}}$ by a factor η. The vacuum mode $\hat{a}_V$ provides a noise contribution obtained from the cross term between its parallel fluctuation component and the amplitude of mode $\hat{d}$. These photon-number noises are respectively

$$\eta N_1^{\frac{1}{2}} \quad \text{and} \quad N_V^{\frac{1}{2}} = \eta^{\frac{1}{2}}(1-\eta)^{\frac{1}{2}}|r||\alpha_L|, \tag{16}$$

and they combine to give a total detected photon-number noise variance

$$N_1^D = \eta^2 N_1 + N_V = \eta|r|^2|\alpha_L|^2\{1-\eta|t|^2+\eta|t|^2\exp(-2s)\}, \tag{17}$$

in agreement with a known result [5]. The achievement of very low detected noise requires a quantum efficiency very close to unity in addition to the conditions outlined in the previous section.

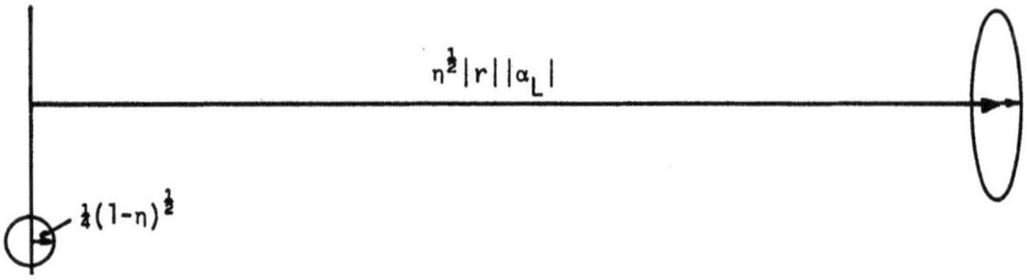

Fig. 3 Representation of the vacuum mode contribution to the noise in photodetection with quantum efficiency η

4. Balanced Homodyne Detection

Now consider the configuration shown on the left of Fig. 1 for conditions in which both output modes are detected, with the detector readings processed to provide the difference between the photon numbers in the two modes. Such four-port balanced homodyne detection has been known for some time to have the advantage over measurement of a single output mode that excess noise from the local oscillator can be cancelled [11-15], and this has been demonstrated experimentally [12,14]. The discussion of the noise given below parallels previous work [10].

The conditions for balanced homodyne detection are different from those for ordinary homodyne detection. Thus it is still assumed that the local oscillator field is much stronger than the signal field, with

$$|\alpha_L| >> |\alpha_S|, \tag{18}$$

but the beam splitter now divides the inputs equally, with

$$r = i/2^{\frac{1}{2}} \quad \text{and} \quad t = 1/2^{\frac{1}{2}}. \tag{19}$$

For the choice of phase angles made in (10), the amplitude of mode $\hat{d}_1$ is determined by the sum of the input amplitudes and that of mode $\hat{d}_2$ is determined by their difference. The situation is represented pictorially in Fig. 4, where the upper part

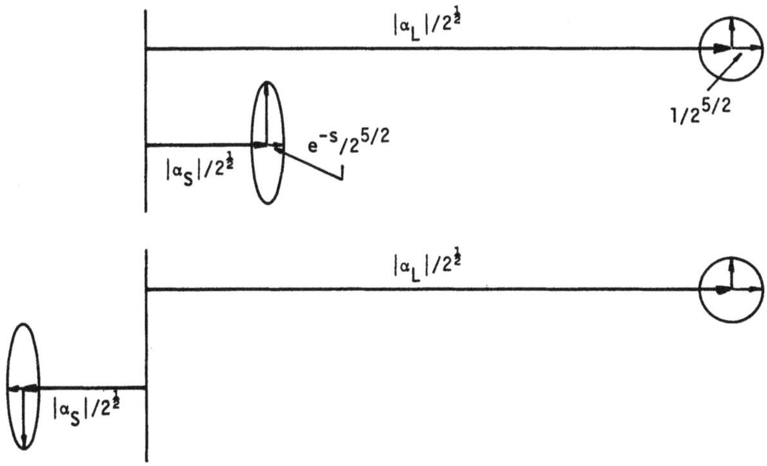

Fig. 4 Local oscillator and signal amplitude and noise diagrams for the two output modes in balanced homodyne detection

refers to mode $\hat{d}_1$ and the lower part to mode $\hat{d}_2$. Note particularly that the sign reversal between the two output modes affects not only the signal amplitude but also its fluctuation components [10]. The contributions to the photon-number uncertainty from the parallel components of the local oscillator and signal field fluctuations, analogous to (12), are respectively

$$\pm\tfrac{1}{4}|\alpha_L| \quad \text{and} \quad \pm\tfrac{1}{4}|\alpha_L|\exp(-s) \tag{20}$$

for output mode $\hat{d}_1$ and

$$\pm\tfrac{1}{4}|\alpha_L| \quad \text{and} \quad \mp\tfrac{1}{4}|\alpha_L|\exp(-s) \tag{21}$$

for output mode $\hat{d}_2$. The corresponding noises on the difference of the photon numbers in the two arms are

$$N_L^{\frac{1}{2}} = 0 \quad \text{and} \quad N_S^{\frac{1}{2}} = |\alpha_L|\exp(-s); \tag{22}$$

the correlated local oscillator fluctuations cancel and the anticorrelated signal field fluctuations add constructively when the difference of the photon-number outputs is formed. The total difference photon-number noise variance is

$$N_{12} = N_L + N_S = |\alpha_L|^2\exp(-2s). \tag{23}$$

These results show that balanced homodyne detection cancels not only the excess noise from the local oscillator but also the shot noise or quantum noise [10,13,15].

The photodetection noise is calculated by the procedure described in the previous section. Thus the noise calculated in (23) is reduced by a factor η^2 but the vacuum mode inputs to the photodetector beam splitters each add noise contributions equivalent to N_V in (16) and (17). The resulting total detected photon-number noise variance is

$$N_{12}^D = \eta|\alpha_L|^2\{1 - \eta + \eta\exp(-2s)\}, \tag{24}$$

in agreement with previous work [13,15].

5. Eight-Port Homodyne Detection

Figure 5 illustrates a more complicated form of homodyne detector with 4 input ports and 4 output ports. Two of the input ports are used for the signal and local oscillator while the remaining two inputs are unexcited. All four output ports are used, the detector readings being processed to provide the differences between the photon counts in adjacent pairs of output arms. The eight-port arrangement has the advantage over the four-port detector that both signal quadratures, or equivalently both signal phase and amplitude, can be measured [16-18], and this has been demonstrated experimentally [19].

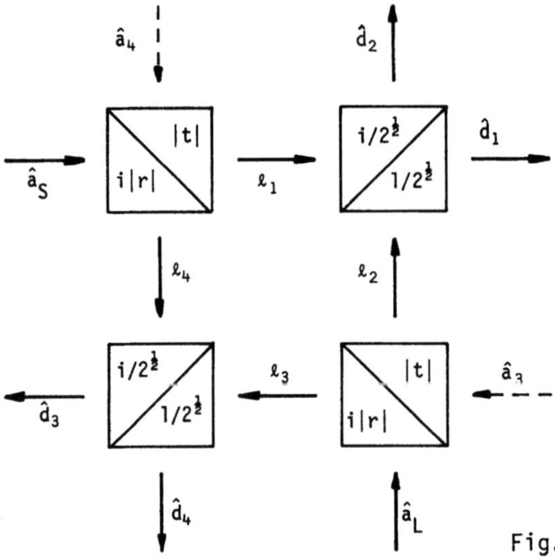

Fig. 5 Eight-port homodyne detector

Diagonally opposite pairs of beam splitters in Fig. 5 are assumed to be identical, with reflection and transmission coefficients as shown, similar respectively to those in (9) and (19). The symmetrical eight-port detector is a special case in which $|r| = |t| = 1/2^{\frac{1}{2}}$, while the arrangement becomes equivalent to the balanced detector of the previous section if either $|r|$ or $|t|$ vanishes. With path lengths chosen so that

$$k(\ell_1-\ell_2) = \tfrac{1}{2}\pi \quad \text{and} \quad k(\ell_4-\ell_3) = \pi, \tag{25}$$

where k is the optical wavevector, two of the output operators are related to the input operators by

$$\hat{d}_1 = i\exp(ik\ell_2)\{t\hat{a}_L + t\hat{a}_S + ir\hat{a}_3 + ir\hat{a}_4\}/2^{\frac{1}{2}}$$

$$\hat{d}_2 = \exp(ik\ell_2)\{t\hat{a}_L - t\hat{a}_S + ir\hat{a}_3 - ir\hat{a}_4\}/2^{\frac{1}{2}}. \tag{26}$$

The operator for the photon-number difference between the two outputs is

$$\hat{d}_1^\dagger\hat{d}_1 - \hat{d}_2^\dagger\hat{d}_2 = |t|^2(\hat{a}_S^\dagger\hat{a}_L + \hat{a}_L^\dagger\hat{a}_S) + \text{other terms}, \tag{27}$$

where the 'other terms' involve either or both of the unexcited input modes. For a coherent local oscillator and a squeezed signal as before, the measured signal is

$$S_{12} = \eta\langle\hat{d}_1^\dagger\hat{d}_1 - \hat{d}_2^\dagger\hat{d}_2\rangle = 2\eta|t|^2|\alpha_L|x_S, \tag{28}$$

where the quantum efficiency η has been inserted and x_S is defined in (5). The other pair of outputs, $\hat{d}_3$ and $\hat{d}_4$ provide a measured signal

$$S_{34} = \eta\langle\hat{d}_3^\dagger\hat{d}_3 - \hat{d}_4^\dagger\hat{d}_4\rangle = 2\eta|r|^2|\alpha_L|y_S. \tag{29}$$

Both signal quadratures can therefore be determined.

The noise contributions to the output modes $\hat{d}_1$ and $\hat{d}_2$ are represented pictorially in the upper and lower parts of Fig. 6 respectively. In each part of the figure the input modes are displayed in the order $\hat{a}_L$, $\hat{a}_S$, $\hat{a}_3$, and $\hat{a}_4$. The signal phase is now assumed to match that of the local oscillator, equivalent to removal of the $\frac{1}{2}\pi$ term from (10). The contributions to the photon-number uncertainty from the four input modes are respectively

$$\pm\tfrac{1}{4}|t|^2|\alpha_L|,\ \pm\tfrac{1}{4}|t|^2|\alpha_L|\exp(-s),\ \pm\tfrac{1}{4}|r||t||\alpha_L|,\ \pm\tfrac{1}{4}|r||t||\alpha_L| \tag{30}$$

for mode $\hat{d}_1$ and

$$\pm\tfrac{1}{4}|t|^2|\alpha_L|,\ \mp\tfrac{1}{4}|t|^2|\alpha_L|\exp(-s),\ \pm\tfrac{1}{4}|r||t||\alpha_L|,\ \mp\tfrac{1}{4}|r||t||\alpha_L| \tag{31}$$

for mode $\hat{d}_2$. The corresponding noises on the difference of the photon numbers in the two outputs are

$$N_L^{\frac{1}{2}} = 0,\ N_S^{\frac{1}{2}} = |t|^2|\alpha_L|\exp(-s),\ N_3^{\frac{1}{2}} = 0,\ N_4^{\frac{1}{2}} = |r||t||\alpha_L|. \tag{32}$$

The total difference photon-number variance is therefore

$$N_{12} = N_S + N_4 = |t|^2|\alpha_L|^2\{|t|^2\exp(-2s) + |r|^2\}. \tag{33}$$

This is quite similar to the expression (14) for the noise in ordinary homodyne detection. However, the eight-port detector also measures the other signal quadrature, and a similar treatment of the photon-number difference noise for outputs $\hat{d}_3$ and $\hat{d}_4$ produces a total variance

$$N_{34} = |r|^2|\alpha_L|^2\{|r|^2\exp(2s) + |t|^2\}. \tag{34}$$

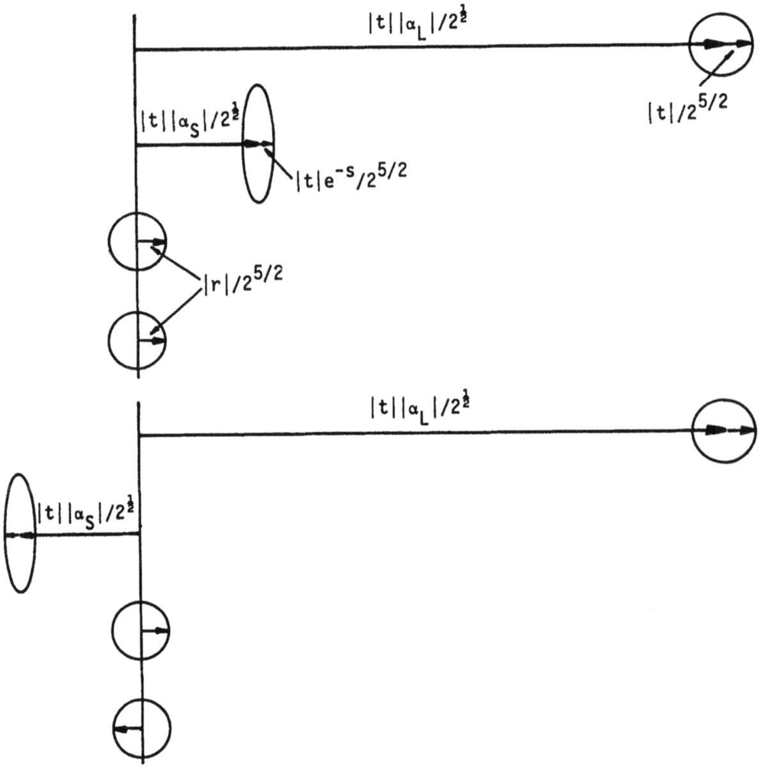

Fig. 6 Amplitude and noise diagrams for the contributions of the four input modes to the output modes $\hat{d}_1$ and $\hat{d}_2$

The noise N_{12} reduces to the four-port result (23) for $|r| = 0$ and $|t| = 1$.

The above results are valid only for the choice of equal local oscillator and signal phase angles but it is not difficult to generalize the calculation to arbitrary angles, when the noises become

$$N_{12} = 4|t|^4|\alpha_L|^2\{(\Delta x)_S^2 + (|r|^2/4|t|^2)\}$$

$$N_{34} = 4|r|^4|\alpha_L|^2\{(\Delta y)_S^2 + (|t|^2/4|r|^2)\}, \qquad (35)$$

where the input signal variances are given in (6). It is seen that each output photon-number noise has a component from the scaled input signal-field noise and a component associated with the homodyne process itself. With the scale factors removed, the variances determined by the homodyne detector are

$$(\Delta x)_H^2 = (\Delta x)_S^2 + (|r|^2/4|t|^2)$$

$$(\Delta y)_H^2 = (\Delta y)_S^2 + (|t|^2/4|r|^2). \qquad (36)$$

These have precisely the forms taken by the variances of conjugate observables in general measurement theory [20]. The additional homodyne contribution can in principle be reduced to zero for a single measurement, as in ordinary or balanced detection, but the uncertainty in a homodyne measurement of the conjugate quadrature then becomes very large. The measured variances satisfy an uncertainty relation

$$(\Delta x)_H^2(\Delta y)_H^2 \geq \tfrac{1}{4}, \tag{37}$$

and the minimum uncertainty product thus exceeds by a factor of 4 the corresponding product (3) of input signal variances.

The homodyne contributions to the noises in (35) and (36) can be interpreted as resulting from the convolution of a detector squeezed-state distribution with the input signal [18]. Thus if a parameter u is defined by

$$\exp(u) = |t|/|r|, \tag{38}$$

where $|t|$ is assumed to be larger than $|r|$, then (36) becomes

$$(\Delta x)_H^2 = (\Delta x)_S^2 + \tfrac{1}{4}\exp(-2u)$$
$$(\Delta y)_H^2 = (\Delta y)_S^2 + \tfrac{1}{4}\exp(2u). \tag{39}$$

The detector squeezed state variance components thus have forms analogous to (7), these simple cases of the more general variance expressions analogous to (6) being selected by the choice of beam splitter phases and relative optical path lengths in the eight-port homodyne detector. The association of a squeezed state with the homodyne detection process itself arises because the measured combinations of operators $\hat{x}$ and $\hat{y}$ defined by (1) have squeezed-state eigenfunctions. Thus the homodyne detection eigenvalue equation analogous to (4) can be written

$$(|t|^2\hat{x}+i|r|^2\hat{y})|\alpha;u,2\arg\alpha_L\rangle = (|t|^2x+i|r|^2y)|\alpha;u,2\arg\alpha_L\rangle \tag{40}$$

where α is arbitrary and

$$x + iy = \alpha\exp(-i\arg\alpha_L). \tag{41}$$

The probability distribution for the homodyne measurements is determined by the matrix element of the signal density operator with respect to the detector squeezed state [18].

The photodetection noise in eight-port homodyne detection is included by the procedure used in section 3 and the results are

$$N_{12}^D = \eta|t|^2|\alpha_L|^2\{4\eta|t|^2(\Delta x)_S^2 + \eta|r|^2 + 1 - \eta\}$$
$$N_{34}^D = \eta|r|^2|\alpha_L|^2\{4\eta|r|^2(\Delta y)_S^2 + \eta|t|^2 + 1 - \eta\}. \tag{42}$$

6. Conclusions

The noises (42) in homodyne detection of both quadratures with measured signals given by (28) and (29), contain three contributions. The first contribution is proportional to the field noise on the input signal, with a scaling factor identical to that which relates the measured signal to the input signal field. The second contribution is associated with the detector configuration; it can be attributed to the vacuum modes that are allowed into the detector by the beam splitters needed to divide the inputs into their quadrature components. This contribution is equivalent to the convolution of a detector squeezed-state field distribution with the field uncertainty distribution of the input signal. It can be reduced to zero for measurement of a single quadrature, as in the four-port result (24), but the uncertainty in a measurement of the other quadrature then increases to infinity. The third and final contribution is the photodetection noise; it can in principle be reduced to zero for unit quantum efficiency, but in practice the quantum efficiency is not usually close to unity.

The noise results and the pictorial representations are all based on the assumption of a strong coherent local oscillator. It is noteworthy that the relevant strength criterion (18) for the balanced four-port and eight-port detectors is less stringent than the corresponding criterion (11) for ordinary homodyne detection.

Finally, it should be mentioned that there are interesting parallels between the noise in eight-port homodyne detection and that in ordinary heterodyne detection [6,21].

1. M.D. Levenson, R.M. Shelby, A. Aspect, M.D. Reid and D.F. Walls: Phys. Rev. A 32, 1550 (1985)
2. M.W. Maeda, P. Kumar and J.H. Shapiro: Phys. Rev. A 32, 3803 (1985)
3. R.E. Slusher, L.W. Hollberg, B. Yurke, J.C. Mertz and J.F. Valley: Phys. Rev. Lett. 55, 2409 (1985)
4. H.P. Yuen and J.H. Shapiro: IEEE Trans. Inf. Theory 24, 657 (1978)
5. J.H. Shapiro, H.P. Yuen and J.A. Machado Mata: IEEE Trans. Inf. Theory 25 179 (1979)
6. H.P. Yuen and J.H. Shapiro: IEEE Trans. Inf. Theory 26, 78 (1980)
7. H.P. Yuen: Phys. Rev. A 13, 2226 (1976)
8. C.M. Caves: Phys. Rev. D 23, 1693 (1981)
9. D.F. Walls: Nature 306, 141 (1983)
10. C.M. Caves: in Coherence, Cooperation and Fluctuations (eds. F. Haake, L.M. Narducci and D.F. Walls) (Cambridge University Press 1986) to be published
11. B.M. Oliver: Proc. IRE 49, 1960 (1961)
12. H.R. Carleton and W.T. Maloney: Appl. Optics 7, 1241 (1968)
13. H.P. Yuen and V.W.S. Chan: Optics Lett. 8, 177 (1983)
14. G.L. Abbas, V.W.S. Chan and T.K. Yee: Optics Lett. 8, 419 (1983)
15. B.L. Schumaker: Optics Lett. 9, 189 (1984)
16. B.M. Oliver: Proc. IRE 50, 1545 (1962)
17. N.G. Walker and J.E. Carroll: AGARD Conf. Proc. 362, Digital Optical Circuit Technology 12.1 (1984)
18. N.G. Walker: to be published
19. N.G. Walker and J.E. Carroll: Electron. Lett. 20, 981 (1984)
20. E. Arthurs and J.L. Kelly: Bell Syst. Tech. J. 44, 725 (1965)
21. J.H. Shapiro and S.S. Wagner: IEEE J. Quantum Electron. 20, 803 (1984)

Oversqueezing Via Four-Order Interaction

P. Tombesi

Dipartimento di Física Università di Roma "La Sapienza",
P. le A.Moro 2, I-00185 Roma, Italy

There has been a great deal of interest in the last decade in the search for a new quantum effect produced by the radiation field in nonlinear optical media /1,2,3/. This effect, the squeezing of noise fluctuations of the electromagnetic field below the quantum limit, has been predicted to be generated via a quadratic interaction /2,4,5/ such as in the parametric amplification, four-wave mixing or similar optical devices. Very recently SLUSHER et al. /6/ reported the first experimental observation of the effect in a backward four-wave mixing /7/ experiment in an optical cavity, by linearly combining the output modes in appropriate way as originally proposed by YUEN and SHAPIRO /8/. Despite the very important result of this experiment, the observed reduction of noise fluctuations below the quantum level is only 7%. To be useful in high-sensitivity interferometer devices /9/, the noise reduction should be much higher.

In a recent report by YUEN and the present author /10/ it has been shown that a two-photon system with an added third-order nonlinearity can produce enhanced squeezing by reducing the interaction time needed to observe it. That could be useful in a traveling wave experiment. The analysis was performed with the following effective Hamiltonian

$$H = \hbar\left[\omega a^{+}a + \frac{\Gamma(t)}{2}a^{+2} + \frac{\Gamma^{*}(t)}{2}a^{2}\right] + \frac{\hbar\kappa}{2}a^{+2}a^{2} \tag{1}$$

The terms inside the square brackets represent the well-studied quadratic Hamiltonian /2/ and the fourth-order term is known to describe optical bistability. The c-number $\Gamma(t)$ is proportional to the external coherent pump assumed undepleted and very intense; κ is proportional to the third-order susceptibility tensor of the nonlinear medium and ω is the angular frequency of the signal. This Hamiltonian could describe an optically bistable two-photon medium or it may describe the effect of self-induced phase fluctuations in degenerate four-wave mixing. In the latter case the terms inside the square brackets represent the effective Hamiltonian discussed by YUEN and SHAPIRO /8/ for the generation of two-photon coherent states /2/, the other term represents the effect of self-induced phase fluctuations of the signal mode.The effect of self-induced phase fluctuations of the pump mode has been discussed by MILBURN, WALLS and LEVENSON /11/. In Ref.10 the analysis was performed by introducing a quantum mechanical quasi-linearization in the nonlinear equations of motion for the operators a and a^+, the destruction and creation operator of the signal mode respectively, which satisfy the canonical commutation relation. The result was very promising,since it was shown that in a particular quadrature phase the noise fluctuations were significantly reduced. However the validity of the approximations introduced was not well justified. The present work is devoted to give a better understanding, by means of a different approach, of the validity of the result in Ref.10. This goal will be accomplished by using an approach similar to the one introduced by HAKEN /12/ for the laser theory. A master equation for the density operator ρ is derived using standard techniques /13/. Then the so-called Q-representation is introduced defined as

$$Q\,(\alpha,\ \alpha^*,\ t) = \langle\alpha|\rho(t)|\alpha\rangle \tag{2}$$

i.e. the diagonal coherent states /14, 15/ matrix element for ρ. Antinormally ordered moments are directly computed in this Q-representation,which also has the following properties: it specifies ρ uniquely /15, 16/, it is an entire analytical function /16/ and, by definition, it never becomes negative; it is a true probability density. By means of the Q-representation the density operator equation of motion can be converted into a Fokker-Planck-like equation for Q (α ,α^*, t). The resulting equation becomes

$$i\frac{\partial Q}{\partial t} = \frac{\partial}{\partial\alpha^*}\left[(\omega-2\kappa)\alpha^*+\alpha(\Gamma^*+\kappa\alpha^{*2})\right]Q-\frac{\partial}{\partial\alpha}\left[(\omega-2\kappa)\alpha+\alpha^*(\Gamma+\kappa\alpha^2)\right]Q+$$
$$-\frac{1}{2}\frac{\partial^2}{\partial\alpha^2}(\Gamma+\kappa\alpha^2)Q+\frac{1}{2}\frac{\partial^2}{\partial\alpha^{*2}}(\Gamma^*+\kappa\alpha^{*2})Q. \tag{3}$$

This equation has a non-positive-definite diffusion matrix and for $\kappa=0$ it reduces to the one exactly solved by YUEN /2/. Given the evolution equation for Q one can show /17/ that the average of any smooth function A(α , α^*) of the complex variables (α , α^*), defined by

$$\langle A(\alpha,\alpha^*,t)\rangle=\int\frac{d^2\alpha}{\pi}\,Q(\alpha,\alpha^*,t)\,A(\alpha,\alpha^*) \tag{4}$$

(where $d^2\alpha = d\mathrm{Re}\{\alpha\}\, d\mathrm{Im}\{\alpha\}$), obeys the same differential equation of motion that can be obtained by formally deriving the stochastic differential equations, or Langevin equations, for α and α^* from (3). The result is a mere consequence of Ito's lemma /18/ and from the independent increment property of the Wiener process. In writing the Langevin equations the two complex variables α and α^* have to be considered as independent stochastic processes. They are complex conjugate only with respect to the average. The diffusion coefficient can be complex. This prescription is similar to the one introduced by DRUMMOND and WALLS /19/ and extensively discussed by DRUMMOND and GARDINER /20/ for the P-representation. We use the Langevin equations corresponding to $Q(\alpha, \alpha^*, t)$ partly because the smooth properties of Q guarantee that our development can be rigorously justified. On the other hand, other quasiprobabilities such as the diagonal P-representation can be used formally in the same fashion.

By means of the above procedure, one establishes a correspondence between c-numbers (α, α^*) and operators (a, a^+). Therefore the operator equations of motion are converted into Ito's stochastic differential equations as following:

$$
\begin{aligned}
d\alpha &= \left[-i(\omega-2\kappa)\alpha - i\alpha^*(\Gamma+\kappa\alpha^2)\right]dt + \sqrt{i(\Gamma+\kappa\alpha^2)}\; dW_1(t)\\
d\alpha^* &= \left[i(\omega-2\kappa)\alpha^* + i\alpha(\Gamma^*+\kappa\alpha^{*2})\right]dt + \sqrt{-i(\Gamma^*+\kappa\alpha^{*2})}\; dW_2(t)
\end{aligned}
\tag{5}
$$

where $dW_j(t)$ is the Ito's increment /21/ of the Wiener process satisfying* the relations

$$
\langle dW_j(t)\rangle = o;\quad \langle dW_j(t)\; dW_i(t')\rangle = \delta_{ij} dt dt' \delta(t-t'). \tag{6}
$$

For convenience we choose the external pump to be

$$
\Gamma(t) = i\Gamma_o \exp\left[-2i(\omega-2\kappa)t\right] \tag{7}
$$

with Γ_0 proportional to its amplitude and we introduce:

$$
\overline{\beta} = \overline{\alpha} \exp\left[i(\omega-2\kappa)t\right] \tag{8}
$$

where $\overline{\alpha}$ represents the deterministic part of the process α. We then consider the deterministic equations obtained from (5) without the stochastic terms. By using (7) and (8) we get

$$
\begin{aligned}
\frac{d\overline{\beta}}{dt} &= \overline{\beta}^*(1+i\varepsilon\overline{\beta}^2)\\
\frac{d\overline{\beta}^*}{dt} &= \overline{\beta}\;(1-i\varepsilon\overline{\beta}^{*2})
\end{aligned}
\tag{9}
$$

where we have introduced the dimensionless time

$$t = t\Gamma_o \tag{10}$$

and the dimensionless coupling constant

$$\varepsilon = \kappa/\Gamma_o . \tag{11}$$

Next we approximate (9) by introducing a perturbative expansion with respect to ε . Thus, the deterministic solution up to the first order in ε is

$$\begin{aligned}\bar\beta(t) \sim \beta_o (e^{t}\cos\phi + ie^{-t}\sin\phi) + \varepsilon\beta_o^3\Big[-\sin\phi\,(te^{t}\cos^2\phi + \tfrac{e^{-t}}{2}\sinh 2t\sin^2\phi) + \\ + i\cos\phi\,(te^{-t}\sin^2\phi + \tfrac{e^{t}}{2}\sinh 2t\cos^2\phi)\Big]\end{aligned} \tag{12}$$

and $\bar\beta^*(t) = (\bar\beta(t))^*$ where

$$\bar\beta(o) = \beta_o(\cos\phi + i\sin\phi) \tag{13}$$

represents the initial coherent state of the signal. In the following we will choose

$$\phi = o \text{ or } \pi/2 \tag{14}$$

Next we assume that the spread of the stochastic trajectory with respect to the deterministic path is very small at least for $t \lesssim 1$. Therefore we can linearize (5) with respect to the deterministic solutions (12) by writing

$$\begin{aligned}\alpha &= \bar\beta \exp[i(\omega - 2\kappa)t] + \beta(t) \\ \alpha^* &= \bar\beta^* \exp[-i(\omega - 2\kappa)t] + \beta^*(t)\end{aligned} \tag{15}$$

where $\beta(t)$ and $\beta^*(t)$ are two independent stochastic processes. Finally, by means of (7), (10), (11) and (15) we get the linearized stochastic differential equations

$$\begin{aligned}d\beta &= \{-i[(\omega/\Gamma_o - 2\varepsilon) + 2\varepsilon|\bar\beta|^2]\beta - i\beta^*(i + \varepsilon\bar\beta^2)e^{-2i(\frac{\omega}{\Gamma_o} - 2\varepsilon)t}\}dt + e^{-i(\omega/\Gamma_o - 2\varepsilon)t}\sqrt{i(i+\varepsilon\bar\beta^2)}\, d\tilde W_1(t) \\ d\beta^* &= \{i[(\tfrac{\omega}{\Gamma_o} - 2\varepsilon) + 2\varepsilon|\bar\beta|^2]\beta^* + i\beta(i - \varepsilon\bar\beta^{*2})e^{2i(\frac{\omega}{\Gamma_o} - 2\varepsilon)t}\}dt + e^{i(\omega/\Gamma_o - 2\varepsilon)t}\sqrt{i(i-\varepsilon\bar\beta^{*2})}\, d\tilde W_2(t)\end{aligned} \tag{16}$$

with

$$\langle d\tilde W_i(t)\rangle = o;\ \langle d\tilde W_i(t)\, d\tilde W_j(t')\rangle = \delta_{ij}\, dt\, dt'\, \delta(t - t') \tag{17}$$

i.e. t is the dimensionless time defined in (10). Let's now introduce

$$i(i+\varepsilon\overline{\beta}^2)=|f(t)|e^{-2i\theta(t)} \tag{18}$$

with

$$|f(t)|=\sqrt{1+\varepsilon^2|\overline{\beta}^2(t)|^2} \tag{19}$$

and

$$\theta(t)=\frac{1}{2}\text{ arctg }\frac{\varepsilon\text{Re}(\overline{\beta}^2(t))}{1+\varepsilon\ \text{Im}(\overline{\beta}^2(t))} \tag{20}$$

where Re (...) and Im (...) the real and imaginary part of $\overline{\beta}^2(t)$ obtained from (12). We introduce the new stochastic variables

$$\begin{aligned}\gamma\ (t)&=\beta(t)e^{i\theta(t)}\\ \gamma^*(t)&=\beta^*(t)e^{-i\theta(t)}\end{aligned} \tag{21}$$

thus (16) becomes

$$\begin{aligned}d\gamma&=\left[-i(\tfrac{\omega}{\Gamma_o}-2\varepsilon+2\varepsilon|\overline{\beta}|^2-\dot{\theta})\gamma-|f(t)|\gamma^*e^{-2i(\frac{\omega}{\Gamma_o}-2\varepsilon)t}\right]dt+|f(t)|^{1/2}e^{-i(\frac{\omega}{\Gamma_o}-2\varepsilon)t}d\tilde{W}_1(t)\\ d\gamma^*&=\left[i(\tfrac{\omega}{\Gamma_o}-2\varepsilon+2\varepsilon|\overline{\beta}|^2-\dot{\theta})\gamma^*-|f(t)|\gamma e^{2i(\frac{\omega}{\Gamma_o}-2\varepsilon)t}\right]dt+|f(t)|^{1/2}e^{i(\frac{\omega}{\Gamma_o}-2\varepsilon)t}d\tilde{W}_2(t)\end{aligned} \tag{22}$$

Within the time interval in which we are interested it is easily seen that one can always choose

$$\frac{\omega}{\Gamma_o}-2\varepsilon>>2\varepsilon|\overline{\beta}|^2-\dot{\theta}(t) \tag{23}$$

which can be obtained for particular values of the initial coherent state of the signal, frequency and amplitude of the external pump. Thus, introducing the quadratures

$$\begin{aligned}\gamma_1(t)&=\frac{1}{2}\left(\gamma(t)\exp\left[i(\tfrac{\omega}{\Gamma_o}-2\varepsilon)t\right]+\gamma^*(t)\exp\left[-i(\tfrac{\omega}{\Gamma_o}-2\varepsilon)t\right]\right)\\ \gamma_2(t)&=\frac{1}{2i}\left(\gamma(t)\exp\left[i(\tfrac{\omega}{\Gamma_o}-2\varepsilon)t\right]-\gamma^*(t)\exp\left[-i(\tfrac{\omega}{\Gamma_o}-2\varepsilon)t\right]\right)\end{aligned} \tag{24}$$

we finally get the stochastic differential equations

$$\begin{aligned}d\gamma_1&=-|f(t)|\gamma_1dt+\frac{1}{\sqrt{2}}|f(t)|^{1/2}d\xi_1(t)\\ d\gamma_2&=|f(t)|\gamma_2dt+\frac{i}{\sqrt{2}}|f(t)|^{1/2}d\xi_2(t)\end{aligned} \tag{25}$$

with

$$<d\xi_i(t)>=o;\ <d\xi_i(t)d\xi_j(t')>=\delta_{ij}dtdt'\delta(t-t'). \tag{26}$$

Therefore the solutions of (25) are

$$\gamma_1(t)=\gamma_1(o)e^{-F(t)}+\frac{e^{-F(t)}}{\sqrt{2}}\int_o^t|f(t')|^{1/2}e^{F(t')}d\xi_1(t') \tag{27a}$$

$$\gamma_2(t)=\gamma_2(o)e^{F(t)}+\frac{ie^{F(t)}}{\sqrt{2}}\int_o^t|f(t')|^{1/2}e^{-F(t')}d\xi_2(t') \tag{27b}$$

where

$$F(t)=\int_o^t|f(t')|dt' \tag{28}$$

obtained by using (19) and (12).
Furthermore we can always choose

$$\gamma_1(o)=\gamma_2(o)=o \tag{29}$$

assuming that at t=0 there is no sthochastic contribution to the process α defined in (15). Then by averaging with respect the initial distribution of the coherent state of the signal the variance, in the quadrature γ_1, is easily obtained as

$$<\Delta\gamma_1^2(t)-1/4>=\frac{1}{4}\,e^{-2F(t)} \tag{30}$$

where by definition $<\Delta\gamma_i^2> = <\gamma_i^2> - <\gamma_i>^2$ and the brackets represent the expectation value with respect to the Q-representation which gives the antinormally ordered moments. One can easily convince oneself that $<\Delta\gamma_i^2-1/4>$ is the required quantum average.

In the quadratures defined in (24) the squeezing is governed by the function F(t) given in (28). By using (12) and (19), (28) becomes

$$F_\pm(t)=\int_o^t\sqrt{1+\varepsilon^2\beta_o^4e^{\pm 4t'}(1+\frac{\varepsilon^2\beta_o^4}{4}\sinh^2 2t')^2}\,dt' \tag{31}$$

where the upper/lower sign for $\phi = 0$, $\pi/2$ respectively.
We immediately see that for $\varepsilon=0$ F(t)=t as it should be. In Fig.1 (30) is reported for various values of the initial phase ϕ and $\varepsilon\beta_o^2=1$. We note the squeezing is significantly enhanced for $\varepsilon\neq 0$. Of course we could get more squeezing when $\varepsilon\beta_o^2>>1$. However since β_o^2 roughly represents the number of photons in the initial state of the signal, in such a case not only (12) but the effective Hamiltonian (1) as well becomes questionable. Indeed, the intensities of the pump and signal both are of the same order of magnitude and one

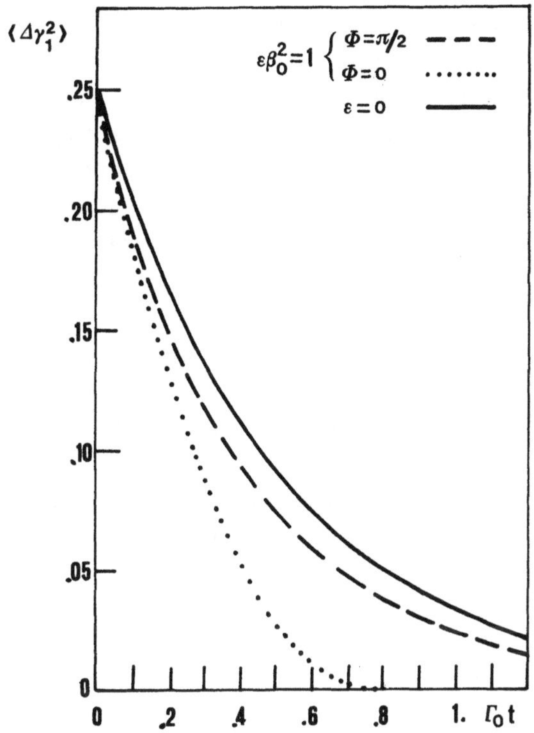

Fig.1: The variance of the quadrature γ_1 is reported for various values of the phase ϕ and $\varepsilon\beta_0{}^2=1$. $\varepsilon=0$ represents the usual squeezing.

should solve the whole quantum problem. The result agrees with Ref.10 where only the case $\phi=0$ was discussed and the deterministic solution was considered up to the zeroth order in ε.

We can now give a rough estimation of the validity of the above procedure of linearization with respect to the deterministic path. We want to emphasize that this procedure is not a perturbative expansion in the small parameter ε, indeed, by solving the linearized equations (16) we get a non-trivial resummation of the ε expansion.

Let us now define

$$s(t)=\left|\frac{<\alpha^2>-<\alpha>^2}{\bar{\alpha}}\right|=\left|\frac{<\beta^2>-<\beta>^2}{\bar{\beta}}\right| \qquad (32)$$

which gives the spread of the stochastic trajectory with respect to the deterministic one. We can thus estimate the time interval within which the spread s(t) is small. This can be obtained by $s(t_0)=1$, therefore as s(t) is monotonic, for $t \ll t_0$ we might be confident of the validity of our procedure. By using (21), (27) and (12) we get

$$s_{\pm}(t)=\frac{1}{2}\;\frac{\sinh 2\,F(t)}{\beta_0^2 e^{\pm 2t}(1+\frac{\varepsilon^2\beta_0^4}{4}\sinh^2 2t)}\;. \qquad (33)$$

A numerical calculation gives $s_{\pm}(1) \sim 10^{\mp 2}\varepsilon$. Within the time interval $t < 1$ (i.e. for the real time $t < \Gamma_o^{-1}$) the stochastic trajectory does not spread very much around the deterministic one, for we can be confident with (16) up to that time at least. The <u>oversqueezing</u> is in the quadrature γ_1 and could be observed by homodyne detection /22/ which involves adding an ordinary coherent optical field to the incoming squeezed field followed by direct photon counting. Of course the phase $\theta(t)$ from (20) and (12) should be compensated by the local oscillator.

A final comment concerning the vacuum initial state is worthwhile. In such a case one should proceed very carefully. In fact by using the c-number correspondence between operators a, a^+ and α, α^* introduced in getting (3), the deterministic equations (12) only give the antinormally ordered part of the quantum correspondent. Insofar as $\beta_0^2 \gg 1$ the classical or quantum values are equal but they become very different for the vacuum. Then, for the vacuum initial state we have to consider different deterministic contributions when we insert (12) in (16).

Aknowledgement - Work partially supported by GNSM and CISM. The financial support of the Italian Foreign Office is greatly appreciated.

1. D.F.Walls; Nature (London) <u>306</u>, 141 (1983), and references therein.
2. H.P.Yuen; Phys.Rev. A<u>13</u>, 2226 (1976).
3. <u>Proceedings of the Seventh International Conference on Laser Spectroscopy</u>, Eds. T.W.Haensch and Y.R.Shen, Series in Optical Science Vol.49 (Springer, Berlin, Heidelberg 1985)
4. D.Stoler; Phys.Rev. D<u>1</u>, 3217 (1970); <u>4</u>, 1925 (1971).
5. E.Y.C. Lu; Nuovo Cimento Lett. <u>2</u>, 1241 (1971); <u>3</u>, 585 (1972).
6. R.E.Slusher, L.W.Hollberg, B.Yurke, J.C.Mertz and J.F.Valley; Phys.Rev.Lett. <u>55</u>, 2409 (1985).
7. A.Yariv and D.Pepper; Opt.Lett. <u>1</u>, 16 (1977).
8. H.P.Yuen and J.H.Shapiro; Opt.Lett. <u>4</u>, 334 (1979).
9. C.M.Caves; Phys.Rev. D<u>23</u>, 1693 (1981).
10. P.Tombesi and H.P.Yuen in <u>Coherence and Quantum Optics V</u>, Eds. L.Mandel and E.Wolf (Plenum, 1984).
11. G.J.Milburn, D.F.Walls and M.D.Levenson; J.Opt.Soc. B<u>1</u>, 390, (1984).
12. H.Haken; <u>Handbuch der Physik</u> (Springer, Berlin 1970) Vol XXV/2C.

13. W.H.Louisell; Quantum Statistical Properties of Radiation (Wiley, New York 1983).
14. R.J.Glauber, in Laser Handbook, Vol.I, Eds. F.T.Arecchi and E.O.Schulz-Dubois (Elsevier, N.Y. 1972).
15. J.R.Klauder and E.C.G.Sudarshan; Fundamentals of Quantum Optics, (Benjamin, N.Y. 1968).
16. C.L.Mehta and E.C.G.Sudarshan; Phys.Rev. 138, B274 (1965).
17. H.P.Yuen and P.Tombesi - Internal Report, Dipartimento di Fisica (1985) Roma (unpublished).
18. I.I.Gihman and A.V.Skorohod; Stochastic Differential Equations (Springer, Berlin 1972).
19. P.D.Drummond and D.F.Walls; Phys.Rev. A23, 2563 (1981).
20. P.D.Drummond and C.W.Gardiner; J.Phys. A13, 2353 (1980).
21. C.W.Gardiner; Handbook of Stochastic Methods for Physics, Chemistry and the Natural Sciences (Springer, Berlin 1983).
22. H.P.Yuen and J.H.Shapiro; IEEE Trans. on Inform.Theory, IT-26, 78 (1980).

Photodetection and Induced Squeezing

P.D. Drummond

Physics Department, University of Auckland, Private Bag,
Auckland, New Zealand

Traditional photodetection theories usually involve either perturbation expansions or the rotating wave approximation. The first type of approximation is necessarily restricted to inefficient photodetection. The second type has the drawback that it ignores corrections due to virtual quanta. A new technique is introduced here in which a canonical transformation is used to transform the Hamiltonian to one which has no counter-rotating terms. As well as allowing an exact treatment of one-mode photodetection, the theory predicts the existence of squeezing induced by the detector coupling.

1. INTRODUCTION

The detection of quanta of the radiation field (or other quantum fields) has an important position in the development of quantum theory. The photo-electric effect is perhaps the earliest example of photodetection, in the sense that this effect was interpreted as resulting from discrete quanta [1]. Of course, at this stage it was not possible to detect individual photons. More recent developments have largely centred around Glauber's treatment of photodetection [2], with its emphasis on photon statistics. This has led to surprising predictions like that of photon antibunching [3,4] and squeezing [5,6], which were later observed in resonance fluorescence experiments.

While Glauber's treatment is intuitively clear, and has a sound foundation in scattering theory, it is also perturbative. This implies that it is less useful in treating <u>efficient</u> photodetection, in which essentially all the photons are observed. An obvious result of efficient photodetection is that two photodetectors placed side by side will have the same count-rate in a plane-wave field; while a photodetector placed behind another one will have a near-zero count-rate. This effect is not allowed for in Glauber photodetector theory, but is true in practice for laboratory photodetectors.

The problem of efficient photodetection or absorption was treated by Mollow [7] using the rotating-wave approximation, and a full nonperturbative treatment of a one-mode model. It is unfortunate that Mollow used an approximation which is not necessary in a scattering theory approach. As is shown in the next section, the rotating-wave approximation has a strong effect in photodetector theory. It essentially removes all of the detector-radiation coupling from the ground state of the interacting system.

In this paper, a technique of unifying the perturbative and absorptive approaches to photodetection is presented. The technique is extremely simple in principle, although nontrivial in practice. It is an alternative to the

Power-Zienau transform used to obtain the multipolar form of a Hamiltonian. In photodetection theory, the Power-Zienau transform is only moderately useful: instead, a transform will be given that alters a one-mode Hamiltonian into a form without counter-rotating terms. This allows the use of the methods pioneered by Mollow, without requiring the rotating wave approximation.

2. PHOTODETECTION PHYSICS

The usual model of photodetection [2,7] is an ionizable quantum system with a dipole coupling to the radiation field. The ionizing transition can be regarded as involving a set of levels, in a discrete approximation. Instead of measuring the field, the measuring process takes place on the ionizing system. Photodetection is reduced to the detection of an electron in the excited states of the detector quantum system, thus allowing an analysis of the detector coupling.

For the present purpose, the existence of many upper levels in the detector system is relatively superfluous. Instead, only one upper level really needs to be taken into account. It is necessary to treat even this case in a physical way. For this reason, as true two-level electric dipole-coupled detectors seem to be non-existent, a harmonic oscillator model will be used. Of course, this does allow multiple photon absorption. The model can be generalised to a large number of weakly-coupled harmonic oscillators in order to reduce multiple excitations.

The Hamiltonian to be treated has the standard minimal-coupling form in the Coulomb gauge of:

$$\hat{H} = \frac{1}{8\pi}\int [\,\hat{\mathbf{E}}_T(\mathbf{x})^2 + \hat{B}(\mathbf{x})^2]d^3\mathbf{x} + \sum_{n=1}^{N}\frac{1}{2m}\left(\hat{\mathbf{p}}_n - \frac{e\hat{\mathbf{A}}(\hat{q}_n)}{c}\right)^2 + V(\hat{q}_1, \,..\, \hat{q}_n) \quad (2.1)$$

This describes N scalar particles of mass m, charge e interacting with the electromagnetic field. For later comparisons, it is noteworthy that $\hat{H}$ can be rewritten in turns of transformed variables $\hat{\mathbf{p}}'$,$\hat{\mathbf{D}}$, using the Power-Zienau transformation. The new operators are still Hermitean (and hence are observables):

$$\begin{aligned}\hat{\mathbf{p}}' &\equiv \hat{S}\hat{\mathbf{p}}\hat{S}^{-1}\\ \hat{\mathbf{D}} &= \hat{S}\hat{\mathbf{E}}_T\hat{S}^{-1}\end{aligned} \quad (2.2)$$

Here $\hat{S}$ is a unitary transformation defined (for a detector with centre-of-mass at $\mathbf{x} = 0$) by:

$$\begin{aligned}\hat{S} &\equiv \exp\left[\frac{i}{c\hbar}\int \hat{P}(\mathbf{x})\hat{A}(\mathbf{x})d^3\mathbf{x}\right]\\ \hat{\mathbf{P}}(\mathbf{x}) &\equiv e\delta^{(3)}(\mathbf{x})\sum_{n=1}^{N}(q_n) + \ldots\end{aligned} \quad (2.3)$$

The higher order terms in the multipolar expansion are omitted intentionally; the end result of the transformation is that:

$$\begin{aligned}\hat{H} &= \frac{1}{8\pi}\int [\hat{D}(\mathbf{x})^2 + \hat{B}(\mathbf{x})^2]d^3\mathbf{x}\, D - \int\hat{\mathbf{P}}(\mathbf{x})\hat{D}(\mathbf{x})d^3\mathbf{x} + 2\pi\int\hat{\mathbf{P}}(\mathbf{x})^2d^3\mathbf{x}\\ &\quad + \frac{1}{2m}\sum_{n=1}^{N}\hat{\mathbf{p}}_n'^2 + V(\hat{q}_1 \,..\hat{q}_n) + \ldots\end{aligned} \quad (2.4)$$

The complete forms (2.1) and (2.4) are equivalent. The usual splitting of the Hamiltonian into 'free' and 'interacting' parts is not equivalent under this transformation. Thus, as is well known, it is possible to obtain results when using approximations, that differ between the Hamiltonians (2.1) and (2.4); even though these describe identical physical systems. The point is that this problem only arises when the approximations used are less than justifiable: it is obviously preferable to write down the Hamiltonian in a form that requires as few approximations as possible to solve it.

3. MEASURING $\hat{\mathbf{D}}$ and $\hat{\mathbf{E}}$

Having written the Hamiltonian in two possible ways, with different field operators, the question clearly arises as to which of these alternatives is relevant to photodetection. Does a photodetector measure $\hat{\mathbf{D}}$ or $\hat{\mathbf{E}}$? In fact $\hat{\mathbf{D}}$ and $\hat{\mathbf{E}}$ are very similar: they only differ by a term equal to $4\pi\hat{\mathbf{P}}(\mathbf{x})$, which is localized at the detector in the dipole approximation. This question is obviously somewhat troublesome: after all, the detector is intended to measure the field at its own location, which is exactly the place where $\hat{\mathbf{D}}$ and $\hat{\mathbf{E}}$ are (nearly) infinitely different from each other.

In scattering theory, a much more relaxed attitude can and does prevail. One simply defines an in-field at a location sufficiently distant from the detector, so that $\hat{\mathbf{D}}$ and $\hat{\mathbf{E}}$ are identical. After this field has scattered from the detector atom, perturbation theory is then able to compute photodetection probabilities. Any problems of interpretation should be nonexistent. A question might still linger about the completely different frequency dependence of the interaction in these two Hamiltonians, which is related to recent controversies on the subject of photon flux [9].

The problem is more serious when non-perturbative techniques are used. In fact calculations exist in the literature which indicate just how troublesome the question really is. Using apparently standard techniques, it is straightforward to calculate completely different photodetection probabilities using $\hat{\mathbf{D}}$ and $\hat{\mathbf{E}}$. Not surprisingly, the differences diverge as the ultraviolet cutoff of the theory is taken to infinity.

In order to illustrate this, a harmonic oscillator example will be used. In the minimal coupling case, using the dipole approximation, the Hamiltonian is written:

$$\hat{H} = \frac{1}{8\pi}\int [\hat{\mathbf{E}}_T(\mathbf{x})^2 + \hat{\mathbf{B}}(\mathbf{x})^2]\, d^3\mathbf{x} + \frac{1}{2}[\hat{\mathbf{p}}^2/m + mv^2\hat{\mathbf{q}}^2] - \frac{e}{mc}\hat{\mathbf{p}}.\hat{\mathbf{A}}(0) + \frac{e^2}{2mc^2}\hat{\mathbf{A}}^2(0) \qquad (3.1)$$

To simplify the problem of a full three-dimensional oscillator as in (3.1), a one-dimensional case can be written, for an oscillation in the direction of a unit vector $\mathbf{u}$ and mode polarisation $\mathbf{e}_K$ as:

$$\hat{H} = \hbar\left(\sum_k \omega_k \hat{a}_k \hat{a}_k + v\hat{b}^+\hat{b} + \sum_k g_k(\hat{b}^+ + \hat{b})(\hat{a}_k + \hat{a}_k)\right) \qquad (3.2)$$

where $g_k = -e\left(\frac{\pi v}{mVck}\right)^{\frac{1}{2}} |\mathbf{e}_k.\mathbf{u}|$

Of course, the general (three-dimensional) situation is more complex, as there are three independent harmonic oscillator operators, with characteristic frequencies that could be different. The Hamiltonian of (3.2) does take into account the different mode polarization directions through $\mathbf{e}_k.\mathbf{u}$: it is obviously no worse than two-level atom theory. The term corresponding to $\hat{\mathbf{A}}^2$ is dropped here, in order to compare directly with calculations in the literature [10].

The normal procedure in published accounts of photodetector theory is to start with the detector in its free ground state, which is then allowed to evolve in time. The probability of photodetection is the excited state population after a defined interaction time. Using the Heisenberg picture, the time evolution equations are:

$$\frac{\partial}{\partial t}\hat{a}_k = -i\omega_k\hat{a}_k - ig_k(\hat{b} + \hat{b}^+)$$

$$\frac{\partial}{\partial t}\hat{b} = -iv\hat{b} - i\Sigma g_k(\hat{a}_k + \hat{a}_k{}^+) \qquad (3.3)$$

At this stage it is clear that the counter-rotating terms have an awkward effect, in that they severely complicate the solution of these equations. A common procedure (which is related to perturbation techniques) is to approximate the time-evolution of a_k by its known free-field behaviour. This gives the result that:

$$\hat{b}(t) = \hat{b}(0)e^{-ivt} - \sum_k \int_0^t ig_k\,[\hat{a}_k^0 e^{-i\omega_k\tau} + \hat{a}_k^{0+} e^{i\omega_k\tau}]\, e^{-iv(t-\tau)}\,d\tau \qquad (3.4)$$

where $\hat{a}_k \approx \hat{a}_k^0 \exp[-i\omega_k t]$.

It is now possible to obtain the photodetection probability, P(T), in a time T. It is useful to define this as the expectation value of the Heisenberg operator $\hat{b}^+(T)\hat{b}(T)$. As a first step, suppose the field was initially in its free vacuum state $|0\rangle_a$; here one might reasonably expect to detect no photons. In fact, the probability of a photon being detected in time T is:

$$\langle\hat{b}^+(T)\hat{b}(T)\rangle = \sum_k \frac{4|g_k|^2}{(v+\omega_k)^2}\sin^2\left[\frac{(v+\omega_k)T}{2}\right] \qquad (3.5)$$

Replacing the sum over discrete k's by an integral, which gives the large mode volume limit, one obtains:

$$P(T) \propto \int_0^{\omega_0} \frac{\omega}{(v+\omega)^2}\sin^2\left[\frac{(v+\omega)T}{2}\right]d\omega \qquad (3.6)$$

Here the (trivial) angular integral is omitted, and use is made of the fact that $g_k{}^2 \propto 1/\omega$. Despite this diminished coupling at large ω, the overall photodetection probability is logarithmically divergent as $\omega_0 \to \infty$. Note that ω_0 has a natural limit: an ultraviolet cut-off at $\omega_0 \approx c/r_0$ is necessary owing to the use of the dipole approximation, where r_0 is an atomic radius.

In summary, not only is there an apparently nonzero photodetection probability, but the calculation is divergent in the large cut-off limit. It is possible to repeat this calculation using the alternate Hamiltonian, (2.4). One now obtains a completely different answer [11], which has an even stronger quadratic divergence:

$$P(T) \propto \int_0^{\omega_0} \frac{\omega^3}{(\nu + \omega)^2} \sin^2 \left[\frac{(\nu + \omega) T}{2} \right] d\omega \tag{3.7}$$

Thus the difference between the two different observables $\hat{\mathbf{D}}$ and $\hat{\mathbf{E}}$ shows itself as a difference in the (divergent) integrals (3.6) and (3.7). The reason that this occurs is that the initial "vacuum" state is in fact different in the two calculations: the free parts of the two Hamiltonians are physically distinct and therefore have completely different ground states.

4. CANONICAL TRANSFORMATIONS

The solution to the paradox outlined in the previous section is simple. It is to use an initial state which does not depend on an artificial splitting into "free" and "interacting" parts. An example is just the ground state of the total Hamiltonian. Since this ground state is an eigenvector of the complete Hamiltonian, it is time-invariant, and does not lead to any time-dependent photodetection probabilities in the case of a vacuum. States of finite photon number should then be defined relative to this true ground state.

Another, less physical solution is also possible. This is to ignore the counter-rotating terms of form $\hat{a}\hat{b}$ or $\hat{a}^+\hat{b}^+$ in the Hamiltonian. With these terms removed, the divergences in fact vanish. This is superficially desirable, but begs the question of motivating the approximation in the first place. Since the rotating-wave approximation (R.W.A.) leads to an infinite (in the limit $\omega_0 \to \infty$) change in the results predicted, it scarcely seems to be a good approximation to use in calculating P(T).

While the R.W.A. has questionable validity in the current calculation, it does provide a guide to solving the problem. It would be advantageous to have a Hamiltonian without counter-rotating terms to begin with. Of course, this could only be achieved through some transformation of the original operators. Once a suitable transformation was found, it is clear that the absence of terms of form $\hat{a}\hat{b}$ or $\hat{a}^+\hat{b}^+$ would greatly simplify the problem. In fact, this provides a simple way to compute the true ground state, as well as handling the time development.

To implement this approach, suppose that a one-mode problem is to be treated. Let:

$$\begin{aligned} \bar{a} &= \hat{U}\,\hat{a}\,\hat{U}^{-1} \\ \bar{b} &= \hat{U}\,\hat{b}\,\hat{U}^{-1} \end{aligned} \tag{4.1}$$

where: $\hat{U}^{-1} = \hat{U}^+$

This unitary transformation leaves the commutation relations invariant, i.e.:

$$\begin{aligned} [\bar{a}, \bar{a}^+] &= [\bar{b}, \bar{b}^+] = 1 \\ [\bar{a}, \bar{b}] &= [\bar{a}, \bar{b}^+] = 0 \end{aligned} \tag{4.2}$$

A suitable transformation can now be evaluated for this case of one mode coupled to a detector, with the requirement of no counter-rotating terms. This implies that the initial and final forms of the Hamiltonian should be (with $\hbar = 1$):

$$\hat{H} = \omega\hat{a}^+\hat{a} + v\hat{b}^+\hat{b} + g(\hat{a} + \hat{a}^+)(\hat{b} + \hat{b}^+) + \frac{g^2}{v}(\hat{a} + \hat{a}^+)^2 \qquad (4.3)$$

and:

$$\hat{H} = \bar{\omega}\bar{a}^+\bar{a} + \bar{v}\bar{b}^+\bar{b} + \bar{g}\bar{a}\bar{b}^+ + \bar{g}\bar{a}^+\bar{b}$$

In order to show how this allows the ground-state to be calculated, (4.3) can be rewritten for the case of $\bar{g} \geq 0$, as:

$$\hat{H} = [\bar{\omega} - \bar{g}]\ \bar{a}^+\bar{a} + [\bar{v} - \bar{g}]\ \bar{b}^+\bar{b} + \bar{g}\{\bar{b} + \bar{a}\}^+\ \{\bar{b} + \bar{a}\} \qquad (4.4)$$

In this form (which has similarities to supersymmetric Hamiltonians) it is clear that $\hat{H}$ is positive semi-definite if $\bar{\omega}$, $\bar{v} \geq \bar{g}$. This follows since each term is individually positive semi-definite:

$$\langle\psi|\hat{H}|\psi\rangle = (\bar{\omega} - \bar{g})\langle\psi|\bar{a}^+\bar{a}|\psi\rangle + (\bar{v} - \bar{g})\langle\psi|\bar{b}^+\bar{b}|\psi\rangle + \bar{g}\langle\psi|(\bar{a} + \bar{b})^+(\bar{a} + \bar{b})|\psi\rangle \qquad (4.5)$$

In particular, the state $|\bar{\psi}_0\rangle = |0\rangle_{\bar{a}}\ |0\rangle_{\bar{b}}$ has a zero eigenvalue. It is annihilated by $\bar{a}$ and by $\bar{b}$, and hence is the true ground state.

It is now necessary to evaluate the transformation U that leaves $\hat{H}$ in the form of (4.4). Just as before (in the multipolar case), the transformation is applied to the operators, and $\hat{H}$ is not altered. Thus this procedure is one of <u>rewriting</u> or rearranging the terms of $\hat{H}$. In particular, it does not alter the time-development or the states of the physical system that $\hat{H}$ describes. Of course, the eigenvectors of $\hat{a}$ are not identical with the eigenvectors of the transformed operator $\bar{a}$.

It is possible to obtain a Hamiltonian like (4.4) by merely diagonalising the corresponding clasical problem: this is in fact a standard way to solve coupled quantum harmonic oscillator problems. The resulting Hamiltonian would trivially be in the form of (4.4), with $\bar{g} = 0$. In doing this, there is a large, resonant mixing between the input (field) and the output (detector) operators. This causes great difficulty in interpreting the resulting dynamics as corresponding to a detection process, since the number states of the $\bar{a}$ and $\bar{b}$ operators do not evolve in time, except trivially due to phase shifts. An alternative procedure that will be utilised here is to require that the transformation causes minimal mixing between the two operators, compatible with the final result having no counter-rotating terms.

A general unitary transformation that leaves commutation relations invariant can be written as:

$$\begin{aligned} \bar{a} &= \mu_0\hat{a} + \mu_1\hat{a}^+ + \mu_2\hat{b}^+ + \mu_3\hat{b} \\ \bar{b} &= v_0\hat{b} + v_1\hat{b}^+ + v_2\hat{a}^+ + v_3\hat{a} \end{aligned} \qquad (4.6)$$

The four equations of (4.2) mean that only four of the above eight coefficients are actually independent. For simplicity only real coefficients will be treated: thus there are four real variables to specify. A reasonable requirement for minimal mixing is that the new $\bar{a}^+$ operator should <u>not</u> create any detector excitation (in the sense of the eigenstates of $\hat{b}^+\hat{b}$). Thus $\mu_3 \approx 0$, and similarly for v_3. As seen previously, the eigenstates of $\hat{b}^+\hat{b}$ themselves depend on the use of minimal coupling or multipolar splittings of "free" and "interacting" parts of the original Hamiltonian. Hence the requirement of $v_3 = \mu_3 = 0$ can be relaxed if necessary in order to keep the necessary final structure of (4.3).

As a first step, the transformations with $v_3 = \mu_3 = 0$ will be investigated. Owing to the four commutation relations, there are only two free variables left, and the general unitary transformation is now written as:

$$\bar{a} = \mu_0 \hat{a} + \mu_1 \hat{a}^+ + \mu_2 \hat{b}^+$$
$$\bar{b} = \mu_0 \hat{b} - \mu_1 \hat{b}^+ + \mu_2 \hat{a}^+ \qquad (4.7)$$

where: $\mu_0{}^2 = 1 + \mu_1{}^2 + \mu_2{}^2$

The transformation defined by (4.7) is necessarily unitary. It is now possible to evaluate $\hat{H}$, using the new operators $\bar{a},\bar{b}$. This just requires the inversion of the unitary transformation. The form of the inverse transformation is:

$$\hat{a} = \mu_0 \bar{a} - \mu_1 \bar{a}^+ - \mu_2 \bar{b}^+$$
$$\hat{b} = \mu_0 \bar{b} + \mu_1 \bar{b}^+ - \mu_2 \bar{a}^+ \qquad (4.8)$$

It is an elementary calculation to substitute the equations for $\hat{a},\hat{b}$ into the minimal coupling Hamiltonian. Having done this, there are terms that appear of form $\bar{a}^2$, $\bar{b}^2$, $\bar{a}\bar{b}$, $\bar{a}^+\bar{b}^+$. In order to eliminate <u>all</u> of these counter-rotating type terms, it is necessary to have more than two independent variables μ_1, μ_2. Nevertheless, one can eliminate them approximately to order (g^2). This calculation gives the result that:

$$\mu_1 = \frac{g^2}{v(v + \omega)}$$
$$\mu_2 = \frac{g}{(v + \omega)} \qquad (4.9)$$

While this calculation is only approximate, it is extremely instructive, indicating that a simple transformation as specified by (4.9) is able to eliminate nearly all the counter-rotating terms from the minimal coupling Hamiltonian. The transformation involves denominators that have the form $1/(v + \omega)$: thus it is <u>nonresonant</u>. This is to be expected, since counter-rotating terms do not have any resonant characteristics. In fact, with slightly greater algebraic complication, it is possible to eliminate the counter-rotating terms exactly. This generalisation will be treated elsewhere. It is worth noting that the transformation defined by (4.9) can be rewritten as:

$$U = \exp[\frac{\alpha_1}{2}[\hat{a}^2 - \hat{a}^{+2} - \hat{b}^2 + \hat{b}^{+2}] + \alpha_2 [\hat{a}\hat{b} - \hat{a}^+\hat{b}^+]] \qquad (4.10)$$

where: $\mu_1 = \frac{\alpha_1}{|\vec{\alpha}|} \sinh|\vec{\alpha}|$, $\mu_2 = \frac{\alpha_2}{|\vec{\alpha}|} \sinh|\vec{\alpha}|$.

Thus, since α_1 is similar to a squeeze parameter, as used to define squeezed states, it might be expected that there is a relationship between the transformation defined here, and the squeezing transformations used to generate the recently discovered squeezed states[5,6].

Finally, note that the new Hamiltonian has an extremely simple time-development for its Heisenberg operators. In fact in the resonant case ($\bar{\omega} = \bar{\nu}$), the exact solution for b is:

$$\bar{b}(t) = [\bar{b}^0 \cos(\bar{g}t) - i\bar{a}^0 \sin(\bar{g}t)]e^{-i\omega t} \tag{4.11}$$

Hence, for an initial state $|\psi\rangle_{\bar{a}}|0\rangle_{\bar{b}}$ the photon detection probability is:

$$\bar{p}(T) = \langle \bar{b}^+(t)\bar{b}(t)\rangle = \sin^2(\bar{g}t)\langle \bar{a}^{o+}\bar{a}^o\rangle \tag{4.12}$$

Thus the detector excitation corresponds in a straightforward way to the input excitation.

5. SQUEEZING

The new Hamiltonian generated by the canonical transformation is of the structure of (4.3). Since the $\bar{a}^+$ operator is only a slight modification (i.e., order g) of the original $\hat{a}^+$ operator, it is reasonable to refer to it as creating photons. This terminology is just as logical as the accepted practice of referring to the field operator $\hat{D}$ in the multipolar Hamiltonian as being an electric field operator. It too differs from the minimal coupling operator $\hat{E}$ by terms of O(g). In fact, one usually regards the lowest energy state of the Hamiltonian as being the state of zero photon number. This is clearly true of $\bar{a}^+\bar{a}$, although not of $\hat{a}^+\hat{a}$, since the <u>bare</u> vacuum state is not an eigenstate of the full Hamiltonian.

While it is simple enough to regard the operators $\bar{a}, \bar{a}^+$ as creating dressed photons, it is necessary to also recall that the definition here of $\bar{a}, \bar{a}^+$ is non-unique. The idea of a dressed photon does, however, provide a useful insight into the inevitable effect that a detector has on the physical system it is measuring. One <u>unique</u> property of <u>any</u> $\bar{a}$ operator as defined here is its ground state: which must be the (unique) ground state of the full Hamiltonian. It is instructive to examine the difference between this and the usual ground state of the free minimal coupling Hamiltonian.

In order to do this, quadrature operators will be defined as:

$$\begin{aligned} \hat{X}_1 &= [\hat{a} + \hat{a}^+] \\ \hat{X}_2 &= i[\hat{a} - \hat{a}^+] \end{aligned} \tag{5.1}$$

In terms of the bare vacuum ($|\psi_0\rangle = |0\rangle_a|0\rangle_b$) one has:

$$\langle\psi_0|\hat{X}_1{}^2|\psi_0\rangle = \langle\psi_0|\hat{X}_2{}^2|\psi_0\rangle \tag{5.2}$$

In the interacting vacuum ($|\bar{\psi}_0\rangle = |0\rangle_{\bar{a}}|0\rangle_{\bar{b}}$) the quadrature fluctuations can be calculated simply on re-expressing $\hat{X}_1, \hat{X}_2$ in terms of the new operators $\bar{a}, \bar{b}$. On doing this, one finds that:

$$\langle\bar{\psi}_0|\hat{X}_1{}^2|\bar{\psi}_0\rangle = \left[1 - \frac{2g^2\omega}{(\nu + \omega)^2\nu}\right]$$

$$\langle\bar{\psi}_0|\hat{X}_2{}^2|\bar{\psi}_0\rangle = \left(1 + \frac{2g^2(\omega + 2v)}{(v + \omega)^2 v}\right)$$

It is notable that the dressed state, $|\bar{\psi}_0\rangle$, is no longer a minimum uncertainty state relative to the $\hat{X}_1, \hat{X}_2$ quadratures. It is, of course, a minimum uncertainty state with equal fluctuations relative to the $\bar{X}_1, \bar{X}_2$ quadratures. With respect to the original quadrature operators, there is a squeezing (uncertainty reduction) in the $\hat{X}_1$ quadrature. This was expected in view of (4.10). The nature of the squeezing is different from that occurring in recent calculations of reduced quadrature fluctations in nonlinear optics, which occur in rotating frame quadrature operators. Here the squeezing is in a laboratory frame quadrature. By comparison, the squeezing calculated here would oscillate at optical frequencies if rotating frame quadrature operators were used. In fact the quadrature with increased fluctuations is an electric field operator, the one with reduced fluctuations a magnetic field operator.

In summary, the new operators defined here allow a natural treatment of photodetection in terms of a Hamiltonian without counter-rotating terms. This removes virtual corrections from the calculation, without any need for the rotating wave approximation. The ground-state fluctuations can also be obtained from the transformation used to generate the operators. The result that the ground state is squeezed is in general agreement with field energy calculations of Persico and Power [12]. The point of interest here is that a simple unitary transformation of the system operators allows these fluctuations to be included in a calculation of photodetection probability.

1. Planck, M., Verh.dt.phys.Ges. 2, 202 & 237 (1900); Einstein, A. Annln Phys. 17, 132 (1905).
2. Glauber, R.J., Phys. Rev. 130, 2529 (1963).
3. H.J. Carmichael & D.F. Walls, J.Phys. 9B, L43 & 1199 (1976).
4. H.J. Kimble, M. Dagenais & L. Mandel, Phys.Rev.Lett. 39, 691 (1977) & Phys.Rev.Lett. 18A, 201 (1978).
5. D.W. Robinson, Comm.Math.Phys.1, 159 (1965).
6. For a review on squeezing, see D.F. Walls, Nature (London) 306, 141 (1983).
7. B.R. Mollow, Phys. Rev. 168, 1896 (1968).
8. E.A. Power & S. Zienau, Phil.Trans.Roy.Soc.Lond. A251, 427 (1959). Also see: W.P Healey, "Non-relativistic Quantum Electrodynamics" (Academic Press, London, 1982).
9. H.J. Kimble & L. Mandel, Phys.Rev.A30, 844 (1984); R.S. Bondurant, Phys.Rev.A32, 2797 (1985).
10. W.H. Louisell, "Quantum Statistical Properties of Radiation" (Wiley, New York, 1973).
11. J.D. Cresser, Physics Reports 94, 47 (1983).
12. F. Persico, E.A. Power, preprint (to appear, Physics Letters).

Generation of Squeezed States: Rydberg Atoms in a High Q Cavity

B.J. Dalton

Physics Department, University of Queensland, St. Lucia, Brisbane, Queensland 4067, Australia

1. INTRODUCTION

Squeezed states [1] of the quantum EM field are characterised by having the fluctuations in one quadrature component of the electric field smaller than that for a coherent state. This is a non classical feature and the theory of the generation of squeezed states has been studied in a wide variety of non linear atom-field processes. Such states have possible applications in spectroscopy, communications, gravitational wave detection and so on. Squeezed states have recently been observed experimentally using the process of non degenerate four-wave mixing in an optical cavity [2]. References to the many papers on squeezed states may be found in the review by WALLS [3] and in the recent paper by REID and WALLS [4].

Recently HEIDMANN, RAIMOND, REYNAUD and ZAGURY [5-7] have studied theoretically squeezing for the radiation field in a high Q millimeter wave cavity with low thermal noise, an effect which is produced when a large number of Rydberg atoms interact with the cavity field. Two types of initial state are considered, both possessing phase information - as is necessary to produced squeezed radiation. The first (spontaneous superradiance) involves the atoms being prepared in an atomic coherent state but with the field mode empty. The second (triggered superfluorescence) has the atoms totally excited and the field prepared in a Glauber coherent state. In both cases large amounts of squeezing occur in the transient radiation field produced when the atoms deexcite. It is suggested that such a system might be suitable for the observation of squeezing.

The present work examines the possibility of squeezing in an extension of the triggered superfluorescence case to a steady-state situation, again in a high Q cavity. An incoherent pumping process from the lower to the upper state of the Rydberg two-level atom (TLA) system is included. As in the work of HEIDMANN et al [5-7], spontaneous emission relaxation is ignored for Rydberg atoms, cavity damping is included and the atom-cavity mode coupling is treated in the rotating wave approximation. For the present thermal noise and collisional relaxation are not taken into account, though both could become important in adversely affecting squeezing. An incident coherent field, coupled to the cavity mode is also included. Finally, only the squeezing in the intracavity field is determined at this stage. The squeezing in particular spectral components of the field external to the cavity can be larger [8-9].

2. CONDITIONS FOR SQUEEZING

Ignoring any polarization and position factors, the positive, negative components E^+, E^- of the electric field for a single-mode cavity of volume V, and whose annihilation, creation operators are a, $a^\dagger$, are given by

$$E^+ = i\lambda a \tag{1a}$$

$$E^- = -i\lambda a^\dagger \tag{1b}$$

with

$$\lambda = (\hbar\omega/2\epsilon_0 V)^{1/2} . \tag{1c}$$

We introduce quadrature components $E_{1\theta}$, $E_{2\theta}$ for frequency ω (which will be chosen equal to that of the incident field) via

$$E_{1\theta} = E^+ e^{i(\omega t+\theta)} + E^- e^{-i(\omega t+\theta)} \tag{2a}$$

$$E_{2\theta} = -i(E^+ e^{i(\omega t+\theta)} - E^- e^{-i(\omega t+\theta)}) \tag{2b}$$

The total field $E = E^+ + E^-$ is then given by

$$E = E_{1\theta} \cos(\omega t+\theta) + E_{2\theta} \sin(\omega t+\theta). \tag{3}$$

From the commutation rules for a, $a^\dagger$ a standard Heisenberg uncertainty principle treatment yields the following condition on the fluctuations $\Delta E_{1\theta}$, $\Delta E_{2\theta}$ of the quadrature components.

$$\frac{\Delta E_{1\theta}}{\lambda} \cdot \frac{\Delta E_{2\theta}}{\lambda} \geq 1 \tag{4}$$

In the case of a coherent state (including the vacuum state) $\Delta E_{1\theta}/\lambda = \Delta E_{2\theta}/\lambda = 1$. In the squeezing situation one component, $E_{1\theta}$ say, has a smaller fluctuation than this, so the criterion for squeezing is given by

$$\frac{\Delta E_{1\theta}}{\lambda} < 1 \tag{5}$$

In idealised two-photon coherent states [1] the minimum uncertainty product condition $\Delta E_{1\theta}/\lambda \cdot \Delta E_{2\theta}/\lambda = 1$ occurs also, though this is not the case for arbitrary squeezed states.

The fluctuation $\Delta E_{1\theta}$ is given by

$$\begin{aligned}\left[\frac{\Delta E_{1\theta}}{\lambda}\right]^2 &\equiv \frac{1}{\lambda^2} \langle (E_{1\theta} - \langle E_{1\theta}\rangle)^2\rangle \\ &= 1 + \Big[-\langle (a - \langle a\rangle)^2\rangle\, e^{2i(\omega t+\theta)} - \langle (a^\dagger - \langle a^\dagger\rangle)^2\rangle\, e^{-2i(\omega t+\theta)} \\ &\qquad + 2\langle (a^\dagger - \langle a^\dagger\rangle)(a - \langle a\rangle)\rangle\Big]\end{aligned} \tag{6}$$

In (6) the quantity in brackets gives the expression for the expectation value of the normally ordered quantity $:(E_{1\theta} - \langle E_{1\theta}\rangle)^2/\lambda^2:$ (annihilation operators to the right of creation operators) which cannot be negative for a classical EM field - hence the non classical nature of squeezing. The expression for the fluctuation $\Delta E_{2\theta}$ of the other quadrature component can be obtained from (6) via $\theta \to \theta - \frac{\pi}{2}$.

3. THE MICROSCOPIC MODEL

The treatment of the problem of N TLA in a cavity (see Fig. 1) is based on a standard master equation for the density operator ρ similar to those used by DRUMMOND and WALLS [10], REID and WALLS [11] for discussing optical bistability and the two photon laser, though of course with appropriate terms neglected. The positive P representation [12] for the quasi distribution function is introduced and a Fokker Planck equation derived via neglecting higher derivatives on the basis of a 1/N scaling argument. Ito stochastic differential equations (SDE) for the atomic and cavity field variables are then obtained. An adiabatic elimination of the atomic variables in terms of the field variables and the Langevin noise terms then follows. SDE's for the field variables containing both deterministic terms and noise terms are then obtained. The deterministic equation for the cavity variables

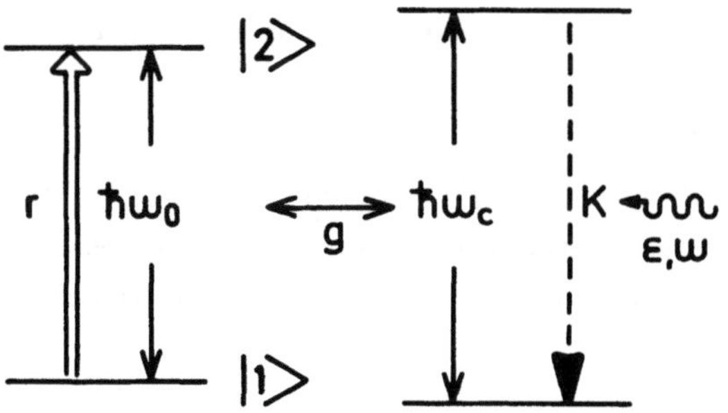

Figure 1. Microscopic model used. N two-level Rydberg atoms (left) interact with a single mode cavity (right). An incident field ϵ of frequency ω is coupled to the cavity mode. K is the cavity damping rate, r is an incoherent pumping rate from the lower to upper level of the two level atom and g is the atom-cavity mode coupling constant.

is then solved to give a steady-state solution, which may be expressed in terms of scaled atomic and cavity detunings, incident and cavity field intensities as well as parameters such as the cooperativity. Linear fluctuation theory (in which the cavity variables are written as a sum of the steady-state solution and a small fluctuation, the deterministic part of the SDE's for the cavity variables expanded about the steady state to the first order in the fluctuations and the coefficients coupling in the fundamental atomic Langevin noise terms evaluated for the steady state) is then used. This leads to SDE's for the cavity-variable fluctuations with both the deterministic and noise terms having constant coefficients. The eigenvalues of the matrix (stability matrix) associated with the deterministic terms are examined to check on the stability of the steady-state solution. Finally the steady-state covariance matrix for the cavity-variable fluctuations, which is related to the quadrature fluctuations $\Delta E_{1\theta}$, $\Delta E_{2\theta}$, can be obtained from the stability matrix, a matrix involved in the coupling of the atomic noise terms and the matrix obtained from the diffusion part of the Fokker Planck equation. The elements of the covariance matrix then can be obtained as solutions for a set of linear equations.

The positive P representation is introduced via the characteristic function $\chi(\underline{\lambda})$ $(\underline{\lambda} = \lambda^+, \eta, \lambda, \beta^+, \beta)$ defined as

$$\chi(\underline{\lambda}) = \mathrm{Tr}\,(e^{i\lambda^+ S^\dagger}\, e^{i\eta S^z}\, e^{i\lambda S}\, e^{i\beta^+ a^\dagger}\, e^{i\beta a}\, \rho) \tag{7}$$

where $S^\dagger = \sum_i (|2\rangle\langle 1|)_i$, $S = \sum_i (|1\rangle\langle 2|)_i$ and $S^z = \sum_i \tfrac{1}{2}(|2\rangle\langle 2| - |1\rangle\langle 1|)_i$ are atomic operators for the N TLA, and $\lambda^+, \ldots, \beta$ are complex quantities.

The distribution function $f(\underline{\alpha})$ $(\underline{\alpha} = v^+, D, v, \alpha^+, \alpha)$ is related to the characteristic function via

$$\chi(\underline{\lambda}) = \int d^2\underline{\alpha}\; e^{i\underline{\lambda}\cdot\underline{\alpha}}\, f(\underline{\alpha}) \tag{8}$$

where $d^2\underline{\alpha} \equiv d^2v^+\, d^2D\, d^2v\, d^2\alpha^+\, d^2\alpha$, $\underline{\lambda}\cdot\alpha = v^+\lambda^+ + D\eta + v\lambda + \alpha^+\beta^+ + \alpha\beta$ and the integral is over the five independent complex planes involved. For convenience we change to final field and atomic variables via $\alpha = \xi\, e^{-i\omega t}$, $\alpha^+ = \xi^+ e^{i\omega t}$, $v = u e^{-i\omega t}$, $v^+ = u^+ e^{i\omega t}$, $D = W/2$ (where ω is the frequency of the incident field) and the Fokker Planck equation is given in terms of the associated distribution function $\Phi(\underline{\xi}) = \tfrac{1}{2} f(\underline{\alpha})$ $(\underline{\xi} = u^+, W, u, \xi^+, \xi)$ normalized so that

$$\int d^2\underline{\xi}\; \Phi(\underline{\xi}) = 1 \tag{9}$$

The importance of the distribution function lies in the fact that the expectation value of any normally ordered product of powers of the operators

$S^\dagger, S^z, S, a^\dagger, a$ (such as those involved in (6)) can be expressed as a integral over the distribution function.

$$\langle (S^\dagger)^p (S^z)^q (S)^r (a^\dagger)^t (a)^u \rangle = e^{i(p-r+t-u)\omega t} \times \int d^2\underline{\xi}\, (u^+)^p (W)^q (u)^r (\xi^+)^t (\xi)^u\, \Phi(\underline{\xi}) \quad (10)$$

Regarding $\underline{\xi}$ as a set of stochastic variables for which a representative ensemble of trajectories is equivalent to the distribution function $\Phi(\underline{\xi})$, the SDE associated with the Fokker Planck equation for Φ are:

$$\begin{aligned}
\dot{\xi} &= -K_c \xi + gu + \epsilon \\
\dot{\xi}^+ &= -K_c^* \xi^+ + gu^+ + \epsilon^* \\
\dot{u} &= -K_r u + gW\xi + \Lambda_u \\
\dot{u}^+ &= -K_r^* u^+ + gW\xi^+ + \Lambda_{u^+} \\
\dot{W} &= -r(-N+W) - 2g(u\xi^+ + u^+\xi) + \Lambda_W
\end{aligned} \quad (11)$$

where

$$\begin{aligned}
\Lambda_u &= d_{uu}\, \Gamma_u + d_{uu^+} \Gamma_{u^+} + d_{uW}\, \Gamma_W \\
\Lambda_{u^+} &= d_{u^+u}\, \Gamma_u + d_{u^+u^+} \Gamma_{u^+} + d_{u^+W}\, \Gamma_W \\
\Lambda_W &= d_{Wu}\, \Gamma_u + d_{Wu^+} \Gamma_{u^+} + d_{WW}\, \Gamma_W .
\end{aligned} \quad (12)$$

The Langevin noise terms satisfy

$$\langle \Gamma_\alpha(t)\, \Gamma_\beta(t') \rangle = \delta_{\alpha\beta}\delta(t-t') \qquad (\alpha,\beta = u, u^+, W) \quad (13)$$

and the coefficients $d_{\alpha\beta}(\underline{\xi})$ are related to the matrix D from the diffusion terms in the Fokker Planck equation via

$$\text{(i)}\quad D_{\mu\nu} = (d\tilde{d})_{\mu\nu} \quad (14a)$$

$$\begin{aligned}
\text{(ii)}\quad D_{uu} &= 2gu\xi \\
D_{u^+u^+} &= 2gu^+\xi^+ \\
D_{uu^+} &= D_{u^+u} = rN \\
D_{uW} &= D_{Wu} = -2ru \\
D_{u^+W} &= D_{Wu^+} = -2ru^+ \\
D_{WW} &= -4g(u\xi^+ + u^+\xi) + 2r(N-W).
\end{aligned} \quad (14b)$$

In the above equations $\delta_c = \omega_c - \omega$ is the cavity detuning, $\delta = \omega_o - \omega$ is the atomic detuning, K is the cavity decay rate, r is the incoherent pump rate, g the atom cavity mode coupling constant (proportional to $1/\sqrt{V}$) and ϵ the amplitude factor for the incident field of frequency ω. $K_c = K + i\delta_c$, and $K_r = \frac{1}{2}r + i\delta$.

After adiabatically eliminating the atomic variables u^+, W, u the following SDE's for the field variables ξ^+, ξ are obtained:

$$\dot{\xi} = -K_c\xi + \frac{g^2N}{K_r A}\xi + \epsilon + G_{\xi u}\Lambda_u + G_{\xi u^+}\Lambda_{u^+} + G_{\xi W}\Lambda_W \quad (15a)$$

$$\dot{\xi}^+ = -K_c^*\xi^+ + \frac{g^2N}{K_r^* A}\xi^+ + \epsilon^* + G_{\xi^+u}\Lambda_u + G_{\xi^+u^+}\Lambda_{u^+} + G_{\xi^+W}\Lambda_W \quad (15b)$$

where

$$\text{(i)}\quad A = 1 + 2g^2\xi^+\xi/(\delta^2 + \frac{1}{4}r^2) \tag{16a}$$

$$\text{(ii)}\quad G_{\xi u} = \frac{g}{K_r}\left[1 - \frac{2g^2\xi^+\xi}{rK_rA}\right]$$

$$G_{\xi u^+} = -\frac{2g^3\xi^2}{r(\delta^2 + \frac{1}{4}r^2)A}$$

$$G_{\xi W} = \frac{g^2\xi}{rK_rA}$$

$$G_{\xi^+u} = -\frac{2g^3\xi^{+2}}{r(\delta^2 + \frac{1}{4}r^2)A}$$

$$G_{\xi^+u^+} = \frac{g}{K_r^*}\left[1 - \frac{2g^2\xi^+\xi}{rK_r^*A}\right]$$

$$G_{\xi^+W} = \frac{g^2\xi^+}{rK_r^*A} \tag{16b}$$

The deterministic equation associated with (15) leads to the steady-state solution ξ_o, $\xi_o^+ = \xi_o^*$ given in terms of the cavity intensity X and phase φ via

$$Y = X\left[\left[1 - \frac{2C}{1 + \Delta^2 + X}\right]^2 + \left[\Delta_C + \frac{2C\,\Delta}{1 + \Delta^2 + X}\right]^2\right] \tag{17a}$$

$$\varphi = \eta + \theta \tag{17b}$$

where

$$\text{(i)}\quad \cos\theta = \left[1 - \frac{2C}{1 + \Delta^2 + X}\right]/S$$

$$\text{(ii)}\quad \sin\theta = \left[\Delta_C + \frac{2C\Delta}{1 + \Delta^2 + X}\right]/S \tag{18a}$$

$$\text{(iii)}\quad S = \sqrt{\left[1 - \frac{2C}{1 + \Delta^2 + X}\right]^2 + \left[\Delta_c + \frac{2C\Delta}{1 + \Delta^2 + X}\right]^2} \tag{18b}$$

For (17), (18) the incident field has been written $\epsilon = |\epsilon|e^{-i\eta}$, the cavity field $\xi_o = |\xi_o|e^{-i\varphi}$ with θ the phase shift. In (17), (18) $\Delta_c = \delta_c/K$ is the scaled cavity detuning, $\Delta = 2\delta/r$ is the scaled atomic detuning and $C = g^2N/rK$ is the cooperativity (which is proportional to the atomic density N/V). $n_o = r^2/8g^2$ is used to define the cavity field intensity X via $X = |\xi_o|^2/n_o$ and the incident field intensity Y via $Y = |\epsilon/K|^2/n_o$.

The intensity equation (17a) is almost the same as that conventionally used to describe optical bistability, differing only in that C is replaced by −C.

Writing ξ as $\xi_o + \delta\xi$ and ξ^+ as $\xi_o^* + \delta\xi^+$ the linear SDE for the fluctuations $\delta\xi$, $\delta\xi^+$ are given as

$$\begin{bmatrix}\dot{\delta\xi}\\ \dot{\delta\xi}^+\end{bmatrix} = \begin{bmatrix}B_{\xi\xi} & B_{\xi\xi^+}\\ B_{\xi^+\xi} & B_{\xi^+\xi^+}\end{bmatrix}\begin{bmatrix}\delta\xi\\ \delta\xi^+\end{bmatrix}$$

$$+ \begin{bmatrix}G_{\xi u}(0) & G_{\xi u^+}(0) & G_{\xi W}(0)\\ G_{\xi^+ u}(0) & G_{\xi^+ u^+}(0) & G_{\xi^+ W}(0)\end{bmatrix}\begin{bmatrix}d_{uu}(0) & d_{uu^+}(0) & d_{uW}(0)\\ d_{u^+u}(0) & d_{u^+u^+}(0) & d_{u^+W}(0)\\ d_{Wu}(0) & d_{Wu^+}(0) & d_{WW}(0)\end{bmatrix}\begin{bmatrix}\Gamma_u\\ \Gamma_{u^+}\\ \Gamma_W\end{bmatrix} \tag{19}$$

where $G_{\alpha\beta}(0)$, $d_{\alpha\beta}(0)$ are the steady-state values of $G_{\alpha\beta}$, $d_{\alpha\beta}$.

In terms of the scaled parameters the stability matrix B in (19) is given by

$$B = -K\begin{bmatrix}a & b\\ b^* & a^*\end{bmatrix} \tag{20}$$

with

(i) $$a = a_R + ia_I \tag{21a}$$

(ii) $$a_R = 1 - \frac{2C(1+\Delta^2)}{(1+\Delta^2+X)^2}$$

$$a_I = \Delta_c + \frac{2C\Delta(1+\Delta^2)}{(1+\Delta^2+X)^2} \tag{21b}$$

(iii) $$b = b_R + ib_I \tag{21c}$$

(iv) $$b_R = \frac{2C\,X}{(1+\Delta^2+X)^2}(\cos 2\varphi - \Delta \sin 2\varphi)$$

$$b_I = \frac{-2C\,X}{(1+\Delta^2+X)^2}(\sin 2\varphi + \Delta\cos 2\varphi). \tag{21d}$$

The stability of the steady-state solution depends on the eigenvalues of B having negative real parts.

The matrix G in (19) can be written as

$$G = \frac{1}{\sqrt{n_o}}\begin{bmatrix}g_{11} & g_{12} & g_{13}\\ g_{12}^* & g_{11}^* & g_{13}^*\end{bmatrix} \tag{22}$$

where

(i) $$g_{11} = \frac{1}{\sqrt{2}\,(1+i\Delta)} - \frac{X\,(1-i\Delta)}{2\sqrt{2}\,(1+i\Delta)(1+\Delta^2+X)} \tag{23a}$$

(ii) $$g_{12} = -\frac{X\,e^{-2i\varphi}}{\sqrt{8}\,(1+\Delta^2+X)} \tag{23b}$$

(iii) $$g_{13} = \frac{\sqrt{X}\,e^{-i\varphi}\,(1-i\Delta)}{4\,(1+\Delta^2+X)}\,. \tag{23c}$$

It is easy to see that $\xi - \langle\xi\rangle = \delta\xi - \langle\delta\xi\rangle$, $\xi^+ - \langle\xi^+\rangle = \delta\xi^+ - \langle\delta\xi^+\rangle$ and we define the covariance matrix $\mathcal{C}$ via the stochastic averages (SA)

$$\mathscr{C}_{\xi\xi} = \langle(\delta\xi - \langle\delta\xi\rangle)^2\rangle_{SA} \tag{24a}$$

$$\mathscr{C}_{\xi^+\xi} = \langle(\delta\xi^+ - \langle\delta\xi^+\rangle)(\delta\xi - \langle\delta\xi\rangle)\rangle_{SA} \equiv \mathscr{C}_{\xi\xi^+} \tag{24b}$$

$$\mathscr{C}_{\xi^+\xi^+} = \langle(\delta\xi^+ - \langle\delta\xi^+\rangle)^2\rangle_{SA} \tag{24c}$$

The covariance matrix at long times $\mathscr{C}(\infty)$ can be obtained (see [13]) from the equation

$$B\mathscr{C}(\infty) + \mathscr{C}(\infty)\,\bar{B} = -\,G(0)\;D(0)\;\bar{G}(0) \tag{25}$$

with

$$D(0) = K\,n_o\,4C \begin{bmatrix} d_{11} & 2 & d_{13} \\ 2 & d_{11}^* & d_{13}^* \\ d_{13} & d_{13}^* & 0 \end{bmatrix} \quad \text{where} \tag{26}$$

$$\text{(i)} \quad d_{11} = \frac{X\,(1 - i\Delta)\,e^{-2i\varphi}}{(1 + \Delta^2 + X)} \tag{27a}$$

$$\text{(ii)} \quad d_{13} = -\,\frac{\sqrt{8X}\,(1 - i\Delta)\,e^{-i\varphi}}{(1 + \Delta^2 + X)}\,. \tag{27b}$$

Thus only the diffusion matrix D(0) (14) is needed along with the coupling matrix G(0) and the stability matrix B(0) in order to determine the long-time covariances.

Equation (25) is the same as the linear equations

$$\begin{bmatrix} 2a & b & b & 0 \\ b^* & 2a_R & 0 & b \\ b^* & 0 & 2a_R & b \\ 0 & b^* & b^* & 2a^* \end{bmatrix} \begin{bmatrix} \mathscr{C}_{\xi\xi} \\ \mathscr{C}_{\xi\xi^+} \\ \mathscr{C}_{\xi^+\xi} \\ \mathscr{C}_{\xi^+\xi^+} \end{bmatrix} = \begin{bmatrix} \mathscr{N}_{\xi\xi} \\ \mathscr{N}_{\xi\xi^+} \\ \mathscr{N}_{\xi^+\xi} \\ \mathscr{N}_{\xi^+\xi^+} \end{bmatrix} \quad \text{where} \tag{28}$$

$$\mathscr{N} = g\,\mathscr{D}\,\bar{g} \tag{29}$$

and g, $\mathscr{D}$ are the matrices associated with (22), (26) but without the factors $1/\sqrt{n_o}$, Kn_o respectively.

4. MINIMUM SQUEEZING

The covariance matrix elements may be related to averages of $(\xi - \langle\xi\rangle)^2$, $(\xi^+ - \langle\xi^+\rangle)(\xi - \langle\xi\rangle)$, $(\xi^+ - \langle\xi^+\rangle)^2$ given in terms of the distribution function $\Phi(\xi)$ via the equivalence of the stochastic process to the latter function. These last averages can be related via (10) to the quantum averages to give

$$\mathscr{C}_{\xi\xi} = \langle(a - \langle a\rangle)^2\rangle\, e^{2i\omega t} \tag{30a}$$

$$\mathscr{C}_{\xi^+\xi} = \langle(a^\dagger - \langle a^\dagger\rangle)(a - \langle a\rangle)\rangle = (\mathscr{C}_{\xi^+\xi})^* \tag{30b}$$

$$\mathscr{C}_{\xi^+\xi^+} = \langle(a^\dagger - \langle a^\dagger\rangle)^2\rangle\, e^{-2i\omega t} = \mathscr{C}_{\xi\xi}^* \tag{30c}$$

Combining (30), (6) and choosing θ to give the minimum and maximum squeezing we obtain finally

$$\left[\frac{\Delta E_{1\theta}}{\lambda}\right]^2_{MIN} = 1 + 2\left[\mathscr{C}_{\xi^+\xi} - |\mathscr{C}_{\xi\xi}|\right] \tag{31a}$$

$$\left[\frac{\Delta E_{1\theta}}{\lambda}\right]^2_{MAX} = 1 + 2\left[\mathcal{C}_{\xi^+\xi} + |\mathcal{C}_{\xi\xi}|\right] \tag{31b}$$

$$= \left[\frac{\Delta E_{2\theta}}{\lambda}\right]^2_{MIN} \tag{31c}$$

with the maximum squeezing corresponding to that for the other quadrature component $E_{2\theta}$ associated with the choice of θ for minimum squeezing.

Examination of the φ dependence of the matrices B, G(0), D(0) shows that the same values for the minimum, maximum values of the fluctuation $\left[\frac{\Delta E_{1\theta}}{\lambda}\right]$ occur if the cavity field phase φ is put equal to zero in (25), though of course the quadrature angle θ will shift. As the phase of the incident field η only appears via φ it follows that the maximum and minimum values of $(\Delta E_{1\theta}/\lambda)$ are not dependent on η.

5. RESULTS

Figure 2 shows the steady-state cavity versus incident field intensity for the absorptive regime with zero detunings, $\Delta = \Delta_c = 0$, and for various cooperativities. The physical situation involved is that of the Rydberg atom laser with injected signal, and is in agreement with the work of Lugiato [14]. Only the upper branches are stable and correspond to laser action, with C = ½ defining the threshold cooperativity.

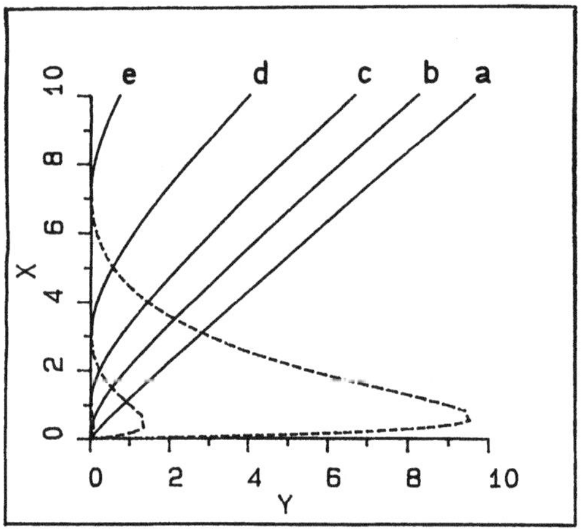

Figure 2. Steady-state cavity field intensity X versus incident field intensity Y for the absorptive regime $\Delta = \Delta_C = 0$ and for various cooperativities C: (a) C = 0.1, (b) C = 0.5, (c) C = 1.0, (d) C = 2.0 and (e) C = 4.0. The solid line indicates a stable branch, the dashed line an unstable branch.

Figure 3 shows the minimum fluctuation $V = (\Delta E_{1\theta}/\lambda)_{MIN}$ (see (31)) as a function of cavity field intensity for the above absorptive case and for the threshold case C = ½. Very little squeezing occurs, and this feature also applies to the other values of C in Fig. 2.

Figure 4 shows the steady-state cavity versus incident field intensity for a dispersive regime ($\Delta \gg 1$, $C \ll \Delta^2$, $\Delta^2 \gg X \gg 1$) situation $C = 2 \times 10^6$, $\Delta = 10^4$ similar to a case examined by REID and WALLS [15] in regard to squeezing via optical bistability in a non pumped medium of TLA. Results for various cavity detunings are shown. Optical bistability occurs with the critical cavity detuning $\Delta_c = -398.4$ here being given by

$$\left|\Delta_c + \frac{2C}{\Delta}\right| = \sqrt{3} \tag{32}$$

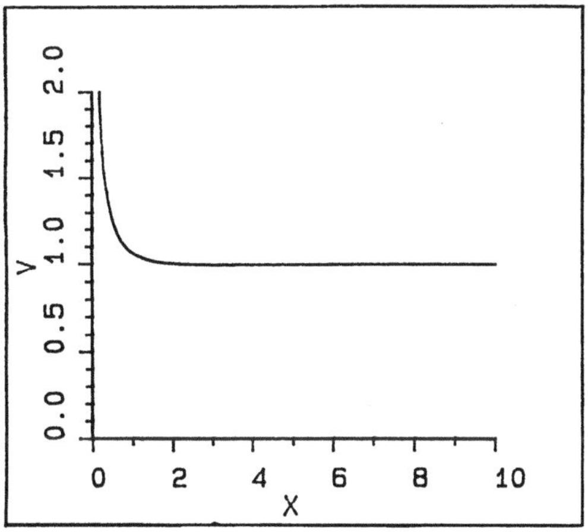

Figure 3. Minimum fluctuation $V = (\Delta E_{1\theta}/\lambda$ in the quadrature component of the cavity field versus cavity field intensity X for the absorptive regime $\Delta = \Delta_C = 0$ and with $C = 0.5$.

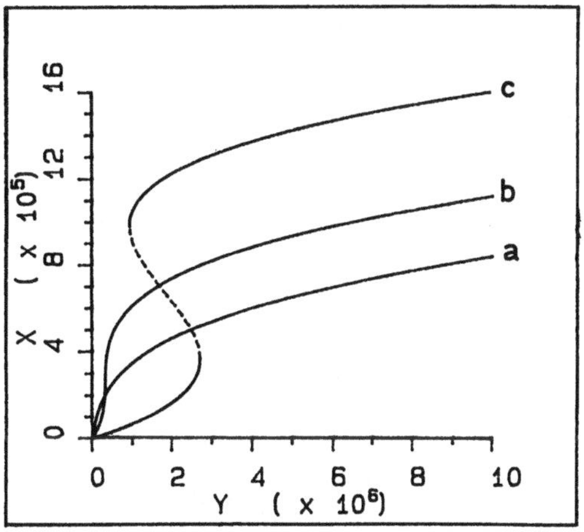

Figure 4. Steady-state cavity field intensity X (in units of 10^5) versus incident field intensity Y (in units of 10^6) for a dispersive regime situation $C = 2 \times 10^6$, $\Delta = 10^4$ and for various cavity detunings Δ_C:

(a) $\Delta_C = -400$, (b) $\Delta_C = -398.4$ and (c) $\Delta_C = -396$. The solid line indicates a stable branch, the dashed line an unstable branch.

This equation has the opposite sign for C to that normally occurring for dispersive optical stability in a passive medium and results in the opposite sign for the critical cavity detuning to that found by REID and WALLS [15]. Relatively little work appears to have been done in the area of optical bistability in a pumped or active medium, see however ARRECHI et al [16].

Figure 5 shows the minimum fluctuation $V = (\Delta E_{1\theta}/\lambda)_{MIN}$ as a function of cavity field intensity for the above dispersive case. Only results for the stable branches are given. The squeezing is reasonably large especially for critical case $\Delta_c = -398.4$, where the value of $\Delta E_{1\theta}/\lambda$ is approximately 0.75 at the critical point, which as in the work of REID and WALLS [15] corresponds to the largest squeezing . The occurrence here of squeezing in an active medium is somewhat unexpected.

6. CONCLUSIONS

A case of optical bistability along with reasonably large squeezing has been found in the dispersive regime for an active medium of pumped two-level Rydberg atoms in a high Q cavity.

ACKNOWLEDGEMENTS

The author is grateful for helpful discussions with Drs. P.L. Knight, S.M. Barnett, H. Carmichael, M.D. Reid and Professor D.F. Walls and for financial support from the British Council.

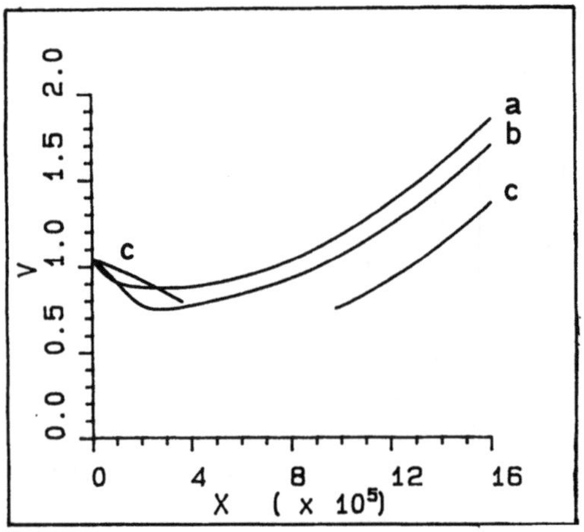

Figure 5. Minimum fluctuation $V = \Delta E_{1\theta}/\lambda$ in the quadrature component of the cavity field versus cavity field intensity X (in units of 10^5) for a dispersive regime situation $C = 2 \times 10^6$, $\Delta = 10^4$ and for various cavity detunings (a) $\Delta_C = -400$, (b) $\Delta_C = -398.4$ and (c) $\Delta_C = -396$. Only stable branches are shown.

REFERENCES

1. H.P. Yuen: Phys. Rev. A13, 2226 (1976)
2. R.E. Slusher, L. Hollberg, B. Yurke, J.C. Mertz and J.F. Valley: Phys. Rev. Lett. 55, 2409 (1985)
3. D.F. Walls: Nature 306, 141 (1983)
4. M.D. Reid and D.F. Walls: Phys. Rev. A31, 1622 (1985)
5. A. Heidmann, J.M. Raimond and S.Reynaud: Phys. Rev. Lett. 54, 326 (1985)
6. A. Heidmann, J.M. Raimond, S. Reynaud and N. Zagury: Opt. Comm. 54, 54 (1985)
7. A. Heidmann, J.M. Raimond, S. Reynaud and N. Zagury: Opt. Comm. 54, 189 (1985)
8. B. Yurke: Phys. Rev. A23, 2563 (1981)
9. C.W. Gardiner and M.J. Collet: Phys. Rev. A31, 3761 (1985)
10. P.D. Drummond and D.F. Walls: Phys. Rev. A23, 2563 (1981)
11. M.D. Reid and D.F. Walls: Phys. Rev. A28, 332 (1983)
12. P.D. Drummond and C.W. Gardiner: J. Phys. A13, 2353 (1980)
13. C.W. Gardiner: Handbook of Stochastic Processes (Academic, New York 1983)
14. L.A. Lugiato: Lett. Nuov. Cim. 23, 609 (1978)
15. M.D. Reid and D.F. Walls: Phys. Rev. A32, 396 (1985)
16. F.T. Arrecchi, G. Giusfredi, E. Petriella and P. Salieri: App. Phys. B29, 79 (1982).

Part II

Optical Bistability and Chaos

Single- and Multimode Instabilities in Lasers

H. Risken

Abteilung für Theoretische Physik, Universität Ulm,
D-7900 Ulm, Fed. Rep. of Germany

1 Introduction

Instabilities play an important role in a large number of fields for example in hydrodynamics, plasma physics, semiconductor physics, ecology and economy. Also in lasers can a great variety of instabilities be found. Compared to hydrodynamics and plasma physics the nonlinear laser equations are, at least for simple ring-laser models, rather simple. Therefore the investigation of the instabilities in lasers is easier to carry out than in most other fields. The main simplification seems to be that for our ring-laser model only one space coordinate needs to be taken into account, whereas for the instabilities in hydrodynamics at least two space coordinates are usually to be considered.

The instabilities in lasers change mainly with increasing pumping strength. The onset of laser action for one mode is the first instability. By increasing further the pumping strength new instabilities occur. Depending on the parameters and on the model, either the intensity of the one mode will start to oscillate or some off-resonance modes will become unstable. The last instability then leads to a build-up of stable light pulses, which travel through the laser cavity. If the pumping strength is increased further, the light pulses themselves are no longer stable. They change their height and their shape in a periodic or even chaotic way. This effect may be termed periodic or chaotic breathing of the pulses. It should be noted that the one-mode laser equations are equivalent to the Lorenz equations. Therefore for appropriate parameters the one-mode laser equations will also show chaotic behaviour.

2 Model and Basic Equations

In our investigations we want to restrict ourselves to the simplest case. We therefore assume that the active material consists of two-level atoms with a homogeneously broadened line. In addition we assume that the cavity losses are homogeneously distributed in the cavity and that we have only one direction of polarization and that the field depends only on the time and on the coordinate x of the laser axis. The **basic equations** [1,2,3] are: 1) the one-dimensional wave equation for the electric field $\tilde{E}(x,t)$ (telegrapher's equation) driven by the polarization $\tilde{P}(x,t)$, i.e.

$$\frac{\partial^2 \tilde{E}}{\partial x^2} - \frac{1}{c^2}\frac{\partial^2 \tilde{E}}{\partial t^2} - \frac{2\kappa}{c^2}\frac{\partial \tilde{E}}{\partial t} = \mu \frac{\partial^2 \tilde{P}}{\partial t^2}\,, \tag{2.1}$$

2) the material equations for polarization $\tilde{P}(x,t)$ and inversion $\tilde{S}(x,t)$ driven by the electric field, i.e.

$$\frac{\partial^2 \tilde{P}}{\partial t^2} + 2\gamma_2 \frac{\partial \tilde{P}}{\partial t} + \omega_0{}^2 \tilde{P} = -2(M^2/\hbar)\,\omega_0 \tilde{E}\tilde{S}$$

$$\frac{\partial \tilde{S}}{\partial t} = \gamma_1 (d_0 - \tilde{S}) + \frac{2}{\hbar\omega_0}\frac{\partial \tilde{P}}{\partial t}\tilde{E} \tag{2.2}$$

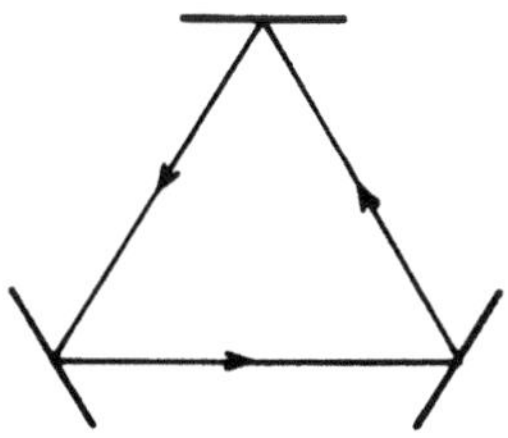

Fig. 1. Ring-laser cavity

and 3) appropriate boundary conditions. In (2.1,2) κ describes the losses, μ is the permeability, $c = 1/\sqrt{\varepsilon\mu}$ the velocity of light in the host material, M the dipole matrix element and $\hbar\omega_0$ the energy separation of the two levels of the active atoms. The constants γ_1 and γ_2 describe the damping of the inversion and polarization respectively and d_0 describes the strength of the pumping. Here we want to consider a ring-laser system, see Fig. 1. Therefore we have the boundary conditions

$$\tilde{\tilde{E}}(x+L,t) = \tilde{\tilde{E}}(x,t), \quad \tilde{\tilde{P}}(x+L,t) = \tilde{\tilde{P}}(x,t), \quad \tilde{\tilde{S}}(x+L,t) = \tilde{\tilde{S}}(x,t)\ , \tag{2.3}$$

where L is the cavity length.

Because the frequency ω_0 is usually much larger than any other time constant involved, (2.1) and (2.2) can be simplified by the **rotating wave approximation.** Assuming only one direction of propagation, we write

$$\tilde{\tilde{E}} = \tilde{E}\exp[i\omega_0(x/c-t)] + \text{c.c.}\ ,$$

$$\tilde{\tilde{P}} = \tilde{P}\exp[i\omega_0(x/c-t)] + \text{c.c.}\ ,$$

$$\tilde{\tilde{S}} = \tilde{S}\ , \tag{2.4}$$

where $\tilde{E}(x,t)$, $\tilde{P}(x,t)$ are slowly varying complex envelopes in x and t and where + c.c means that we have to add the complex conjugate. If higher harmonics are neglected the inversion $\tilde{S}$ will also slowly vary in x and t. Neglecting higher order harmonics, the insertion of (2.4) into (2.1,2) leads to ($\partial/\partial t$ is now denoted by a dot)

$$c(\partial\tilde{E}/\partial x) + \dot{\tilde{E}} + \kappa\tilde{E} = i(\omega_0/2\varepsilon)\tilde{P}$$

$$\dot{\tilde{P}} + \gamma_2\tilde{P} = -i(M^2/\hbar)\tilde{E}\tilde{S}$$

$$\dot{\tilde{S}} + \gamma_1\tilde{S} = \gamma_1 d_0 + (2i/\hbar)(\tilde{E}\tilde{P}^* - \tilde{E}^*\tilde{P})\ . \tag{2.5}$$

If we assume that **one cavity mode is in resonance** with the atomic frequency ω_0, i.e. $\exp(i\omega_0 L/c)=1$, the boundary conditions for the slowly varying fields $\tilde{E}$, $\tilde{P}$, $\tilde{S}$ will be the same as those in (2.3) for the rapidly varying fields.

For $d_0 < \tilde{S}_{cw}$ where $\tilde{S}_{cw}$ is defined in (2.6) we have only the trivial solution $\tilde{P} = \tilde{E} = 0$, $\tilde{S} = d_0$. This trivial solution is stable for $d_0 < \tilde{S}_{cw}$ but unstable for $d_0 > \tilde{S}_{cw}$.

The equations (2.5) have a nontrivial solution for d_0 larger than $\tilde{S}_{cw}$, which is independent of x and t. This cw solution is given by

$$\tilde{S}_{cw} = (2\varepsilon\hbar\kappa\gamma_2)/(\omega_0 M^2)$$

$$\tilde{E}_{cw} = \sqrt{(d_0-\tilde{S}_{cw})\gamma_1\hbar\omega_0/(2\kappa\varepsilon)/2}$$

$$\tilde{P}_{cw} = -i(2\varepsilon\kappa/\omega_0)\tilde{E}_{cw} \ . \tag{2.6}$$

It will be useful to introduce the following **normalized fields**

$$E = \tilde{E}/\tilde{E}_{cw}, \quad P = \tilde{P}/\tilde{P}_{cw}, \quad \sigma = \tilde{S}/\tilde{S}_{cw}. \tag{2.7}$$

The equations (2.5) then reduce to the following normalized form [4]

$$c(\partial E/\partial x) + \dot{E} + \kappa E = \kappa P$$

$$\dot{P} + \gamma_2 P = \gamma_2 E\sigma$$

$$\dot{\sigma} + \gamma_1\sigma = \gamma_1[\lambda + 1 - \lambda(E^*P + EP^*)/2] \ , \tag{2.8}$$

where the normalized pump parameter λ is given by

$$\lambda = (d_0 - \tilde{S}_{cw})/\tilde{S}_{cw} \ . \tag{2.9}$$

3 Single-Mode Instability

Now we want to investigate the stability of the cw solution (2.6) for the resonant mode. Therefore the field envelope does not depend on x. Furthermore we consider only real values of E and P (i.e. we assume that the phases as given by (2.6) are constant). Equations (2.8) then reduce to

$$\dot{E} + \kappa E = \kappa P$$

$$\dot{P} + \gamma_2 P = \gamma_2 E\sigma$$

$$\dot{\sigma} + \gamma_1\sigma = \gamma_1(\lambda + 1 - \lambda EP) \ . \tag{3.1}$$

The main trick in investigating the stability of a solution of nonlinear equations like (3.1) is to introduce additional small deviations to the known solution. Because of their smallness one can neglect products of the small deviations and thus obtain linear equations which are much easier to solve than the original nonlinear ones. Thus introducing the small deviations e, p, δ from the cw solution ($E = P = \sigma = 1$), we obtain the **linearized equations** for e, p, δ

$$\dot{e} + \kappa(e - p) = 0$$

$$\dot{p} + \gamma_2(p - e - \delta) = 0$$

$$\dot{\delta} + \gamma_1[\delta + \lambda(e+p)] = 0 \ . \tag{3.2}$$

The ansatz

$$\begin{pmatrix} e \\ p \\ \delta \end{pmatrix} = \begin{pmatrix} e_0 \\ p_0 \\ \delta_0 \end{pmatrix} \cdot \exp(\beta t) \tag{3.3}$$

then leads to the cubic equation

$$\beta^3 + (\gamma_1+\gamma_2+\kappa)\beta^2 + \gamma_1(\gamma_2+\kappa+\gamma_2\lambda)\beta + 2\gamma_1\gamma_2\kappa\lambda = 0 \ . \tag{3.4}$$

It follows from the Hurwitz criterion that the solutions (3.3) are stable, i.e. have a negative real part, if

either $\quad \gamma_1 + \gamma_2 \geq \kappa$

or $\quad \gamma_1 + \gamma_2 < \kappa; \quad \lambda \leq \lambda_s \ . \qquad (3.5)$

Here λ_s is given by [5]

$$\lambda_s = (\gamma_1+\gamma_2+\kappa)(\gamma_2+\kappa)/[(\kappa-\gamma_1-\gamma_2)\gamma_2] . \tag{3.6}$$

The solutions are unstable if

$$\gamma_1 + \gamma_2 < \kappa, \quad \lambda > \lambda_s . \tag{3.7}$$

Thus in order to get unstable solutions for the one-mode case the losses of the cavity must exceed the sum of the losses of the polarization and the inversion (**bad cavity case**). In the **good cavity case** $\kappa<\gamma_1+\gamma_2$, the cw one-mode solution is always stable.

As it was pointed out by HAKEN [3,6] the equations (3.1) are equivalent to the Lorenz equations. Introducing

$$\tau = \gamma_2 t; \quad X = \sqrt{\lambda\gamma_1/\gamma_2}E; \quad Y = \sqrt{\lambda\gamma_1/\gamma_2}P;$$

$$Z = \lambda + 1 - \sigma \tag{3.8}$$

(3.1) are transformed to

$$dX/d\tau = a(Y - X)$$

$$dY/d\tau = X(r - Z) - Y$$

$$dZ/d\tau = XY - bZ , \tag{3.9}$$

which are identical to the LORENZ equations [7]. Here a, b, r are given by

$$a = \kappa/\gamma_2, \quad b = \gamma_1/\gamma_2, \quad r = \lambda + 1 .$$

Therefore, in addition to a limit cycle solution, chaotic solutions are also possible. It should be noted that the condition (3.6,7) (in the notation of the Lorenz equation (3.9)) has been derived by LORENZ himself [7].

4 Instability of Nonresonant Modes

In the good cavity case a single resonant mode is always stable. In the presence of the resonant mode other nonresonant modes, however, may become unstable. To investigate this instability we again introduce **small deviations** e, p, σ, ϕ, ψ from the cw solutions, i.e.

$$E = (1 + e)\cdot\exp(i\phi); \quad P = (1 + p)\cdot\exp(i\psi); \quad \sigma = 1 + \delta . \tag{4.1}$$

The nonlinear system (2.8) thus reduces to the linear system

$$c(\partial e/\partial x) + \dot{e} + \kappa(e-p) = 0$$

$$\dot{p} + \gamma_2(p-e-\delta) = 0$$

$$\dot{\delta} + \gamma_1[\delta + \lambda(e+p)] = 0$$

$$c(\partial\phi/\partial x) + \dot{\phi} + \kappa(\phi-\psi) = 0$$

$$\dot{\psi} + \gamma_2(\psi-\phi) = 0 . \tag{4.2}$$

The ansatz

$$\begin{pmatrix} e \\ p \\ \delta \\ \phi \\ \psi \end{pmatrix} = \begin{pmatrix} e_0 \\ p_0 \\ \delta_0 \\ \phi_0 \\ \psi_0 \end{pmatrix} \exp[i(\alpha/c)x + \beta t] + c.c. \tag{4.3}$$

leads to the characteristic equations

$$\beta^3 + (\gamma_1+\gamma_2+\kappa+i\alpha)\beta^2 + [\gamma_1(\gamma_2+\kappa+\gamma_2\lambda)+(\gamma_1+\gamma_2)i\alpha]\beta + 2\gamma_1\gamma_2\kappa\lambda + \gamma_1\gamma_2(1+\lambda)i\alpha = 0 \tag{4.4}$$

$$\beta^2 + (\kappa+\gamma_2+i\alpha)\beta + i\gamma_2\alpha = 0 . \tag{4.5}$$

Because of the boundary condition (2.3) the wave number α/c should have the values

$$\alpha/c = m\cdot 2\pi/L; \quad m=0, \pm1, \pm2, \dots . \tag{4.6}$$

The values for $m\neq0$ correspond to the different off-resonant modes. It can be derived from (4.4) that the real part of β can become positive if the pump parameter exceeds a critical value [4,8,9]

$$\lambda > \lambda_c = 4 + 3(\gamma_1/\gamma_2) + 2\sqrt{4 + 6(\gamma_1/\gamma_2) + 2(\gamma_1/\gamma_2)^2} . \tag{4.7}$$

The dispersion relations $\beta(\alpha)$ for two typical cases with $\lambda>\lambda_c$ are shown in Figs. 2 and 3. Thus if the discrete α values lie in the appropriate region, these off-resonances become unstable. For the lower part on Fig. 2 and 3 the corresponding region for α reads [4]

$$\alpha_{min} < \alpha < \alpha_{max} ;$$

$$\alpha_{\substack{max \\ min}} = \sqrt{\gamma_1(3\gamma_2\lambda - \gamma_1 \pm R)/2}\cdot\{1 - 2\kappa/[\gamma_2(\lambda-2)-\gamma_1 \pm R]\} ;$$

$$R = \sqrt{\lambda^2\gamma_2^2 - 2(3\gamma_1+4\gamma_2)\lambda\gamma_2 + \gamma_1^2} . \tag{4.8}$$

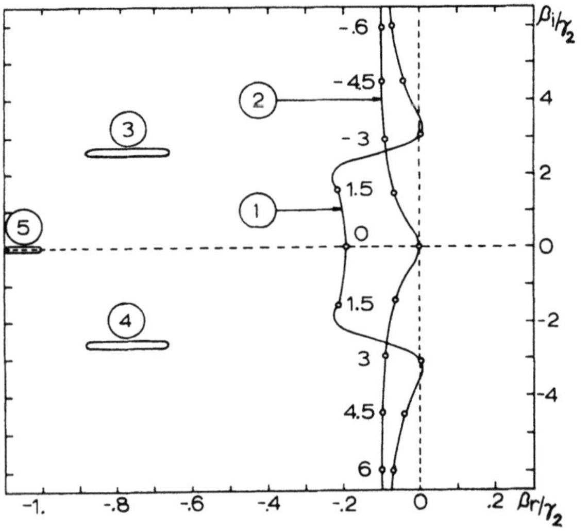

Fig. 2. Roots of the characteristic equations (4.4) and (4.5) with α varying from $-\infty$ to $+\infty$. The fixed parameters are $\gamma_1=\gamma_2/2$, $\kappa=\gamma_2/10$ and $\lambda=15$. ①, ③ and ④ are the roots corresponding to the amplitudes, ② and ⑤ are the roots corresponding to the phases. The instable region of ① is $2.9\gamma_2<|\alpha|<3.6\gamma_2$. The discrete values α_n are shown for $2\pi c/L=1.5\gamma_2$ for root loci ① and ②. Two instable modes are $\alpha_2=-3\gamma_2$ and $\alpha_{-2}=3\gamma_2$

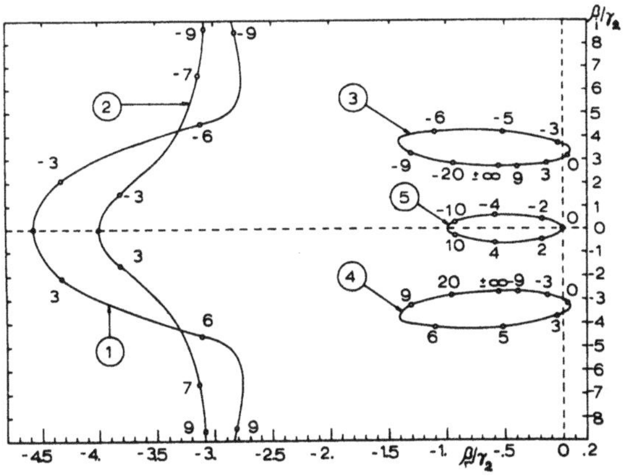

Fig. 3. Roots of the characteristic equations (4.4) and (4.5) with α varying from $-\infty$ to $+\infty$ as marked on the curves. The fixed parameters are $\gamma_1=\gamma_2/2$, $\kappa=3\gamma_2$ and $\lambda=15$. ①, ③ and ④ are the roots corresponding to the amplitudes, ② and ⑤ are the roots corresponding to the phases. The instable regions of ③ and ④ are $-2.5\gamma_2<\alpha<0.9\gamma_2$ and $-0.9\gamma_2<\alpha<2.5\gamma_2$, respectively

One remarkable feature of (4.7) is that it is independent of κ. It can be shown that the critical value (3.6) for varying $\kappa>\gamma_1+\gamma_2$ is larger than or equal to the above critical value ($\lambda_s \geq \lambda_c$). The minimum $\lambda_s=\lambda_c$ occurs for $\kappa=\kappa_{min} = \gamma_1+\gamma_2[1+\sqrt{4+6(\gamma_1/\gamma_2)+2(\gamma_1/\gamma_2)^2}]$. Therefore λ_c is also a lower bound of λ for the instability of a single mode in the bad cavity case. (It may be shown that for $\lambda>\lambda_s\geq\lambda_c$ α_{min} becomes negative in agreement with the instability results of the previous chapter for a single mode, i.e. for $\alpha=0$.)

If one of the resonant modes is unstable, the deviations e, p, and δ grow to infinitely large values. Thus the linearized equations (4.2) cannot be used for larger times. Therefore one has to solve the original nonlinear equations (2.8). There are two main methods to integrate these nonlinear equations numerically.

In the first method one uses a discretisation of the variables x and t. If the differential quotients are replaced by appropriate difference quotients one gets difference equations which can then be solved numerically. When using this procedure one must be careful to obtain stable difference equations, i.e. those where the differences of physical quantities belonging to neighbouring space points do not blow up, see for instance [10].

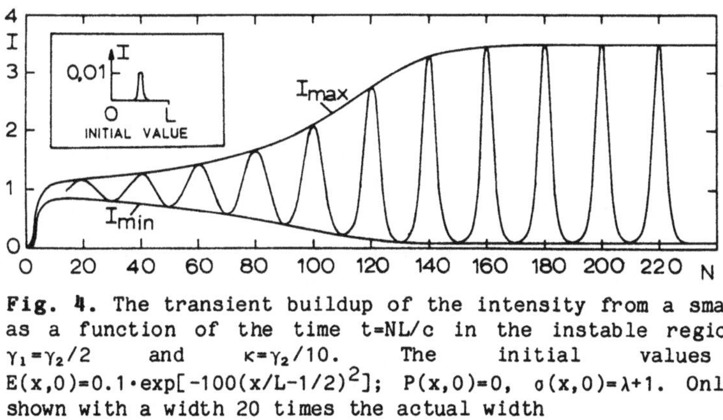

Fig. 4. The transient buildup of the intensity from a small Gaussian disturbance as a function of the time $t=NL/c$ in the instable region $\lambda=15$, $L=2\pi c/(3.2\gamma_2)$, $\gamma_1=\gamma_2/2$ and $\kappa=\gamma_2/10$. The initial values are given by $E(x,0)=0.1\cdot\exp[-100(x/L-1/2)^2]$; $P(x,0)=0$, $\sigma(x,0)=\lambda+1$. Only every 20th pulse is shown with a width 20 times the actual width

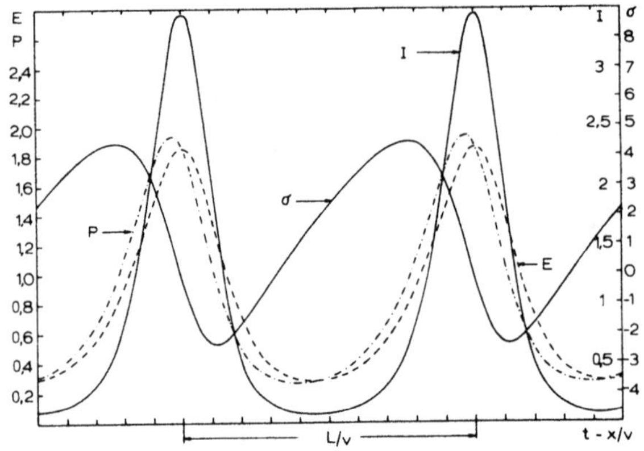

Fig. 5. Steady-state fields E, P (dotted) and σ, $I=E^2$ (solid) vs. t-x/v for $\gamma_1=\gamma_2/2$, $\lambda=15$, $L/v=2\pi/(3.47\gamma_2)$. Direction of propagation is from right to left along the x axis. The one-period average intensity is $\langle I\rangle=0.973$, the average inversion $\langle\sigma\rangle=1.40$

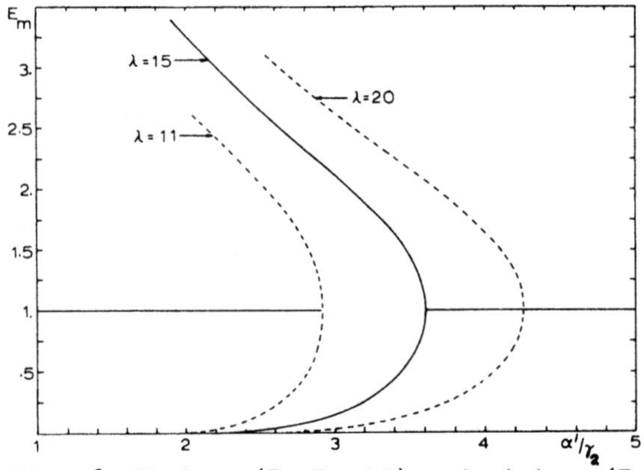

Fig. 6. Maximum ($E_m=E_{max}>1$) and minimum ($E_m=E_{min}<1$) electric field pulse amplitudes as a function of $\alpha'=2\pi v/L$ for $\gamma_1=\gamma_2/2$ and for three pump parameters $\lambda=20$ (dotted line), $\lambda=15$ (solid line), $\lambda=11\approx\lambda_c$ (dotted line); v is the velocity of the corresponding pulse and is approximately equal to the velocity of light in the host material for $\kappa\ll\gamma_2$. The instable region of the cw solution for $\lambda=15$ is marked by a blank space between $2.92\gamma_2<\alpha'<3.62\gamma_2$. Since for $\alpha'>\alpha_{max}'=\alpha_{max}/(1-\epsilon\kappa)$ a velocity v is not defined, we may use for $\alpha'\geq\alpha_{max}'$, $v=v(\alpha_{max}')$

In the second method one expands the fields into the modes $\exp[i(\alpha/c)x]$ with α given by (4.6) and solves the system of truncated nonlinear ordinary differential equations for the expansion coefficients numerically. As shown in [4], by the first method a pulse builds up if one of the nonresonant modes is unstable, see Fig. 4. The steady-state pulse, which travels with constant velocity $v>c$ through the cavity depends on x and t only in the combination x-vt. The values of E, P and σ for a steady state pulse are shown in Fig. 5. It also seems to be of interest that for certain values of L and $\lambda>\lambda_c$ both the cw solution and the pulse solution can be stable (bistable operation), see Fig. 6.

In addition to the tuned cw solution $E=P=\sigma=1$, detuned cw solutions of the system (2.8) are also possible. These solutions are given by ($\Omega^2<\lambda$) [11]

$$E = \sqrt{1-\Omega^2/\lambda}\cdot\exp\{i\Omega[(\gamma_2+\kappa)x/c-\gamma_2 t]\}$$

$$P = E(1 + i\Omega)$$

$$\sigma = 1 + \Omega^2 \ . \qquad (4.9)$$

GERBER and BÜTTIKER have shown [11] that the solution (4.9) becomes unstable if the parameter Ω is large enough. Because of $\lambda>\Omega^2$ the pump parameter λ must then also be large. In this case the phase-branch in Fig. 2 crosses the imaginary axis.

5 Instability of Pulses

If the pump parameter is increased still further the pulses themselves will change. The maximum value of the pulse, its minimum value and its shape will alter in time while it is travelling through the cavity. This behaviour has been termed breathing of the pulse. Breathing of the pulse can occur in a periodic or in a chaotic manner. The maximum intensity of the pulses as a function of time and its Fourier transform are shown in Figs. 7 and 8 for the case of periodic and chaotic breathing [12].

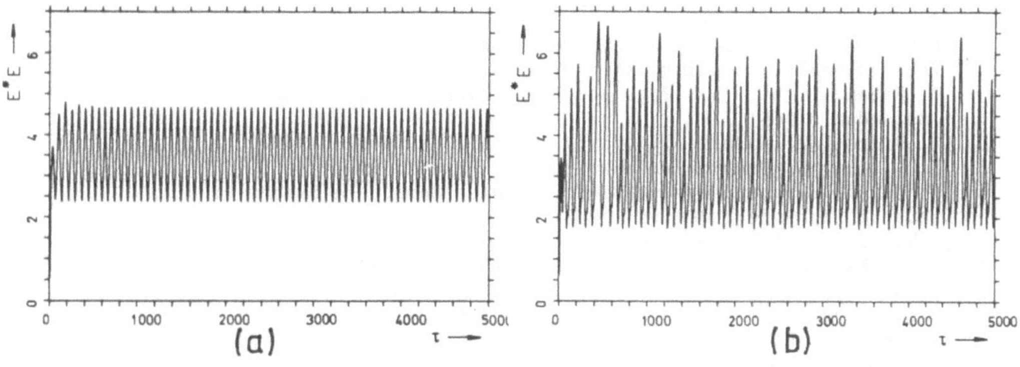

Fig. 7. The maximum intensity E^*E in the cavity as a function of the normalized time $\tau=\gamma_2 t$ as obtained by a numerical integration of the coupled equations for the expansion coefficients. a) Periodic breathing. Parameters are $\lambda=20$, $\Omega^2=9$, $\gamma_1=\gamma_2/2$, $\kappa=\gamma_2/10$. b) Chaotic breathing. Parameters are $\lambda=25$, $\Omega^2=10$, $\gamma_1=\gamma_2/2$, $\kappa=\gamma_2/10$. The initial values of both cases are the Gerber-Büttiker solutions where the Fourier coefficients of the inversion have been perturbed by adding 0.01 to all coefficients

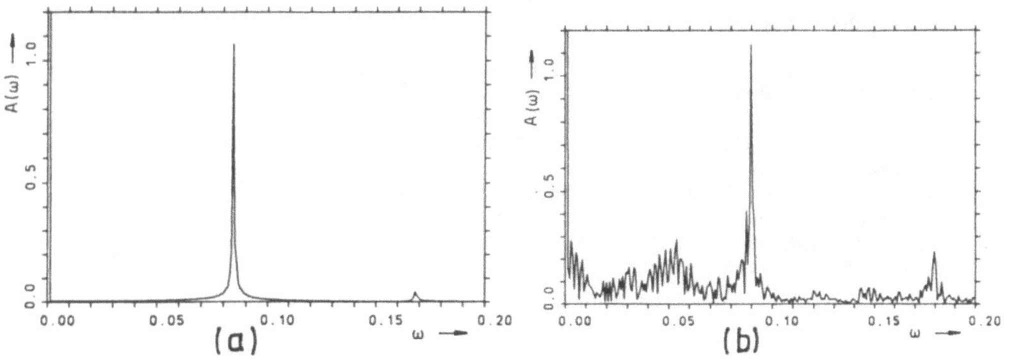

Fig. 8. Fourier spectrum of the maximum intensity corresponding to Fig. 7. The amplitude $A(\omega)$ is plotted versus ω

6 Conclusion

We have seen that the equations for a simple laser model (ring-laser with one direction of propagation , spatially homogeneously distributed loss and gain, two-level atoms with a homogeneously broadened line, one direction of polarization) show a great variety of instabilities. The simple model treated so far can be extended in several ways. One may in addition to the active material include a passive material [13-17]. The extension to an inhomogeneously broadened line was given by CASPERSON [18]. A recent paper by LUGIATO et al. [19] treats the case where the loss of the electric field in the cavity occurs at one of the mirrors. Selfpulsing in ring-lasers with bistable absorption was discovered some time ago, see for instance [20,21] for reviews. One should also mention the IKEDA instability [22], where the essential reason for its occurrence seems to be a delay mechanism. In an experiment done with a ring-laser system having a homogeneously broadened line, a number of instabilities have been observed [23]. As far as I understand, a quantitative theoretical explanation of all the observed instabilities has not yet been given.

References

1 H.Haken, Laser Theory, in Encyclopedia of Physics, Vol. XXV/2c, Springer, Berlin (1970)
2 M.Sargent, M.O. Scully, W.E. Lamb, Laser Physics, Addison-Wesley, Reading, MA (1974)
3 H. Haken, Light, Vol. 2, Laser Light Dynamics, North Holland, Amsterdam (1985)
4 H. Risken, K. Nummedal, J. Appl. Physics 39, 4662 (1968)
5 H. Risken, C. Schmid, W. Weidlich, Z. Physik, 194, 337 (1966)
6 H. Haken, Advanced Synergetics, Springer Series in Synergetics, Vol. 20, Springer, Berlin (1983)
7 E.N. Lorenz, J. Atmos. Sci. 20, 130 (1963)
8 H. Risken, K. Nummedal, Phys. Lett. 26A, 275 (1968)
9 R. Graham, H. Haken, Z. Physik 213, 420 (1968)
10 G.E. Forsythe, W.R. Wasow: Finite-Difference Methods for Partial Differential Equations, Wiley, New York (1967)
11 P.R. Gerber, M. Büttiker, Z. Physik B33, 219 (1979)
12 M. Mayr, H. Risken, H.D. Vollmer, Optics Comm. 36, 480 (1981)
13 A.P. Kazantsev, S.G. Rautian, G.I. Surdutovich, Sov. Phys. JETP 27, 756 (1968)
14 L.A. Lugiato, P. Mandel, S.T. Dembinski, A. Kossakowski, Phys. Rev. A18, 238 (1978)
15 H. Knapp, H. Risken, H.D. Vollmer, Appl. Phys. 15, 265 (1978)
16 J.C. Antoranz, M.G. Velarde, Optics Comm. 38, 61 (1981)
17 E. Arimondo, F.Casagrande, L.A. Lugiato, P. Glorieux, Appl. Phys. B30, 57 (1983)
18 L.W. Casperson, J. Opt. Soc. Am. B2, 993 (1985)
19 L.A. Lugiato, L.M. Narduci, E.V. Eschenazi, D.K. Bandy, N.B. Abraham, Phys. Rev. A32, 1563 (1985)
20 R. Bonifacio (ed.), Dissipative Systems in Quantum Optics - Resonance Fluorescence, Optical Bistability, Superfluorescence, Springer, Berlin (1982)
21 L.A. Lugiato in: Progress in Optics XXI, p.69, Ed. E. Wolf, Elsevier Science Publisher, Amsterdam (1984)
22 K. Ikeda, Optics Comm. 30, 257 (1979)
23 L.W. Hillman, J. Krasinski, R.W. Boyd, C.R. Stroud Jr., Phys. Rev. Lett. 52, 1605 (1984)

Ordered and Chaotic Response of a Modulated or Driven NMR Laser

E. Brun, B. Derighetti, M. Ravani, G. Broggi, P. Meier, and R. Stoop
Physics Institute, University of Zürich, CH-8001 Zürich, Switzerland

R. Badii
Institute of Theoretical Physics, University of Zürich, CH-8001 Zürich, Switzerland

1 Introduction

The *Feature Issue on Instabilities in Active Optical Media* of the Journal of the Optical Society of America B [1] shows the increasing interest in nonlinear phenomena in laser systems. The ruby nuclear-magnetic-resonance (NMR) laser [2], among others, is an excellent test system with which to study the nature of instabilities and the routes to chaos of lasing systems. In our contribution [3] we reviewed theoretical and experimental investigations of the ruby NMR laser that operates in the rf regime and has its counterpart in the homogeneously broadened single-mode ring laser. Two topics were stressed in particular: the transition from ordered to chaotic response of the parametrically modulated laser (PML), and the transition to injection-locking of a laser with an injected signal (LIS). In both cases a control parameter of the system was slowly swept in time. For the PML it was either the amplitude or the frequency of the parametric modulation signal $A_p \sin\omega_p t$; for the LIS it was the amplitude of the injected signal $V_0 \sin\omega t$ which was detuned by $\Delta\omega_a = \omega_a - \omega$ from the NMR frequency ω_a. In scanning the control parameter we observed in both cases sequences of instabilities leading to either regular or irregular response.

We found that the PML can follow a route to chaos via a sequence of period-doubling bifurcations for ω_p about ω_{relax}, the frequency of the relaxation oscillation of the free-running NMR laser. The chaotic regime includes merging chaotic bands and periodic windows with period-doubling bifurcations therein. Computer simulations with a low-dimensional NMR laser model are in good qualitative agreement with the various experiments. The calculations are consistent with the scaling properties of the Feigenbaum scenario [4]. We also observed intermittency in the PML. But the question, whether the observed intermittent behavior is that of the Pomeau-Manneville [5] scenario, still requires an answer.

In setting $\Delta\omega_a \approx \omega_{relax}$, we detected with the LIS a wealth of interesting nonlinear effects along its route to injection-locking where the laser signal and injection signal become synchronized. In the low-injection domain, the LIS revealed normal beating of two signals of different frequencies. Our preliminary experiments [3] indicated a pushing of the laser frequency with respect to ω for $\Delta\omega_a > \omega_{relax}$. With

$\Delta\omega_a$ close to ω_{relax}, the response of the LIS was more complicated. We observed a variety of quasi-stable NMR laser states: normal beating of frequency ω_b, periodic self-pulsations of approximately fractional beat frequencies ω_b/n, $n = 2,3,4,\ldots$, and irregular spiking near the injection-locking threshold.

In the present paper, we describe new investigations with the NMR laser, which complement our former work with the LIS and PML. After a brief introduction to the physical properties of our system and its theoretical model in Sec.2, we treat in Sec.3 the frequency pushing or pulling of the LIS in the low-injection limit. We then present measurements of the frequency pushing up to injection-locking and illustrate the fractional self-pulsation region together with the irregular spiking region with time plots and online Fourier spectra. In Sec.4 we discuss the PML as a parametric small-signal detector that operates near the critical point of a period-doubling bifurcation.

2. Ruby NMR laser and its model

The active entities of the NMR laser are the gyromagnetic ^{27}Al nuclei in ruby which perform the rf-forced precession about a static magnetic field B_0. Single-mode spins flip between adjacent Zeeman states $|m\rangle$ and $|m+1\rangle$ which determine the NMR frequency $\omega_a = gB_0$ (g the gyromagnetic ratio of ^{27}Al with $I = 5/2$). The spins thus are treated as members of a fictitious spin-1/2 two-level system.

An LC-circuit of the frequency $\omega_c = 1/\sqrt{LC}$ forms the resonant structure (cavity) in which the spins are embedded. The NMR coil L supports the radiation field which interacts with the magnetic moments of ^{27}Al. A microwave source pumps the nuclear spins to an inverted Zeeman population by means of the dynamic nuclear polarization (DNP) technique.

To describe the NMR laser, a rotating reference frame (u,v,z) is introduced. Its rotation frequency is usually set equal to the frequency ω of the injected signal. The interaction of the nuclear moments with the radiation field can be translated into the nonlinear Bloch-Kirchhoff (BK) equations given in [3]:

$$\begin{aligned} \dot{m} &= -(\gamma_\perp + i\Delta\omega_a)m + isgbM_z \\ \dot{M}_z &= -\gamma_\parallel(M_z - M_e) + \frac{ig}{2}(mb^* - m^*b) \\ \dot{b} &= -(k + i\Delta\omega_c)(b - b_d) + i\chi m \;. \end{aligned} \tag{1}$$

The macroscopic laser variables $m = M_u + iM_v$, M_z, and $b = B_u + iB_v$ represent the precessing nuclear magnetization, the longitudinal nuclear magnetization or spin polarization, and the relevant rotating components of the magnetic radiation field inside the NMR coil, respectively. M_e is the longitudinal nuclear magnetization to which M_z would be pumped by DNP in the absence of the radiation field. For the NMR laser, M_e is negative and thus a measure of the population inversion for the two Zeeman states involved in the laser action. The quantity

$$b_d = B_u^d + iB_v^d \tag{2}$$

reflects the driving field which stems from the injected rf-signal of frequency ω. If $b_d=0$, the laser is free-running. In analogy to the NMR detuning, $\Delta\omega_a=\omega_a-\omega$, we call $\Delta\omega_c = \omega_c-\omega$ the coil (cavity) detuning. The constants $k=\omega_c/2Q$ and $\gamma_\perp=1/T_2$ characterize the damping of the coil and the transverse spin damping, respectively. The constant $\gamma_\parallel=1/T_e$ describes the speed of the DNP pump. Hence, T_e is called the pumping time. It plays the role of an effective spin-lattice relaxation time T_1. The spin factor s is proportional to the matrix elements $|\langle m|I_\pm|m+1\rangle|^2$ of the angular momentum operators (I=5/2) between the two fictitious spin-1/2 states $|m\rangle$ and $|m+1\rangle$. The coupling constant

$$\chi = \frac{\mu_0 \eta Q k}{2} \tag{3}$$

characterizes the feedback of the precessing spins to the radiation field. Here, η is the filling factor of the coil and $\mu_0=4\pi\times10^{-7}$Vs/Am.

Since our NMR laser is governed by a hierarchical set of damping constants:

$$k \cong 4\times10^5 s^{-1} \gg \gamma_\perp \cong 3\times10^4 s^{-1} \ggg \gamma_\parallel = 10 s^{-1}, \tag{4}$$

adiabatic elimination of macroscopic variables becomes feasible. Thus the BK model (1) can be reduced to fewer dimensions. Use of this procedure will be made throughout the forthcoming paragraphs.

3. NMR laser with injected signal (LIS)

We limit ourselves to the case where the NMR coil is perfectly tuned to the frequency $\omega_a= gB_0$ of the free spin precession. The injected signal of frequency ω thus is equally detuned from the NMR frequency as from the cavity frequency; that is $\Delta\omega_a=\Delta\omega_c=\Delta\omega$.

3.1 LIS models

Adiabatic elimination of the field variables in the 5-equation BK model (1) leads to the 3-equation model:

$$\dot{P}_u = -\gamma_\perp [(1+2CD)P_u - \frac{\Delta\omega}{\gamma_\perp} P_v - \frac{\Delta\omega}{k} DY_d]$$

$$\dot{P}_v = -\gamma_\perp \, [(1+2CD)P_v + \frac{\Delta\omega}{\gamma_\perp} P_u - DY_d] \tag{5}$$

$$\dot{D} = -\gamma_\parallel \, [D+1-2C(P_u^2+P_v^2)+Y_d(P_v + \frac{\Delta\omega}{k} P_u)] \; .$$

The dimensionless variables

$$P_u = \frac{M_u}{|M_e|} \left(\frac{\gamma_\perp}{s\gamma_\parallel} \right)^{1/2}, \quad P_v = \frac{M_v}{|M_e|} \left(\frac{\gamma_\perp}{s\gamma_\parallel} \right)^{1/2}, \quad D = \frac{M_z}{|M_e|} \tag{6}$$

describe the nuclear magnetization in the rotating frame which rotates with the frequency ω of the injected field B_u^d about B_o.

$$Y_d = gB_u^d \left(\frac{s}{\gamma_\perp \gamma_\parallel}\right)^{1/2}, \quad 2C = p = \frac{|M_e|}{|M_{th}|}, \quad |M_{th}| = \frac{2\gamma_\perp}{\mu_o s \eta g Q} \tag{7}$$

are the dimensionless driving field, the pumping parameter as introduced by Lugiato [6], and the laser threshold, respectively. Typical values of the adjustable system parameters are

$$\Delta\omega = 280 s^{-1}; \quad Y_d \approx -10^{-3}; \quad p \cong 1.5 \quad . \tag{8}$$

Keeping (4) and (8) in mind, the following rescaling is indicated:

$$\begin{aligned} & x = pP_u, \quad y = pP_v, \quad z = 1+pD, \\ & \sigma = p-1, \quad \Delta = \frac{\Delta\omega}{\gamma_\perp} \approx 10^{-2}, \quad t = \Delta \cdot \gamma_\perp t_{exp} , \\ & K = -\frac{Y_d}{\Lambda} \approx 0.1, \quad \beta = \frac{\gamma_\parallel}{\gamma_\perp \Lambda} \cong 3\times10^{-2}, \quad \theta = \frac{\Delta\omega}{k} \cong 7\times10^{-4} . \end{aligned} \tag{9}$$

Thus, we obtain

$$\begin{aligned} \dot{x} &= -\frac{1}{\Delta} xz + y \\ \dot{y} &= -\frac{1}{\Delta} yz - x - K(z-1) \\ \dot{z} &= -\beta(z+\sigma-x^2-y^2- K\Delta y). \end{aligned} \tag{10}$$

In (10) all the terms containing θ have been dropped. Due to their smallness this approximation is numerically valid in the whole accessible parameter range.

In order to better elucidate the symmetries of the LIS model (10), we introduce polar coordinates $x=r\cos\phi$, $y=r\sin\phi$ and obtain

$$\dot{r} = -\frac{1}{\Delta} rz - K(z-1)\sin\phi$$

$$\dot{\phi} = -1 - \frac{1}{r} K(z-1)\cos\phi \tag{11}$$

$$\dot{z} = - \beta\ (z+\sigma-r^2-K\Delta r\ \sin\phi).$$

In the zero-injection limit, K=0, eqs. (11) reduce to

$$\dot{r} = - \frac{1}{\Delta} rz$$

$$\dot{\phi} = - 1 \qquad (12)$$

$$\dot{z} = - \beta(z+\sigma-r^2).$$

The 3-equation model (12) describes two uncoupled oscillators: (i) the free-running NMR laser through the variables r (amplitude) and z (spin inversion), and (ii) a free rotator of constant amplitude and angular frequency $\dot{\phi}= - 1$ (in units of $\Delta\omega$).

The free NMR laser relaxation frequency $\omega_o = \omega_{relax}/\Delta\omega$ can be found by setting $q=\ell nr$ which fulfills the dynamic equation

$$\ddot{q} + \beta\dot{q} = F(q) = \frac{1}{\Delta}\beta(\sigma-e^{2q}) = - \frac{\partial V}{\partial q}. \qquad (13)$$

V is the associated Toda-like potential [7]. Linearizing around the potential minimum leads to

$$\omega_o = (\frac{2\beta\sigma}{\Delta})^{1/2} \qquad (14)$$

The asymptotic limit-cycle solution of (12) thus becomes

$$r = R_o = \sigma^{1/2}$$

$$\phi = \phi_o = - t \qquad (15)$$

$$z = z_o = 0$$

In phase space, therefore, we have a circular orbit of radius R_o with its center at the origin.

In the strong-injection case (11) has a fixed-point solution which reflects injection-locking. The critical value of K for which locking sets in is approximately

$$K_c = \sigma^{1/2}. \qquad (16)$$

3.2 Frequency pushing or pulling at low injection

The low-injection domain, $0<K<<K_c$, is accessible to an analytical treatment. The observed normal beating is considered as a perturbation of the exact limit-cycle solution (15). Following the method suggested by Oppo and Politi [8], we assume that the perturbed orbit in phase space is still approximately circular, however, with a slightly different averaged angular frequency $<\dot{\phi}>$. This allows the use of the Lindstedt-Poincaré technique [9] to evaluate corrections to the unperturbed frequency $\dot{\phi}_o = - 1$. Therefore, we expand the variables r, ϕ and z around the zeroth-order solution (15) as follows:

$$r(t) = R_o + \varepsilon r_1(t) + \varepsilon^2 r_2(t) + \dots \tag{17}$$
$$\phi(t) = \phi_o + \varepsilon\phi_1(t) + \varepsilon^2\phi_2(t) + \dots + \varepsilon\omega_1 t + \varepsilon^2\omega_2 t$$
$$z(t) = \varepsilon z_1(t) + \varepsilon^2 z_2(t)$$

with $\varepsilon = K$.

The terms ϕ_1 and ϕ_2 in the expansion of $\phi(t)$ are nonsecular, that is, they remain finite in the limit $t \to \infty$. The only secular terms are $\phi_o = -t$ and the two explicit corrections $\varepsilon\omega_1 t$ and $\varepsilon^2\omega_2 t$ [10]. The first-order equations become

$$\dot{r}_1 = -\frac{R_o}{\Delta} z_1 - \sin t$$
$$\dot{\phi}_1 = \frac{1}{R_o}\cos t - \omega_1 \tag{18}$$
$$\dot{z}_1 = -\beta(z_1 - 2R_o r_1 + R_o\Delta \sin t).$$

The solutions of (18) are substituted in the 2nd-order equation for ϕ:

$$\dot{\phi}_2 = -\omega_2 - \frac{1}{R_o} z_1\cos t - \frac{1}{\sigma} r_1\cos t + \frac{1}{\sigma}\sin^2 t. \tag{19}$$

The coefficients ω_1 and ω_2 must be chosen in such a way that all secular terms vanish. Therefore $\omega_1 = 0$, while other secular terms originate from the time-integration of constants arising from the terms in $\cos^2 t$ and $\sin^2 t$. The requested condition leads to

$$\omega_2 \cong \frac{\alpha}{\alpha^2+\Delta^2(\omega_o^2-1)^2}\left(\omega_o^2-1-\frac{\alpha}{2}\right) \tag{20}$$

with $\alpha = \beta\Delta << 1$.

The average frequency $\langle\dot{\phi}\rangle$ in the low-injection domain can now be expressed as

$$\langle\dot{\phi}\rangle = -1 + \varepsilon^2\omega_2 \tag{21}$$

Thus, if $\omega_2 > 0$, the NMR laser frequency is pulled towards the frequency ω of the injection signal. For $\omega_2 < 0$, there is frequency pushing.

Multiplying (21) by $\Delta\cdot\gamma_\perp = \Delta\omega$ brings us back to physical units, and the beat frequency in s^{-1} is

$$\omega_b = \Delta\omega(\varepsilon) = \Delta\omega(1-\varepsilon^2\omega_2). \tag{22}$$

In other words we have "pushing" or "pulling" whenever the detuning $\Delta\omega$ is larger or smaller than ω_{relax}, respectively. A similar result was reported in [8] for a different laser type.

3.3 Experimental frequency pushing and beyond

The perturbative approach to the beating phenomenon in the low-injection domain stimmulated us to investigate experimentally the full range below injection-locking. With

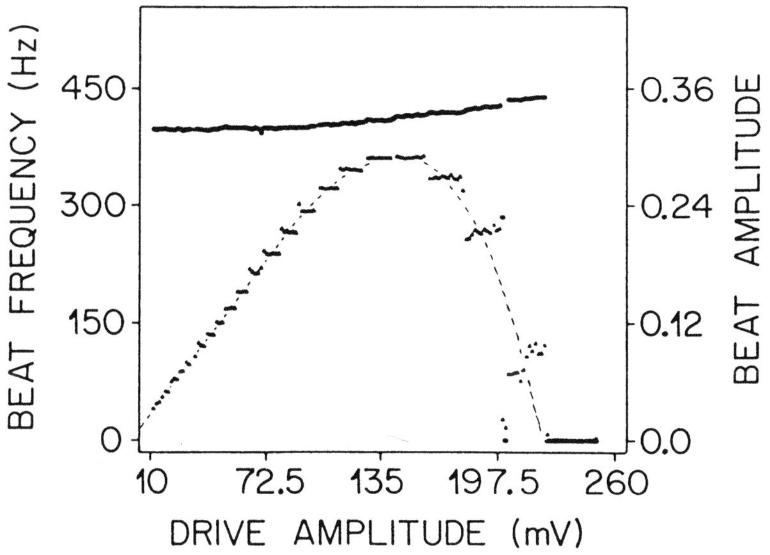

Fig. 1 Measured beat frequency f_b (heavy line) and beat amplitude (dashed line for visual help; the steps in the data are instrumental) of the LIS which is driven with a detuned rf-field of about 12 MHz. The detuning was $\Delta f_a \cong 400$ Hz. The drive amplitude was slowly swept up to the region of injection-locking (≥ 200 mV)

$\Delta\omega$ appreciably greater than ω_{relax}, we were able to obtain reproducible results. As a function of time, we measured both the beat frequency and the beat amplitude for a set of fixed detunings $\Delta\omega = 2\pi\Delta f$. A typical result is shown in Fig.1. The solid line depicts the observed beat frequency $f_b = \omega_b/2\pi$ with its characteristic jump to zero at the locking-threshold K_c. The broken curve, with a hump below K_c, reflects the beat amplitude which goes smoothly to zero at K_c. The appearance of the finite beat frequency together with the continuous decrease of the beat amplitude near K_c indicates that the limit-cycle behavior for $K<K_c$ and fixed-point behavior for $K>K_c$ is the effect of a Hopf bifurcation at K_c. The overall increase of f_b shows that there is frequency pushing. This result finds further support from the data of Fig.2. The measured beat frequency f_b is plotted up to the injection-locking threshold for various fixed detunings $\Delta f = \Delta\omega/2\pi$ as a function of K. With $f_{relax} = \omega_{relax}/2\pi \cong 60$Hz, the LIS responds in the given detuning range with frequency pushing. This observation is in agreement with numerical calculations using the 3-equation LIS model (5).

For smaller detunings the response of the LIS changes dramatically. Normal beat patterns are still seen in the low-injection region of K. However, pushing or pulling is too weak to be measured. With increasing K the normal beating becomes unstable. The LIS then shows an almost regular, pulsating output of a frequency which is approximately a fraction 1/n, n = 2,3,4,5,..., of the detuning frequency $\Delta\omega$. Narrow regions of such fractional pulsations alternate with regions of normal beating when

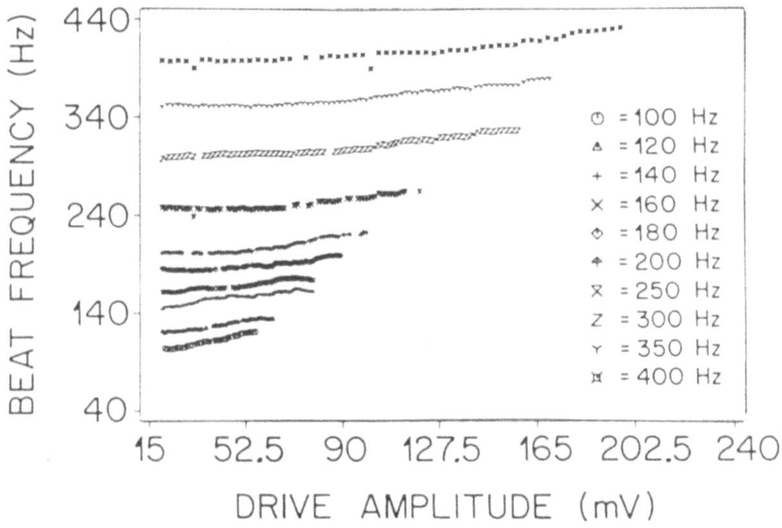

Fig. 2 Experimental data for frequency pushing of the LIS with injected rf for a series of detunings (100 Hz ÷ 400 Hz). The data points for each run end at the injection-locking threshold

the driving signal is slowly swept to larger K values. Near injection-locking the LIS output becomes irregular. Unstable patterns of strongly peaked pulsations of variable shape, amplitude and period are commonly found. Finally, the slowing-down of the spiking rate heralds injection-locking with its highly stable output. By reversing the sweep in the locked state, the LIS returns to the unstable behavior. However, the up and down scans have a strong hysteresis, as is shown in Fig.3. We measured the LIS output stroboscopically with equal time intervals t_s and stored the digitized values as a function of the drive amplitude V_0 in a computer. The figure depicts the recorded values for the up scan (t=0 to 24.5 s) and the subsequent down scan. The LIS responds with normal beating for small V_0, regular or irregular

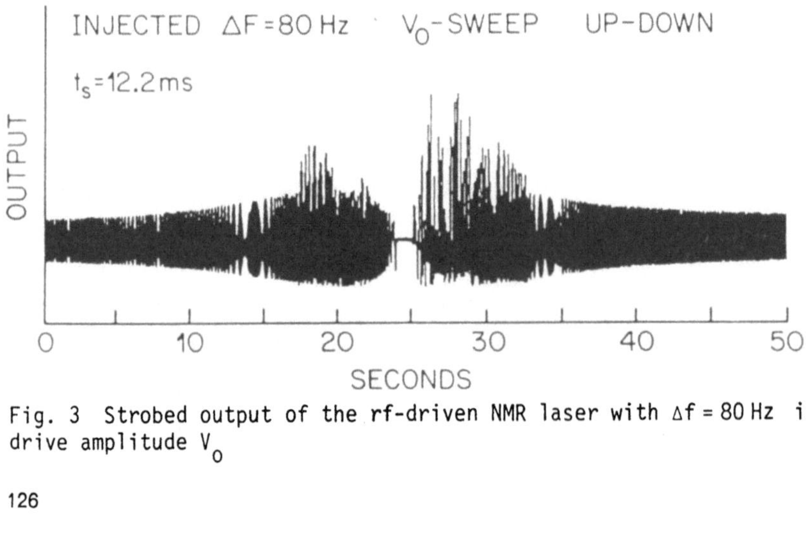

Fig. 3 Strobed output of the rf-driven NMR laser with $\Delta f = 80$ Hz in function of the drive amplitude V_0

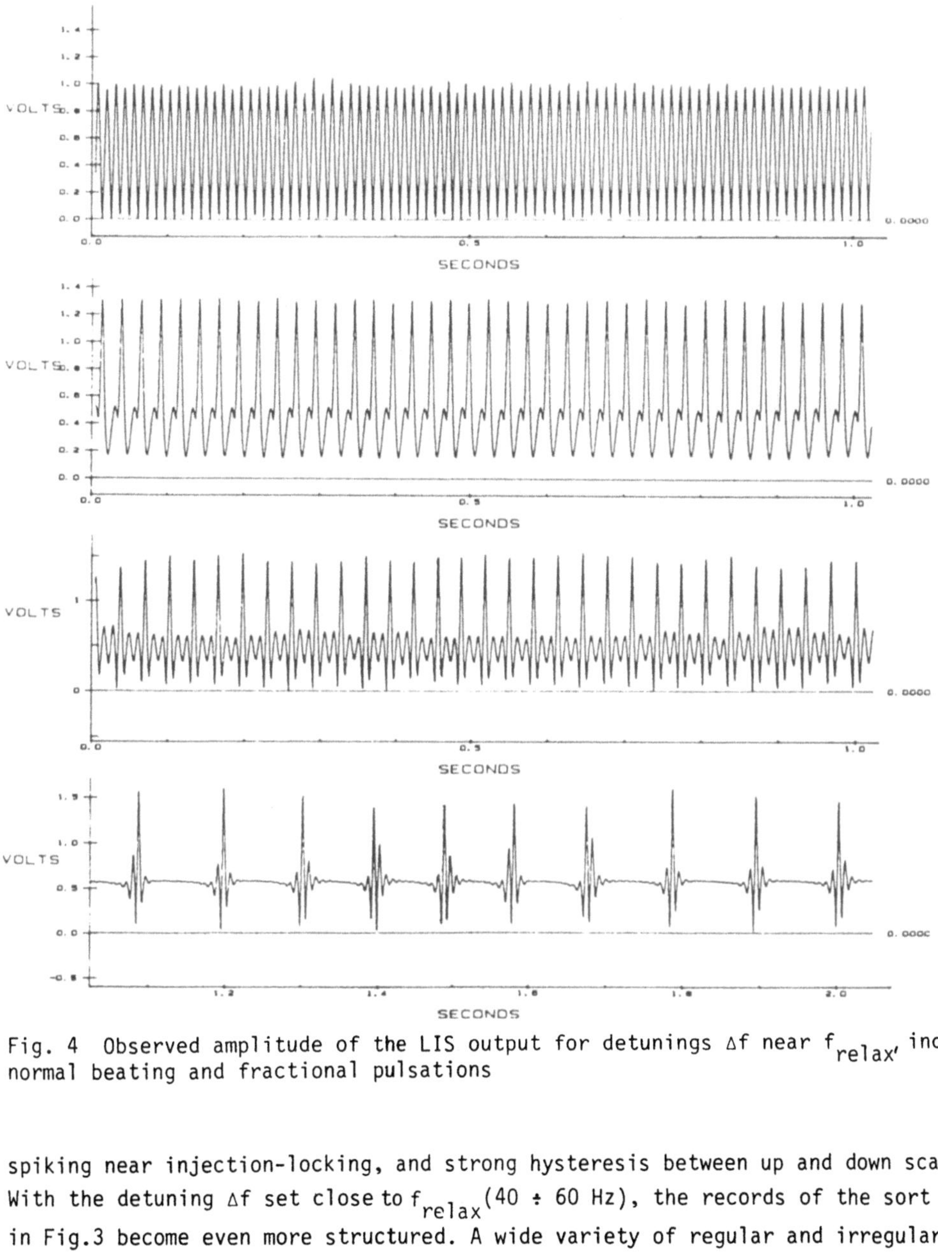

Fig. 4 Observed amplitude of the LIS output for detunings Δf near f_{relax}, indicating normal beating and fractional pulsations

spiking near injection-locking, and strong hysteresis between up and down scans. With the detuning Δf set close to f_{relax} (40 ÷ 60 Hz), the records of the sort given in Fig.3 become even more structured. A wide variety of regular and irregular sections alternate. Strong hysteresis effects make it difficult to reproduce the results. However, one can always find basins of attraction of the above-mentioned LIS behavior. Some examples are given for illustration now. Figure 4 shows time plots of the LIS output. The four continuous time records demonstrate quasi-stable behavior. From top to bottom we have the normal beating in the low-injection domain and the spiking pulsations of almost fractional beat frequencies for higher driving signals. Fig.5 shows two time pictures for the LIS output together with the corres-

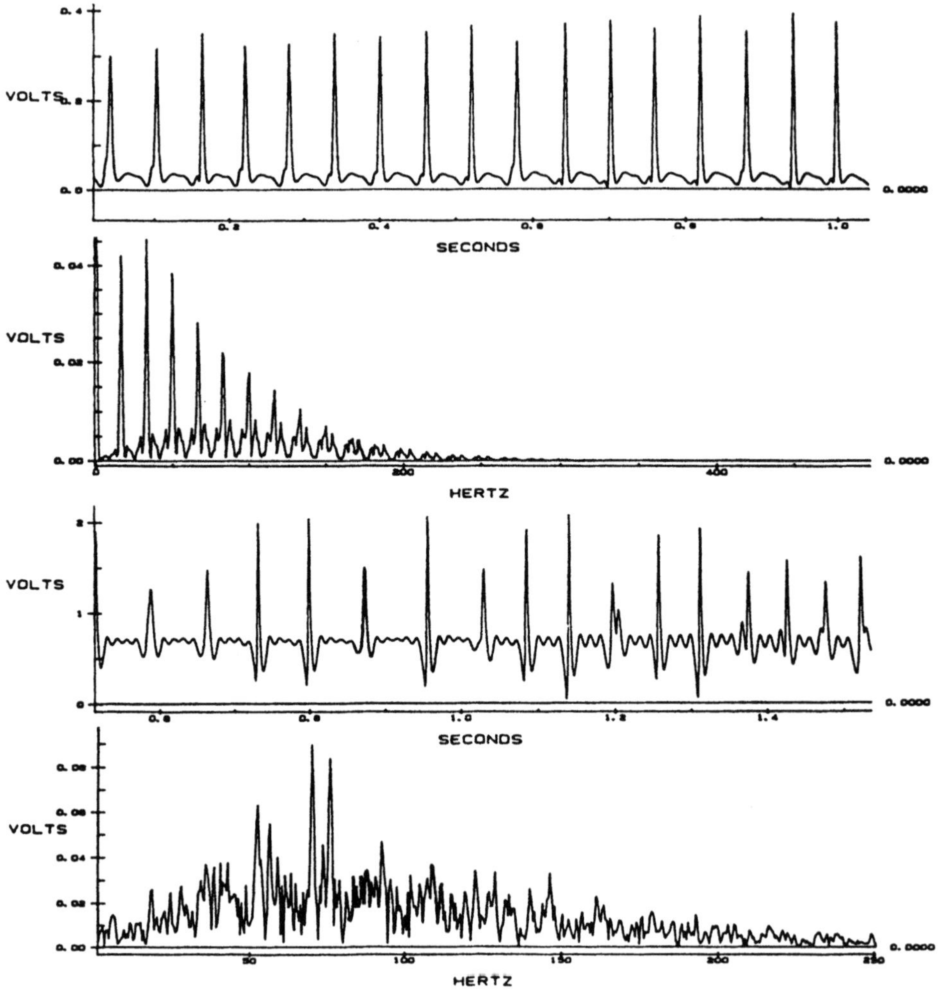

Fig. 5 Observed amplitude of the LIS output and online Fourier spectra for a detuning Δf near f_{relax} ($\Delta f \cong 62\,Hz$)

ponding Fourier spectra. The records of the two runs depict the quasi-periodic fractional pulsation and the irregular spiking near the injection-locking threshold. In order to simulate the stationary LIS behavior we performed also numerical calculations with the 3-equation BK model (5). Many of the experimental observations could thus be confirmed. Good qualitative agreement with the measured frequency pushing was obtained with $\Delta\omega > \omega_{relax}$ up to the injection-locking threshold. For $\Delta\omega \cong \omega_{relax}$ we could numerically reproduce the scenario depicted in Fig.4, where stable beating alternates with pulsations of lower and lower fractional frequencies as the injected signal amplitude is increased. However, in contrast to Fig.5b, no chaotic spiking has been found yet with the 3-equation model.

In summing up, one may say that our experimental and numerical results show many of the features of the LIS that were predicted and partly observed in the optical regime [11,12,13,14,15].

4. NMR laser with parametric double-modulation (PDML)

Recently, we described first investigations [16] with the tuned NMR laser to demonstrate a new method [17] which allows us to enhance small signals by a nonlinear system showing period-doubling bifurcations. We started from the PML where a signal $S_p = A_p \sin\omega_p t$ modulates one of the system parameters. Then, we supplied a second, small signal $s_i = a_i \sin\omega_i t$ to the PML in order to generate an additional frequency in the system. For selected values of A_p, ω_p and ω_i, we observed a strong nonlinear coupling of s_i with S_p thus providing a mechanism for a frequency-selective small-signal detection scheme. We called S_p and s_i the parametric pump and input signals of this parametric double-modulation laser (PDML).

In the PML mode ($s_i = 0$), A_p and ω_p are adjusted such that a sequence of period-doubling bifurcations occurs. In keeping ω_p fixed, A_p is set close to, but below, one of the observed critical values $A_c^{(n)}$, $n = 1,2,3,\ldots$, for the instabilities. The PML output oscillates under the forcing signal S_p with constant amplitude and the fundamental frequency $\omega_p/2^{n-1}$, as it is shown in Fig.6a for n = 1. Thus, we have a stable limit-cycle behavior. When s_i is switched on, the output will change periodically its amplitude, as is depicted in Fig.6b.

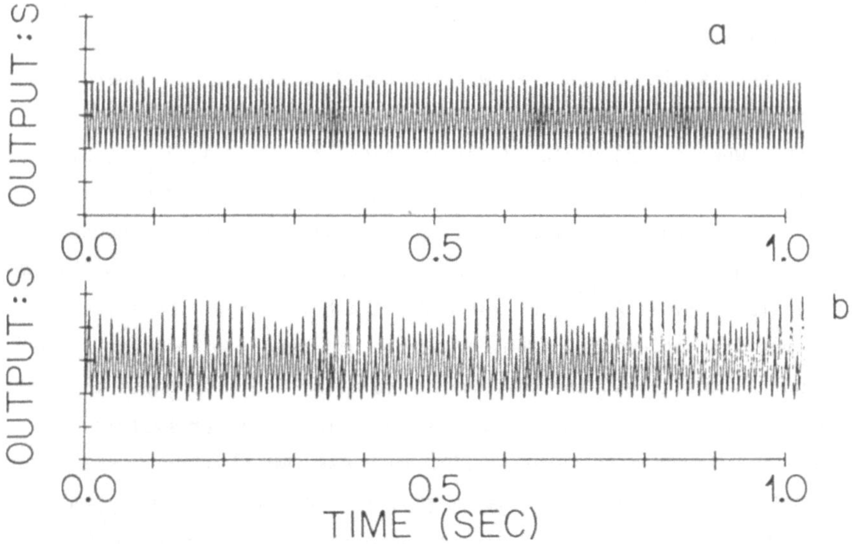

Fig. 6 Observed oscillations of the NMR laser output S: (a) with a single parametric signal (pump) $S_p = A_p \sin\omega_p t$ (PML), (b) with a second parametric signal (input) $s_i = a_i \sin\omega_i t$ (PDML). We have set $A_p \lesssim A_c^{(1)}$, ω_i near $\omega_p/2$. Note, the constant amplitude of the oscillation in (a), and the dynamic precursor of the period-doubling bifurcation in (b) with the slow variation of the height of the even or odd numbered peaks

The characteristic form of the time behavior (b) suggests the following procedure to define operationally a response function for the PDML as a small-signal detector near a period-doubling bifurcation: Firstly, measure the differences between the heights of consecutive peaks (high-low); secondly, determine the average amplitude ΔS of these time-dependent differences. Since ΔS is obviously a measure of the strength of the dynamical precursor of the nearby bifurcation ($\omega_p \to \omega_p/2$), one may call ΔS the experimental sensitivity of the system to the input $s_i = a_i \sin \omega_i t$ and $\Delta S(\omega_i)$ the corresponding experimental response function. That definition then can readily be extended to higher bifurcations ($\omega_p/2^{n-1} \to \omega_p/2^n$), where ΔS refers to peaks which differ in time by $2^{n-1}T_p$, with T_p the period of the pump signal $S_p = A_p \sin\omega_p t$. Typically, $\Delta S(\omega_i)$ is peaked at certain frequencies, the lowest being the fractional frequency $\omega_p/2^n$. A strong peaking, and therefore a high detection sensitivity for s_i, requires a pump frequency ω_p of the order of ω_{relax} and a pump amplitude A_p as close as possible to one of the instability points $A_c^{(n)}$.

To demonstrate PDML characteristics, we measured $\Delta S(\omega_i)$ by means of a strobing technique for the proper testing of the PDML output. There, the signal is stroboscopically sampled at times that differ by $2^{n-1}T_p$. The phase of the discrete timing is adjusted to render maximum sensitivity for $\omega_i = \omega_p/2^n$. The discrete output values are digitalized and stored in a computer; ΔS is then calculated and displayed. In a run, the computer is used also to control the scan of ω_i in small quasi-adiabatic steps. In Fig.7 we depict the outcome of two typical runs (a) and (b) to measure $\Delta S(\omega_i)$. The two cases correspond to the frequency setting $\omega_p \approx 2\omega_{relax}$. In run (a), A_p was set below the instability point $A_c^{(1)}$ where the first experimental period-doubling ($\omega_p \to \omega_p/2$) occurs. In run (b), A_p was increased to almost $A_c^{(2)}$ where the second period-doubling ($\omega_p/2 \to \omega_p/4$) sets in. There is a remarkable difference between the runs (a) and (b). Whereas (b) yields distinct peaks at $\omega_i \to \omega_p/4$, $3\omega_p/4$, $5\omega_p/4$,.., (a) yields a single peak at $\omega_i \to \omega_p/2$ only. The results of run (b) indicate the presence of a true period-doubling bifurcation at $A_c^{(2)}$ with a typical response function $\Delta S(\omega_i)$ as predicted by [17]. In contrast to (b), run (a) leads to the conclusion that the instability at $A_c^{(1)}$ does not reflect a bifurcation. More likely, the PML at $A_c^{(1)}$ jumps discontinuously from one stable attractor to another which has twice that period. In this view, the different results of Fig. 7 suggest the use of the PDML scheme to study also the nature of instabilities in nonlinear systems.

To substantiate the observed features of PDML response, we also performed numerical calculations based on the 3-equation system (5). The detuning was set $\Delta\omega_a = \Delta\omega_c = 0$, and two modulation terms, $A_p \sin\omega_p t$ and $a_i \sin\omega_i t$, were introduced. In a specific example, we adjusted A_p and ω_p to get a limit-cycle solution of frequency $\omega_p/2$ in PML-mode. That was done in analogy to the experimental conditions of run (b) in Fig.7.

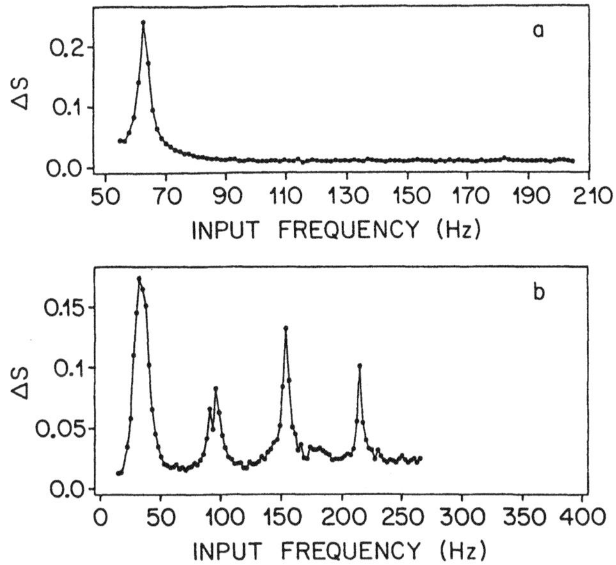

Fig. 7 Stroboscopically measured small-signal detection sensitivity $\Delta S(f_i = \omega_i/2\pi)$: (a) with $A_p \lesssim A_c^{(1)}$, $f_p = \omega_p/2\pi \cong 130$Hz. (b) with $A_p \lessapprox A_c^{(2)}$, $f_p \cong 120$Hz. In both cases, a_i was adjusted to warrant a linear behavior of $\Delta S(a_i)$ at $\omega_i = \omega_p/2$ in (a) and $\omega_i = \omega_p/4$ in (b)

The dynamical evolution was calculated both for the PML ($a_i = 0$) and the PDML ($a_i \sin\omega_i t$). The amplitude m of the oscillating difference between the two solutions was then determined and plotted as a function of ω_i. The result is shown in Fig.8 for several values of A_p. Since $m(\omega_i)$ measures the overall distortion of the original limit-cycle, the various runs in Fig.8 demonstrate the increasing

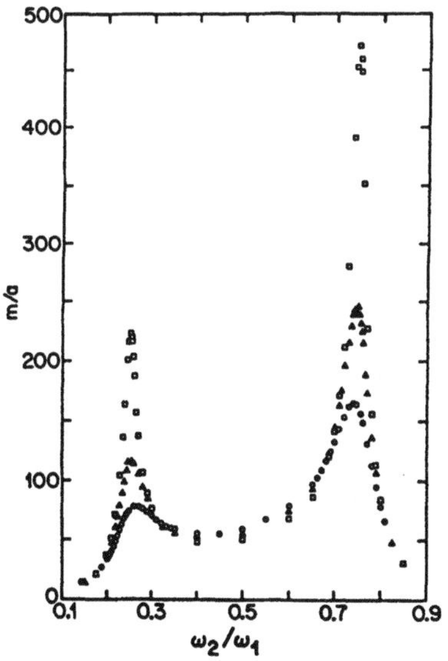

Fig. 8 Scaled amplitude m/a_i of the difference of PDML and PML solutions versus the relative input frequency ω_i/ω_p for $A_p = 0.0355$ (circles), 0.0360 (triangles), and 0.0365 (squares). The PML bifurcates from period 2 to period 4 at $A_c^{(2)} = 0.03705$

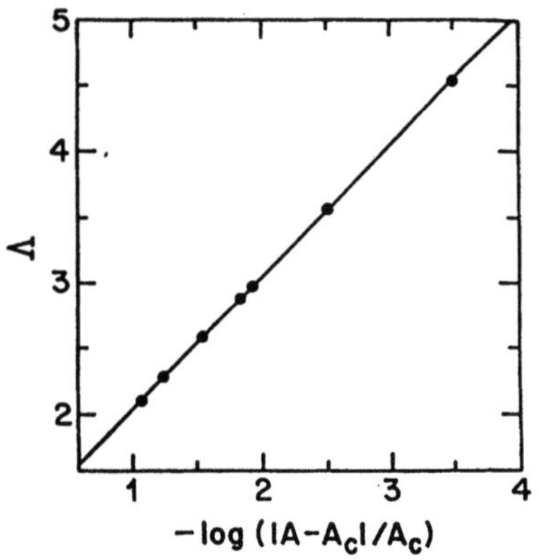

Fig. 9 Logarithm Λ of $(\sigma(\omega_p/4))^{1/2}$ evaluated for the input frequency $\omega_p/4$ versus $-\log\ \varepsilon$ where $\varepsilon = |A_p - A_c^{(2)}|/A_c^{(2)}$

influence of the input signal for ω_i near $\omega_p/4$ and $3\omega_p/4$ as A_p approaches the critical value $A_c^{(2)}$ where the PML undergoes a bifurcation from period 2 to period 4. $m(\omega_i)$ has been found to be proportional to a_i in a wide range ($10^{-3}>a_i>10^{-7}$).

We computed also the Fourier power spectrum of the oscillating difference and determined, in particular, its spectral component σ at the frequency $\omega_p/4$. The square root of $\sigma(\omega_p/4)$ increases inversely proportional to $|A_p - A_c^{(2)}|/A_c^{(2)}$ as is shown in Fig.9. The same scaling behavior holds also for m at $\omega_i = \omega_p/4$. The presence of the Fourier component $\sigma(\omega_p/4)$ indicates the existence of a dynamical precursor of the period-doubling bifurcation $\omega_p/2 \rightarrow \omega_p/4$. Hence, $\sigma(\omega_p/4)$ is directly related to the experimental sensitivity ΔS. Indeed, calculations of $\sigma(\omega_p/4)$ as a function of ω_i give results in good agreement with the ones of Fig.7b. Preliminary measurements of $\Delta S(\omega_p/4)$ as a function of a_i agree with the stated power law above.

5. Acknowledgements

The work reported here has been very much a team effort of experimentalists and theoreticians, therefore the many useful discussions with colleagues of both species are gratefully acknowledged. In particular, we would like to express our gratitude to N.B. Abraham, J.R. Tredicce and D. Walls, L.A. Lugiato. L.M. Narducci, A. Politi who have strongly influenced our work. We thank also the Swiss NSF for support.

References

1. Feature Issue of Instabilities in Active Optical Media, J. Opt. Soc. Am. B2 (1985).
2. P. Bösiger, E. Brun, D. Meier, Phys. Rev. Lett. 38 (1977) 671, Phys. Rev. A18 (1978) 671, Phys. Rev. A20 (1979) 1073.
3. E. Brun, B. Derighetti, D. Meier, R. Holzner, M. Ravani, J. Opt. Soc. Am. B2 (1985) 156.
4. M.J. Feigenbaum, J. Stat. Phys. 19 (1978) 25.
5. Y. Pomeau, P. Manneville, Commun. Math. Phys. 74 (1980) 189.

6. L.A. Lugiato, Lett. Nuovo Cimento 23 (1978) 609.

7. G.L. Oppo, A. Politi, Z. Phys. B59 (1985) 111.

8. G.L. Oppo, A. Politi, G.L. Lippi, F.T. Arecchi, in the Technical Digest of the International Meeting on Instabilities and Dynamics of Lasers and Nonlinear Optical Systems, Rochester, USA, June 1985.

9. A.H. Nayfeh, "Perturbation Methods", J. Wiley & Sons, New York (1973).

10. We acknowledge a useful discussion on this particular point with A. Politi.

11. L.A. Lugiato, L.M. Narducci, D.K. Bandy, C.A. Pennise, Opt. Commun. 46 (1985) 64.

12. D.K. Bandy, L.M. Narducci, L.A. Lugiato, J. Opt. Soc. Am. B2 (1985) 148.

13. J.R. Tredicce, F.T. Arecchi, G.L. Lippi, G.P. Puccioni, ibid. (1985) 173.

14. J.R. Tredicce, private communication.

15. Van Lerberghe, P. Cottin, J.L. Boulnois, F.T. Arecchi, G.P. Puccioni, in the Technical Digest of the International Meeting on Instabilities and Dynamics of Lasers and Nonlinear Optical Systems, Rochester, June 1985.

16. B. Derighetti, M. Ravani, R. Stoop, P.F. Meier, E. Brun, R. Badii, Phys. Rev. Lett. 55 (1985) 1746, ibid. 55 (1985) 2739(E).

17. K. Wiesenfeld and B. McNamara, Phys. Rev. Lett. 55 (1985) 13.

Bistable Operation of an Infrared Dye Laser

J.D. Harvey[1], *A. Seilmeier*[2], *and H. Roskos*[2]

[1]Physics Department, University of Auckland, Private Bag, Auckland, New Zealand

[2]Physik Department E11, Technische Universität München, Arcistrasse 21, D-8000 München 2, Fed. Rep. of Germany

1. INTRODUCTION

In recent years, the range of operating wavelengths of synchronously pumped mode locked dye lasers has been steadily broadened. Operation in the near infrared has been obtained with mode locked ion lasers as the pump source and recently, operation in the 1.3μ region has been reported [1] using direct pumping by a mode locked Nd-Yag laser. The geometry of the dye laser cavity for infrared operation is very similar to that of a mode locked Rhodamine 6G dye laser operating at visible wavelengths. However, the quantum efficiency of infrared dyes is several orders of magnitude smaller ($Q=10^{-3}-10^{-4}$)resulting in somewhat different operating characteristics [1,3].

We have adapted a previously developed [2] numerical model of the synchronously pumped Rhodamine 6G dye laser and applied it to the Nd-Yag pumped infrared dye laser, in order to determine whether a simple rate equation model is adequate to explain the more complex behaviour of this system [3]. The results indicate that the operation of the infrared dye laser can be understood on the basis of this simple model. In particular the model predicts a novel type of bistable operation which is in agreement with experimental observations. In contrast, such bistable operation is neither observed nor predicted in Rhodamine lasers with standing wave cavities.

2. NUMERICAL MODEL

The model used here has been described previously [2], and simulates the operation of a synchronously pumped dye laser in a simple three-mirror standing wave cavity. The active medium is modelled as a two-level system in which the excited and ground state population densities are given by N_1 and N_0 respectively. The gain at the dye jet is given by

$$G(t) = N_1(t)\sigma_e - N_0(t)\sigma_a$$

where σ_e and σ_a are the cross-sections for stimulated emission and absorption respectively. The coupled rate equations for the system are

$$\frac{dN_1(t)}{dt} = N_0(t)\sigma_p I_p(t) - G(t)\left[I_d(t) + I_d(t-\tau) \right] - \frac{N_1(t)}{\tau_d}$$

$$\frac{d\,I_d(t)}{dt} = c\,G(t)\,I_d(t) + \frac{cQ\varepsilon\,N_1(t)}{\tau_d}$$

$$\frac{d\, I_d(t - \tau)}{dt} = cG(t)\ I_d(t - \tau) + \frac{cQ\varepsilon\ N_1(t)}{\tau_d}$$

In these equations τ is the elapsed time between passes through the dye jet, τ_d is the spontaneous decay lifetime for the excited state, c is the speed of light, Q is the fluorescence quantum efficiency and ε the proportion of spontaneous fluorescence contributing to the dye laser intensity $I_d(t)$. These equations are integrated numerically using a chosen step size Δt and a computational "window" of 800 ps around the time of arrival of the pump pulse. Total cavity losses are treated as the sum of the output coupler loss and the lumped cavity losses. This approach allows the absolute efficiency of operation of the laser to be calculated. The effect of a birefringent filter is included numerically by convolving the amplitude of the dye laser pulse with the transform of the filter function, after stepping the entire time window through the dye jet (in both directions). The process is then repeated for each round trip. Convergence of the solution was monitored by computing the normalised sum of the squares of the differences between pulse intensities of consecutive round trips.

3. EXPERIMENTAL

The experimental system has also been described in detail elsewhere [1]. The dye laser cavity consisted of a similar three-mirror arrangement to that of the Rhodamine laser previously analysed. Operation of the infrared laser has been investigated experimentally with one two and three-plate birefringent filters, but in the bulk of the numerical results reported here a three-plate filter was assumed. Output of the infrared dye laser pumped by a modelocked Nd-Yag laser was monitored using an autocorrelator and a high-speed Ge photodiode. Narrow pulse widths are only obtained close to threshold with this system,and the laser operating characteristics were accordingly investigated in this region.

4. RESULTS

The most dramatic differences between the operation of a Rhodamine laser and the operation of an IR dye laser are in the overall efficiency (which is much lower in the infra red) and in the cavity detuning for optimal operation. The length adjustment of the Rhodamine laser is much less critical than that of the infrared laser, and the Rhodamine system will operate with cavity detunings of several picoseconds. In contrast the IR laser will only operate over a range of cavity detunings an order of magnitude smaller.

The rate equation model correctly predicts this behaviour. The low efficiency can be readily understood on the basis of the short fluorescence lifetime of the infrared dye, while the short cavity detuning range close to threshold is a consequence of the low fluorescence quantum efficiency and small gain per pass coupled with the fact that the fluorescence lifetime of the IR dye is much less than the pump pulse width. The model calculations for the IR laser also differ from the Rhodamine results in two other ways. Firstly the calculations take an order of magnitude longer to converge (requiring typically thousands of round trips instead of hundreds), and secondly the model often predicts an output pulse which circulates within the cavity arriving at the dye jet travelling in the same direction as the pump pulse. In contrast, the Rhodamine simulations invariably predict that the dye pulse collides with the pump pulse at the dye jet. This behaviour has also been experimentally verified for Rhodamine [2]. Physically, this difference in operating characteristics has its origin in the very short fluorescence lifetime of the infrared dye (~11ps) which ensures that no significant population remains in the excited state after passage of the pulse to the near high reflecting end mirror and back (which takes 330ps in the laser used here). In contrast, the excited states of Rhodamine 6G do not decay significantly in this time, and the returning pulse experiences further gain and pulse shaping on its second pass through the dye jet.

A particularly interesting feature of these results is that for long cavity detunings (physical cavity length greater than that of the pump laser) the model yields one stable solution with two pulses which collide in the dye jet in synchrony with the pump pulse arrival (see Fig 1a). For short cavity detunings (physical cavity length less than that of the pump laser), the model yields two stable solutions in which one or the other of these two pulses dominates (see Fig 1b). Numerical studies show that in this (bistable) regime the system can be tripped from one stable solution to the other by means of a suitable perturbation. The double pulse regime, however, is stable against such perturbations.

The experimental laser was found to exhibit much the same behaviour. Depending on the detuning, the laser either generated (for long cavities) double pulses with pulse spacing equal to the round trip time to the near mirror or (for short cavities) single pulses. The single pulses were unstable, however, and the position of the pulse jumped randomly between the two time positions simultaneously occupied in the case of double pulse operation. This can be understood on the basis of the model calculations as bistable operation, with the transition being caused by some perturbation in the laser system (either in the pump pulse or in the dye jet or resonator).

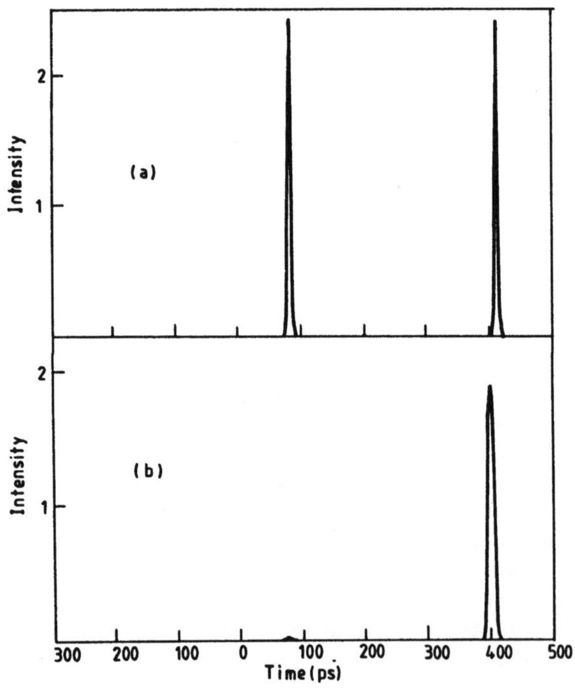

Figure 1 Typical pulse intensities predicted by the numerical model. (a) Intensity versus time for the intracavity pulse with a long cavity (detuning=15fs), (b) Intracavity pulse with a short cavity (detuning=-10fs). In these calculations, the peak of the pump pulse occurs at t=400ps. Other parameters:- $Q=10^{-3}$, $\tau_d=11ps$, $\sigma_e=10^{-16}cm^2$, $\sigma_a=10^{-18}cm^2$, $\sigma_p=3.5x10^{-16}cm^2$, pump pulse duration=70ps.

In common with many other systems exhibiting a bifurcation to bistable operation (see eg [4]), the calculations exhibit critical slowing down behaviour in the neighbourhood of the critical point, and for a small region of detunings around this point, convergence was not obtained after ten thousand round trips. The behaviour of the solutions as a function of cavity detuning is shown in Figure 2. It should be noted that convergent solutions are obtained for both positive and negative absolute detunings with the bifurcation point occurring at a small positive detuning (at approximately 1.5μ or 5fs for the parameters chosen for the simulations leading to Fig 2). In contrast, stable solutions have only been obtained for the Rhodamine laser at substantial positive absolute detunings [2].

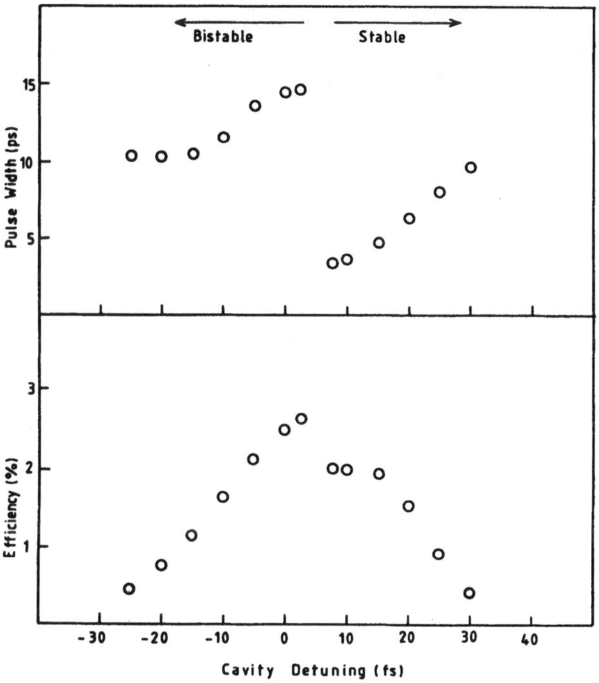

Figure 2 Variation of pulse width and power output with cavity detuning for the IR dye laser.
In the bistable regime, the output consists of one dominant pulse (see eg Fig 1b) in one of two positions.
In the stable regime, the output consists of a double pulse (see eg Fig 1a) both components of which have the same shape and intensity.

5. DISCUSSION

The simple rate equation approach used previously for the Rhodamine laser [2] has been shown to be capable of explaining the much richer behaviour of the synchronously pumped IR dye laser. The model successfully explains the very short cavity detuning range of the IR dye laser and its two operating regimes. The rate equation approach could also be applied to the study of colliding pulse ring dye lasers, and to lasers containing saturable absorbers. It is reasonable to suppose that such systems could also exhibit bistable operation under some circumstances, by analogy with the behaviour of the infra red dye laser [5]. In this case stable unidirectional circulation of a single dominant pulse might be obtained without the use of optical elements to suppress one of the directional modes of operation [6].

The cavity detuning range shown in Fig 2 is in good agreement with experiment, but experimental determinations of the pulse width as a function of detuning do not exhibit the pronounced minimum at small positive detunings shown in the figure.

The most likely explanation for this is that the pump laser exhibits a small phase jitter. A shift of only 2 ps in the position of the (70 ps FWHM) pump pulse in a microsecond, for example, is equivalent to a cavity length detuning of 0.02 ps or 6μ. Random phase jitter of this order would average the values of the pulse width over this range of detunings, since the system takes considerably longer than a microsecond (in which the pulse makes only 100 round trips) to evolve to the converged results shown in Fig 2.

The fact that the IR laser operates with negative absolute detunings is a consequence of the low gain per pass and very strong pumping needed to lift the laser above threshold. In such a situation, the cavity length detuning is compensated each round trip by a pulse-shaping mechanism which is anomalous. For short cavity lengths the circulating pulse has its leading edge attenuated and its trailing edge amplified by the increasing population inversion during its passage through the dye jet. In the long cavity regime (where the Rhodamine laser operates) the pulse is shifted forward by amplification of the leading edge and attenuation of the trailing edge. This situation is easily understood, since the population inversion is strongly depleted by stimulated emission during the passage of the narrow pulse.

The observed bistable operation of the IR dye laser could be stabilised if one of the two possible pulses were given an advantage in some way. Alternatively, for investigation of processes occurring on a time scale of less than 100 ps the laser can be used in the stable double pulse regime where it generates the shortest pulses.

6. REFERENCES

1. A. Seilmeier, W. Kaiser, B. Sens. & K.H. Drexhage: Optics Lett. 8 205 (1983).

2. L.M. Davis, J.D. Harvey and J.M. Peart: Optics Comm. 50 49 (1984).

3. H. Roskos, S. Opitz, A. Seilmeier, W. Kaiser: to be published, I.E.E.E. J. Quant. Electron. (1986).

4. Ma, Shang-Keng: "Modern Theory of Critical Phenomena" Frontiers in Physics 46 W.A. Benjamin, Inc. (1976).

5. A.M. Johnson, W.M. Simpson: Optics Lett. 8 554 (1983).

6. J. Dobler, H.H. Schulz, W.Zinth: to be published, Optics Comm. (1986).

Intrinsic Optical Bistability

C.M. Bowden

Research Directorate, AMSMI-RD-RE-QP Research, Development, and Engineering Center, U.S. Army Missile Command, Redstone Arsenal, AL 35898, USA

I. Introduction

The phenomenon of optical bistability (OB) is the nonlinear response of a material to incident light, resulting in bistability and reversible hysteresis of the output as a function of the input intensity. The first theoretical prediction of OB was by Szoke et al. [1] for a saturable absorber placed in a cavity to provide feedback and is called "absorptive" OB. Later, McCall [2] showed by numerical integration of the Maxwell-Bloch equations with boundary conditions for a Fabry-Perot cavity, that under suitable conditions such a system can exhibit differential gain and transistor action. These predictions led to the experiments of Gibbs, McCall and Venkatesan [3] in which they used Na vapor in a Fabry-Perot cavity. Although the main predictions of McCall were realized in the experiment, detailed analysis of the data indicated that the observed OB was dispersive in character rather than absorptive, and this was subsequently explained in terms of a simple phenomenological model.

The results of the earlier work stimulated much theoretical and experimental activity [4-7]. Of particular significance is the contribution to the fundamental understanding of absorptive OB by Bonifacio and Lugiato [6,8,9] who showed that OB can be regarded as a first order phase transition far from thermodynamic equilibrium. The case for mixed absorptive and dispersive OB was also treated by them [6,9] and the general conditions for OB in collections of two-level atoms in a high-Q optical cavity have been predicted [6,10], and studied experimentally [11,12].

Although most of the theoretical treatments have used the ring cavity configuration [4,6], for convenience, the experimental work has used predominantly the Fabry-Perot arrangement [3-5,7] and has been dispersive in nature. However, the theoretical predictions have been largely borne out in experiments; of particular note is the singular work of Kimble [5,7,11] and coworkers, who have verified through experiments using atomic beams the quantitative predictions of Lugiato and Bonifacio [6,8,9] based upon a simple model for two-level atoms. Most of the current effort focuses upon the utilization of semiconductor optical nonlinearities [5,7,12] due to the importance in leading to device applications in optical computing. More recently, the nature of various types of instabilities in OB has become a very lively area of activity [5,6,7,12]. This was sparked by the early predictions of Ikeda [13] and the experiments of Gibbs, Hopf, Kaplan, and Shoemaker [14] in which they studied Feigenbaum-like bifurcations to chaos [15] in a ring cavity configuration, and the work of Lugiato [6] and coworkers who predicted conditions for self-pulsing in absorptive OB and chaos in mixed absorptive and dispersive OB, and treated switching in terms of passage time statistics [16].

In all of these treatments, the essential ingredients for OB have been: (a) the illumination of an optically nonlinear material, and (b) which is contained in a high-Q cavity to provide feedback. It was pointed out early by Bowden and Sung [17] and by Bowden [18] who used a fully quantized, many-body model in the mean field limit, that OB may occur for a system comprised of a collection of atoms interacting with the electromagnetic field and driven by an externally applied coherent field, without external feedback, such as is provided by the mirrors of a cavity. Certainly, the material boundaries are expected to produce reflection, but this is usually insufficient to cause OB, although it can affect the hysteresis cycle. In this case, the role of the "feedback" is provided by the induced coherent dipole-dipole interaction among the atoms of the material. For sufficiently low values of the incident field, the strength of the induced interatomic interaction increases with the applied field. This produces a reaction field opposing the externally applied field and thus reduces the internal field. For sufficiently large values of the incident field, the individual atoms' interaction with the electromagnetic field is dominated much more strongly by the externally applied field than that due to the interatomic interaction. This overwhelms the interatomic interaction via the electromagnetic field, and thus the contribution to the reaction field due to the coherent dipole-dipole interaction is largely destroyed. Bowden and Sung [17] and Bowden [18] showed that this occurs as a first order phase transition, far from thermodynamic equilibrium. Since this is a genuine first order phase transition, completely analogous to the Meisner effect in the theory of superconductivity [17-20], we refer to this kind of OB as intrinsic OB (IOB).

Later, Flytzanis and Tang [21] pointed out that the intrinsically bistable nature of a collection of particles which interact anharmonically with the electromagnetic field can cause critical behavior in the intensity of the reflected phase-conjugate field when a weak-probe signal is introduced. Later Goldstone and Garmire [22] treated the IOB of an anharmonic optically nonlinear medium in terms of propagational effects and obtained bistability and hysteresis in the reflected intensity as well as switching of the relative phase between the incident and reflected field amplitudes. They showed that reflectivity was induced due to a discrete, spacially dependent change in the nonlinear index of refraction in the material. Ritchie and Bowden [23] studied the switching dynamics of the intrinsically bistable Duffing oscillator and made comparison with an anharmonic quantized version [24], whereas, Yuan, et al. [25] have studied instabilities in essentially the same quantized model.

Renewed interest in IOB stimulated the work of Hopf, Bowden and Louisell [26] who re-examined the case for two-level atoms interacting with the electromagnetic field, relaxing the mean field approximation made earlier [17,18]. In this treatment, the macroscopic Maxwell-Bloch formulation was used and modified by introducing the local field correction. The correction to the macroscopic theory was justified according to the work of Van Kronendonk and Sipe [27] who trace its origin to the reinclusion of an atom's self-field into the field that couples to the atom in passage from microscopic to macroscopic electrodynamics. Elimination of the local field in the macroscopic Bloch equations using the local field correction results in a coupled set of equations which are nonlinear in the atomic variables. It was shown that IOB can exist under steady-state conditions,and that the length of the material in the propagation direction must be much less than an atomic wavelength in order that the system not be opaque to the incident light. This is due to the threshold requirement that the system be strongly absorptive.

To test the results of the steady-state macroscopic theory, Hopf and Bowden [28] performed a heuristic treatment of the semiclassical coupled Maxwell-Bloch formulation within the framework of microscopic electrodynamics [28]. In this model, there is no "correction" necessary for the local field, and fluctuations were introduced into the dynamical evolution of the system composed of a finite number of spacially extended two-level atoms. The problem was treated statistically by introduction of temporal fluctuations, and this gave rise to passage time statistics. In the limit of infinite average passage times, the heuristic results of the semiclassical, microscopic, statistical model confirm precisely the results, under corresponding conditions, of the macroscopic as well as microscopic semiclassical deterministic treatments [28].

The first detailed experimental study of IOB was conducted almost simultaneously by Hajto and Janossy [29] using amorphous $GeSe_2$, who interpreted their results as due to temperature-dependent induced optical absorption in the material, and Bohnert, Kalt and Klingshirn, who used CdS [30], and also Rossmann, Henneberger and Voigt using the same material [31]. The process in the first case depends upon absorption due to temperature variation induced by the incident field, whereas the latter cases depend upon saturation of absorption due to the generation of carriers in the material, and the IOB has been interpreted [32] as due to bandgap renormalization. Later, Miller et al. [33] observed IOB in a GaAs/GaAlAs multiple-quantum-well semiconductor. This was interpreted by them as due to shrinkage of the bandgap due to heating of the material by the incident light. Shortly thereafter, Dagenais and Sharfin [34] reported IOB in CdS when the incident laser frequency was tuned below the absorption peak of the bound I_2 exciton resonance. They interpreted their results in terms of shift of the absorption peak due to the temperature distribution in the material induced by the incident field.

Theoretically, Haus, Sung, Bowden and Cook [35] predicted multiple IOB in CdS, due to increasing as well as decreasing absorption, using the bound I_2 exciton optical nonlinearity and exhibited conditions for multiple, opposite sense, IOB hysteresis cycles explicitly in terms of the heat conductivity boundary conditions. Quite recently, Li, Sung, Haus, and Bowden [36] have treated the case for the temperature dependence of the bandgap due to induced absorption in a GaAs/GaAlAs quantum well structure, and predict similar behavior and showed how the hysteresis cycles can be engineered by control of the heat conductivity boundary conditions. Longitudinal spacial inhomogeneities in IOB in CdS as well as GaAs/GaAlAs quantum well structures were recently demonstrated in calculations by Bowden, Haus, Sung and Cook [37]. These results show that switching of the output intensity, I_0 , as a function of the input intensity I_I , in IOB in CdS and GaAs/GaAlAs quantum well structures corresponds to "switching" of the induced absorption peak or the induced absorption edge, respectively, across a material boundary due to the heat conductivity boundary conditions. Thus, the two boundary conditions at each end of a cylindrical sample of large radial extent compared with the incident intensity spot size, give rise to the possibility for two distinct hysteresis cycles, which correspond to the two, distinct, heat conductivity boundary conditions, and hence multiple IOB due to induced absorption.

Of particular note is the prediction of Schmidt et. al [32] of IOB in CdS due to bandgap renormalization in the electron-hole-electromagnetic field interaction, and explains the experimental results of Bohnert, et al. [30]. Most recently, Ben-Aryeh and Bowden [38] have obtained results predicting the conditions for IOB from a fully quantized model for a collection of spacially distributed two-level atoms interacting with the electromagnetic field and driven by an externally applied coherent field. The results qualitatively confirm the early results of Bowden and Sung [17] and Bowden [18] who used a similar model, but in the mean field approximation. The results quantitatively confirm the recent results of Hopf and Bowden [28] based upon a corresponding semiclassical heuristic model with fluctuations. The results agree in the limit of very long average passage times.

The purpose of this article is threefold. First, to present a review of the main contributions to IOB; second, to exhibit the conditions for IOB and the major distinguishing characteristics of the main categories of systems that show IOB behavior; and third, to present the sufficient conditions for IOB which underlie all categories of the behavior. The latter will be discussed in terms of a simple "prototype model". The first point has been covered. The other points will be brought to bear in the next section.

II. Models and Conditions for Intrinsic Optical Bistability

A. Anharmonic Oscillator

It is well known that the driven Duffing oscillator is intrinsically bistable. There have been a number of recent studies of its properties in connection with IOB [21-25]. A very enlightening treatment of the statistical properties in the bistable regime of a quantized version of the anharmonic oscillator was given by Drummond and Walls [39]. Following previous work we shall use the quantized version of an anharmonic oscillator driven by a coherent field and coupled to a reservoir of oscillators. It will be shown that this model is the simplest prototype which exhibits all of the essential characteristics for IOB which have been studied.

The Hamiltonian is

$$H = \Delta a^{+}a - X(a^{+}a)^{2} + \sum_{m=o}^{\infty} \Delta_{m} b_{m}^{+} b_{m} + \Omega^{*} a^{*} + \Omega a + \sum_{m=o}^{\infty} \beta_{m} [a^{+} b_{m} + b_{m}^{+} a], \qquad \text{(A-1)}$$

where Δ and Δ_m are the detunings of the laser frequency ω_L from the oscillator frequency ω_0 and the reservoir frequency ω_m, respectively, i.e., $\Delta = \omega_0 - \omega_L$, $\Delta_m = \omega_m - \omega_L$. The shift operators for the anharmonic mode with anharmonic strength, X, are (a,a^{+}), and operators for the bath are (b,b_m^{+}), and each set of operators obeys Bose commutation rules. The amplitude of the laser driving field envelope is Ω. After adiabatic elimination of the reservoir variables, the equations of motion for the amplitude, a, and exciton $N \equiv a^{+}a$, are

$$\dot{a} = -i[\Delta' - 2XN - \tfrac{i}{2}\gamma]\, a - i\Omega^{*} \qquad \text{(A-2)}$$

$$\dot{N} = i\Omega a - i\Omega^{*} a^{+} - \gamma N \qquad \text{(A-3)}$$

where $\gamma = 2\pi\rho\beta^{2}$ where ρ is the density of states of the bath and β is the coupling between the oscillator and the bath, and $\Delta' = \Delta - X$. If we use the factorization $\langle a^{+}aa\rangle = \langle a^{+}a\rangle \langle a\rangle$, then (A-2) and (A-3) become equations for the factorized complex amplitude $\langle a\rangle \equiv a$ and excitation $\langle N\rangle \equiv N$. It is to be noted, then, that the effect of the anharmonic contribution of strength, X, is to cause an excitation, N, dependent renormalization of the bare frequency Δ', where the excitation, N, is driven by the interaction.

For steady-state, (A-2) and (A-3) give the equation of state

$$|\Omega|^{2} = N\,[1/4\gamma^{2} + (\Delta' - 2XN)^{2}] \;, \qquad \text{(A-4)}$$

which is cubic in N. This equation has two real roots provided $|\Delta'|^{2}/\gamma^{2} > 3/4$, and $\Delta' > 0$. When these conditions are satisfied, N as a function of $|\Omega|^{2}$ exhibits the familiar "s-curve" characteristic of OB [4-7, 23, 25].

The essential characteristics which cause the IOB are explicit in (A-2) and (A-3) and they are: (a) an excitation-dependent renormalization of the bare frequency Δ', (A-1), where the excitation is driven by the interaction, (A-2). (b) The sign of $\Delta' > 0$ causes a kind of positive feedback with regard to the nonlinearity in (A-2) caused by the anharmonic strength X. This simplest model forms a prototype model for the qualitative features of each representative model treated below.

B. Intrinsic Optical Bistability Due to Induced Absorption

We consider as a representative case, the saturable optical nonlinearity associated with the bound I_2 exciton resonance in CdS [34,35]. Since induced nonlinear absorption together with temperature-dependent shift of the absorption peak plays the major role, propagation of the field and heat conduction specify the required set of equations for the model. A similar analysis has been conducted for GaAs/GaAlAs quantum well structures [36].

For the forward- and backward-propagating fields, E_F^+ and E_B^+, respectively, we have, in steady-state [35-37]

$$\partial E_F^+/\partial x + \eta/2 \, E_F^+ = 0 \quad , \tag{B-1}$$

$$\partial E_B^+/\partial x - \eta/2 \, E_B^+ = 0 \quad , \tag{B-2}$$

where $\eta = \alpha + i\beta$ and α is the absorption function and β is the corresponding dispersion which are connected by the Kramers-Kronig relations. Empirically, [35]

$$\eta(x) = \frac{\alpha_0[T(x)]}{1 + i\gamma\{\omega-\omega_0\,[T(x)]\}} \tag{B-3}$$

where T(x) is the absolute temperature in the medium. Equations (B-1) and (B-2) are connected with the heat conduction equation in steady state,

$$\kappa \frac{\partial^2 T}{\partial x^2} = I\,\alpha \quad , \tag{B-4}$$

where $I = |E_F^+|^2 + |E_B^+|^2$, and κ is the heat conductivity, here assumed constant for simplicity. The coupled equations (B-1), (B-2), and (B-4) are nonlinear through the nonlinear induced absorption $\alpha[T(x)]$, which is given empirically as [35]

$$\alpha = \frac{1}{10} \; \frac{[3 + 0.04\,(T-T_b)^2]}{\{1 + [\omega-\omega_0(T)]^2\,(0.4)^2\}} \tag{B-5}$$

in units of reciprocal micrometers, and $(\omega-\omega_0)$ is in units of reciprocal centimeters. Here, T_b is the ambient temperature in Kelvins, which we take to be $T_b = 7°K$. The temperature dependence of the resonance peak is found to be given approximately by the empirical relation [35],

$$\omega_0(T) = \omega_0(0) - AT^2 \quad . \tag{B-6}$$

Here, $A = 1/80\ cm^{-1}\ °K^{-2}$. The existence of IOB depends crucially upon the detuning Δ as well as the boundary conditions associated with (B-4). IOB is observed when the laser frequency ω is tuned just below the absorption resonance at low field values.

For simplicity, we adopt the "Newton flow" form for the boundary conditions associated with the heat conduction equation,

$$H_x[T(x)-T_b] = I(x)\,\alpha(x)] \begin{cases} x = 0 \\ x = L \end{cases} \quad . \tag{B-7}$$

where H_x is the "heat resistance" of the boundary. Thus, using (B-5) and (B-6) in (B-7) the heat conductivity boundary conditions are nonunique, thus the output intensity I_0 as a function of the input intensity I_I can be nonunique, i.e., bistable. We find that for the two boundary conditions corresponding to the opposite faces of the sample in the propagation direction, where the spot size is much smaller than the width of the sample, at most two separate hysteresis cycles can be obtained [35].

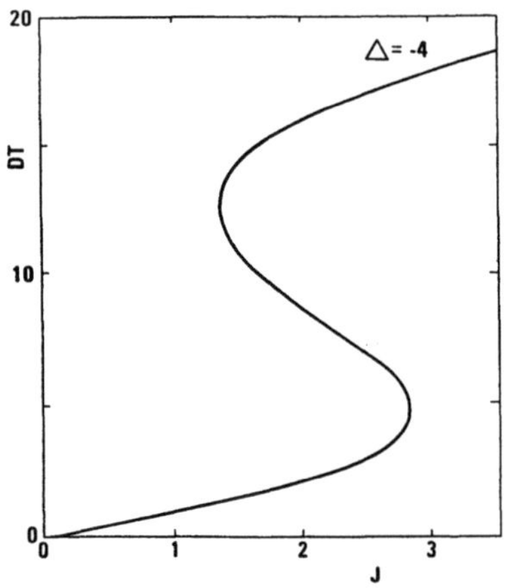

Fig. 1. The temperature difference across the boundary, DT ≡ $T-T_b$, vs. the normalized intensity at the boundary, J ≡ I/H, (see (B-7)), for the input boundary, is shown for a detuning of $\Delta = -4$, where Δ is normalized to γ (see (B-3))

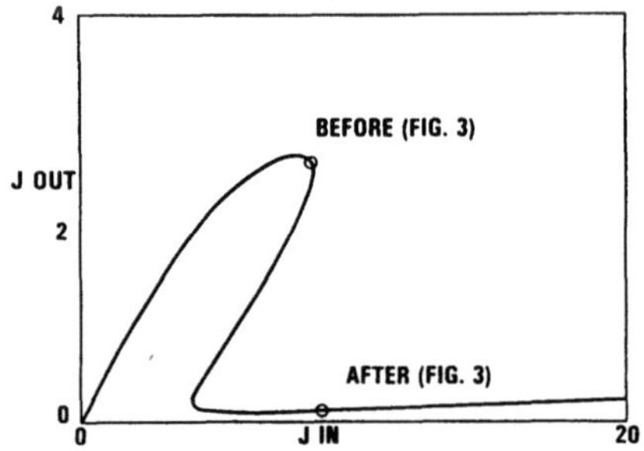

Fig. 2. Scaled intensities in the medium at the input and output faces for the boundary conditions shown in Fig. 1 and $H_0 = 3H_L$. The two circles indicate the conditions corresponding to the respective absorption distributions shown in Fig. 3

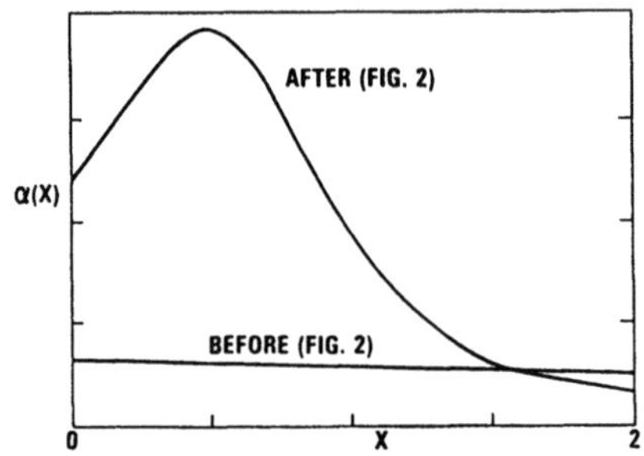

Fig. 3. Absorption $\alpha(x)$ as a function of propagation distance into the medium x corresponding to the conditions labeled in Fig. 2, before and after switching in the output intensity

A typical set of boundary conditions is portrayed in Fig. 1 and the corresponding hysteresis cycle of the output I_0 vs. the input I_I is shown in Fig. 2. An interesting spacial dependence in switching from one state to the other in the output for the same input is shown in Fig. 3. Here it is seen that the peak of the absorption maximum "switches" across the left-hand boundary of the sample, i.e., the peak of the absorption spacial distribution sweeps into the material across the boundary [37].

With the laser frequency ω tuned just below the absorption peak ω_0 at low field, as the incident intensity I_I is increased, the internal intensity increases,causing an increase in the local temperature in the medium. This causes a shift of the absorption resonance $\omega_0(T)$ to lower values, (B-6). Thus the resonance peak $\omega_0(T)$ is shifted nearer the field frequency ω which causes the field to be coupled into a larger density of states with respect to the absorption curve, and so the nonlinearity together with the detuning cause a "feedback", where the sign of the initial detuning Δ determines whether the feedback is positive or negative. These general attributes are representative qualitatively, in prototype, by the model discussed in Section II, A.

The interesting spacial switching $\alpha(x)$, depicted in Fig. 3 is descriptive of the connection between the local IOB in the material and the boundary conditions. After switching, the material near the left-hand boundary has been shifted in absorption resonance past the field frequency ω , to varying degrees, thus causing high material absorption and low intensity input. Thus, the IOB is due to increasing absorption. As the incident intensity is increased, the peak of the absorption distribution $\alpha(x)$ shifts further into the material and eventually near the right-hand boundary. It is possible for the boundary conditions at the right-hand boundary to be such that, at a higher value for the input intensity, the absorption peak "switches" across that boundary and out of the material. This causes a hysteresis cycle in the opposite sense of the former and is due to decreasing absorption [35,36].

C. Intrinsic Optical Bistability Due to the Local Field Correction: Semiclassical Macroscopic Theory

The local field correction [27] was used in the Maxwell-Bloch formulation [26] to show under what circumstances the local field effect gives IOB. This work drew heavily upon the work of Van Kronendonk and Sipe [27] who trace the origin of the local field correction in macroscopic electrodynamics, to the error commonly made in passage from the microscopic electrodynamic formulation to the macroscopic formulation by reinclusion of the atom's self-field into the field that drives the atom. The local field correction is, in this sense, simply a matter of subtracting out the self-field in macroscopic theory.

The usual Bloch equations for a system of two-level atoms having polarization P and inversion W are written as

$$dP/dt = -[i\Delta+\gamma]P + u/2\hbar\, E_L W \tag{C-1}$$

$$dW/dt = -\gamma[i+W] + u/\hbar\, [E_L^* P + E_L P^*] \tag{C-2}$$

where γ is dephasing and relaxation, u is the matrix element of the transition dipole moment and E_L is the field that drives an atom. This field consists of the externally applied field and the reaction fields from all the dipoles, but not the self-field of the dipoles. The Maxwell field on the other hand contains all fields present, including the self-field contributions. To render (C-1) and (C-2) consistent with Maxwell's equation

$$c^2\nabla^2 E - \frac{\partial^2 E}{\partial t^2} = 4\pi \frac{\partial^2 P}{\partial t^2} \tag{C-3}$$

the local field E_L must be eliminated in terms of the Maxwell field E . For this, we use the local field correction [27]

$$E_L = E + \sigma u N P \tag{C-4}$$

where the last term on the right is the correction to the Maxwell field to remove the atoms' self-field to give the field E_L which drives an atom. Here, N is the density of atoms, and σ is a complex structure factor which depends upon the local geometry of discrete dipoles.

When (C-4) is used in (C-1) and (C-2), we get

$$dW/dt = -\gamma(1+W) + u/\hbar\,\{E^*P+EP^* - 2u^2N/\hbar\,(\mathrm{Re}\,\sigma)P^*P\}\ . \tag{C-5}$$

$$dP/dt = -i[\Delta - u^2/2\hbar\ NW\ \mathrm{Im}\,\sigma]P - [\gamma + u^2/2\hbar\ N\ (\mathrm{Re}\,\sigma)W] + u/2\hbar\ EW. \qquad \text{(C-6)}$$

It is to be noted that (C-5) and (C-6) bear a remarkable resemblance to (A-2) and (A-3) if the contribution of Re σ is ignored. Usually Re σ makes a negligible contribution [26,28] and in the notation used here for the Bloch equations the more widely used form for (C-4) is with σ purely imaginary. In this case, the effect of the local field correction is to leave the equation for the inversion, (C-5), invariant in form, but results in a term nonlinear in the atomic variables in the equation for the polarization, (C-6). The effect of the additional nonlinear term, as for (A-2), is to cause an excitation-dependent shift to the bare frequency, and the excitation is driven by the atom-field interaction, (C-5). The sign of the detuning Δ causes positive or negative feedback.

Thus, based upon Sec. A, wė expect IOB. Equations (C-5) and (C-6) are coupled to Maxwell's equation (C-3), and, the problem, solved in steady state with the appropriate boundary conditions, is quite similar to that of the previous section. It was found [26] that for IOB to exist, the thickness of the sample in the propagation direction must be much shorter than a resonant wavelength. This is due to the threshold requirement on the density of the dipoles. This presented a unique boundary value problem [26] in solving the coupled set of equations, (C-3), (C-5), and (C-6).

D. Intrinsic Optical Bistability Due to Coherent Dipole-Dipole Interaction: Semiclassical Microscopic Theory

The case for two spacially separated two-level atoms interacting, in the semiclassical approximation, by coherent dipole-dipole interaction and driven by an externally applied coherent field was treated by Hopf and Bowden [26,28]. If the atoms are located well within a resonance wavelength from each other, and if it is stipulated that their induced dipole moments lie perpendicular to a common plane and along the line of the polarization of the incident field, and if the atoms are considered indistinguishable, the equations of motion for the expectation value for the inversion per atom w and the polarization per atom p are given by

$$dW/dt = u/\hbar\,\{E_I^* p + E_I p^* - 2u^2/\hbar\ \ p^* p\} - \gamma(1+w) \qquad \text{(D-1)}$$

$$dp/dt = -i[\Delta - u^2/2\hbar\ \alpha w]p - [\gamma + u^2/2\hbar\ \beta w]p + u/2\hbar\ E_I w\ , \qquad \text{(D-2)}$$

where $E = E_I + iuAp$ is the field which drives each atom, E_I is the externally applied field and $iuAp$ is the dipole field of one dipole onto the other. Within the approximations mentioned $A = \alpha + i\beta$, where α and β are real, $\alpha \approx \frac{1}{2}kr[1-2/(kr)^2]$, $\beta \approx 5/6i$, and r and k are the interatomic separation and mode k-vector, respectively (see ref. [28]).

Immediately, it is seen that (D-1) and (D-2) are identical in form with (C-5) and (C-6), which is not at all surprising. Again, the collective contribution to the relaxation, which depends upon β, does not depend upon r in this approximation and does not affect the existence of OB.

Hopf and Bowden [28] performed an heuristic calculation from a model consisting of a collection of spacially distributed two-level atoms obeying decayless Bloch equations and driven by an externally applied field and interacting with each other by coherent dipole-dipole interaction. Temporal fluctuations were introduced into the calculation by imposing relaxation with regard to each member of the collection and statistically determined for each atom at each time step of the calculation. It was found that bimodality in the results from the heuristic calculation corresponds quantitatively to bistability in the corresponding deterministic results obtained by replacing the statistics of interruption relaxation by decay rates in the equations. Quasi-stationary results associated with hysteresis loops occur in the heuristic results when the average passage times become sufficiently long. The average passage times are dependent upon the number of interacting atoms as well as the strength of the coherent dipole-dipole interaction.

E. Quantum Mechanical Treatment of Intrinsic Optical Bistability

The first quantum mechanical treatment for IOB was conducted by Bowden and Sung [17] and by Bowden [18] in which a medium consisting of a large number of two-level atoms interacting with the electromagnetic field and driven by an externally applied coherent field was considered in the mean-field approximation and in the bad cavity limit. A first-order phase transition was shown to occur in the steady state due to interatomic interaction via the electromagnetic field provided a mechanism was included for breaking the J^2-symmetry intrinsic to the Dicke Hamiltonian [40] for a small volume. It was shown that IOB is caused by interaction-dependent nonlinear dressing of the atomic frequency produced by the interatomic interaction. Furthermore, a second-order phase transition was shown to exist from the model under certain conditions [41] in the limit of vanishing externally applied field. This model suffered from two major defects: (a) the mean-field approximation was made, thus propagation and retardation effects were altogether ignored; (b) the effects of fluctuations were neglected.

To relax these approximations in the quantum mechanical approach, Ben-Aryeh and Bowden [38,42] recently reconsidered this problem for a spacially distributed system of two-level atoms. The quantum electrodynamical case for two, two-level atoms was considered where the atoms are separated by a distance r much smaller than a resonance wavelength, and driven by a coherent, linearly-polarized field. For adiabatic elimination of the field variables in the bad cavity limit and using normal ordering of the operators, i.e., field operators containing creation photon operators always appear to the left of atomic operators, the Heisenberg equations of motion for the atomic inversion $\sigma_3^{\binom{p}{q}}$ and atomic shift operators $\sigma_+^{\binom{p}{q}}$ for each atom labeled by p and q, respectively, are

$$\frac{d\sigma_3^{\binom{p}{q}}(t)}{dt} = -\gamma[1+\sigma_3^{\binom{p}{q}}(t)] - \Big\{\sigma_+^{\binom{p}{q}}(t)\Big[\alpha(r)\,\sigma_-^{\binom{q}{p}}(t \mp r/c) + \omega_R\, e^{-i(\omega_L t - \underline{k}_L\cdot\underline{r})}\Big] + 2\sigma_+^{\binom{p}{q}}(t)\, f^{(+)}(t) + \text{h.c.}\Big\} \qquad \text{(E-1)}$$

$$\frac{d\sigma_+^{\binom{p}{q}}(t)}{dt} = \frac{i}{2}[\varepsilon+i\gamma]\,\sigma_+^{\binom{p}{q}}(t) + \frac{1}{2}\Big[\alpha(r)\,\sigma_+^{\binom{q}{p}}(t \mp r/c) + \omega_R\, e^{-i(\omega_L t - \underline{k}_L\cdot\underline{r})}\Big]\sigma_3^{\binom{p}{q}}(t) + f^{(-)}(t)\;\sigma_3^{\binom{p}{q}}(t). \qquad \text{(E-2)}$$

Here, σ_3 and σ_+ are the usual Pauli operators for a two-level system, γ is the natural relaxation rate, ε is the atomic level separation, ω_R the externally applied field envelope with carrier frequency for the applied field ω_L and wave vector K_L, and $\alpha(r)$ is a c-number function of the atomic separation and we have made the relative dipole approximation. The terms involving the operators $f^{(\pm)}(t)$

stem from vacuum fluctuations and comprise Langevin source terms and can be ignored for the present discussion.

The hierarchy of moments equations of motion generated according to (E-1) and (E-2) do not close within a finite set due to the presence of operators with retarded time arguments. However, if the atomic separation r is such that $r K_L \ll 1$, then, at least for steady state, retardation effects are entirely negligible [43]. It has been shown by Ben-Aryeh and Bowden [38,42] that thus ignoring retardation in the arguments of the operators, the hierarchy of moments equations close through their second moments, and therefore the set of fifteen independent equations, comprise a linear set and therefore cannot give rise to bistability [38,42]. So, strong interatomic correlation due to coherent dipole-dipole interaction via the electromagnetic field cannot, in itself, cause bistability, a fact also noted recently by Senitzky [44]. On the other hand, since the work of Carmichael and Walls [45] and Hassan and Walls [46], it is well established that simple factorization of the equations (E-1) and (E-2) in steady-state results in a cubic nonlinearity and bistability in the parameter space. In fact, the equations (E-1) and (E-2) are then, of course, closed and become identical in form to (D-1) and (D-2) when Langevin source terms are ignored. Thus, the strong implication is that strong interatomic interaction by coherent dipole-dipole interaction alone is insufficient to cause IOB; mechanisms must be present to validate factorization of products of different atom atomic operators, at least to some sufficient degree. This point was also noted recently by Senitzky [44].

Motivated by the observations discussed above and the results of previous work [17,18,28], Ben-Aryeh and Bowden [38,42] considered the quantum mechanical treatment for a collection of a large number of spacially distributed interacting, driven two-level atoms, and treated the problem from the many-body standpoint. The intrinsic many-body statistics tend to allow for factorization and decorrelation, at least to varying degrees of justification, even within a small volume containing many atoms. Two cases were treated [38,42]: (a) the thin sample geometry with propagation direction smaller than a resonance wavelength and infinite transverse dimension, and (b) the extended sample geometry where propagation effects are important.

It was shown [38,42] that for the thin sample geometry, IOB occurs in steady state for $|\Gamma| \gg \gamma$, $|\Gamma| \gg \Delta$ where Γ is the spacial average of the coherent dipole-dipole interaction, γ is the dephasing rate and Δ is the incident field frequency detuning from the undressed atomic frequency. Under these conditions, switching from the low to the high transmission state occurs for the incident field intensity [38,42], $|\omega_R|^2$,

$$|\omega_R|^2/|\Gamma|^2 = 2/3 \qquad \text{(E-3)}$$

with the equation of state for the inversion per atom $\langle\sigma_3\rangle$ as a function of the incident field intensity $|\omega_R|^2$ obtained as [38,42]

$$\gamma[\langle\sigma_3\rangle + 1] \{[\langle\sigma_3\rangle \, \mathrm{Re}\, \Gamma - \gamma]^2 + [\langle\sigma_3\rangle \, \mathrm{Im}\, \Gamma + \Delta]^2\} = -\gamma |\omega_R|^2 \langle\sigma_3\rangle, \qquad \text{(E-4)}$$

which is obviously cubic in $\langle\sigma_3\rangle$.

In the case for the extended medium, retardation and propagation effects are important. Propagation is conveniently handled, and retardation effects are precisely taken into account by a transformation to the retarded time frame for the forward propagation direction of the incident field, and noting that the backward propagating wave is uncoupled to a good approximation and therefore does not contribute to the expectation values of the forward propagating, slowly varying components of the atomic operators. It was shown [38,42] that interatomic correlation is built up in the direction of propagation of the incident field in the medium, a phenomenon which was noted and utilized previously in the analysis of the spectrum of transmitted fluorescent light in a strongly coherently pumped three-level system [47].

The results show that for cubic or spherical symmetry for the relative location of the atoms of the medium, the major contribution to the reaction field is provided by the "distant" dipoles, whereas in the near region for which $|r-r'| > \lambda$, where λ is a resonance wavelength, the net contributions vanish, a result obtained from classical models [27,48]. These contributions do not produce nonlinear terms in the local atomic operations in the equations of motion and do not lead to IOB. However, an additional local contribution to the reaction field was obtained, corresponding to integration over the region $|r-r'| < \lambda$ and is of the form

$$\varepsilon = -i\lambda^3 \gamma\rho \langle\sigma_+(r)\rangle \tag{E-5}$$

where $\langle\sigma_+(r)\rangle$ is a continuous variable of the spacial coordinates, and ρ is the density of atoms per cubic wavelength. Equation (E-5) causes inversion-dependent renormalization Δ', of the detuning from resonance Δ , given by

$$\Delta' = \Delta - \gamma\lambda^3\rho \langle\sigma_3(r)\rangle . \tag{E-6}$$

Thus, by analogy with the results discussed in the previous sections, the contribution from (E-5) can cause IOB provided the density ρ , (actually, the number of atoms per cubic wavelength), and/or the transition dipole moment $|u|$ is sufficiently large. These results for IOB are formally equivalent to those for the thin sample, (E-3) and (E-4), and indeed, the results for the long sample reduce to those for the thin sample in the limit. The contribution (E-5) is consistent with the contribution which arises from the local field correction in the Maxwell-Bloch formulation [26], and stems from the same origin, i.e., the passage from the discrete, microscopic representation to the macroscopic continuous representation.

III. Conclusion

Intrinsic optical bistability, IOB, has been discussed in the previous sections for representative systems. Each case discussed is characterized by the same essential local characteristics to produce IOB. As enunciated previously in Sec. II-A, they are: (1) an inversion or excitation-dependent renormalization of the resonance frequency for the nonlinear optical material; (2) the excitation or inversion is dependent upon the field-medium interaction; (3) the sign of the detuning of the incident field frequency from the resonance frequency of the medium determines a positive or a negative "feedback". These are the essential properties that cause IOB for the anharmonic oscillator as shown in Sec. II-A. That model can thus be considered as a simple prototype model for IOB.

1. A. Szoke, V. Danen, J. Goldhar and N. A. Kurnit, Appl. Phys. Lett. 15, 376 376 (1969)
2. S. L. McCall, Phys. Rev. A9, 1515 (1974)
3. H. M. Gibbs, S. L. McCall and T. N. C. Venkatesan, Phys. Rev. Lett. 36, 1135 (1976)
4. Optical Bistability, edited by C. M. Bowden, M. Ciftan and H. R. Robl (Plenum, New York, 1981)
5. Optical Bistability 2, edited by C. M. Bowden, H. M. Gibbs and S. L. McCall (Plenum, New York, 1984)
6. L. A. Lugiato, Progress in Optics XXI, edited by E. Wolf (North-Holland, 1984), p. 89-216
7. Optical Bistability: Controlling Light With Light, H. M. Gibbs, (Academic Press, 1985)
8. R. Bonifacio and L. A. Lugiato, Opt. Comm. 19, 172 (1976); Phys. Rev. A18, 1129 (1978)
9. R. Bonifacio and L. A. Lugiato, Lett. Nuovo. Cimento 21, 517 (1978)
10. S. S. Hassan, P. D. Drummond and D. F. Walls, Opt. Comm. 27, 480 (1978)

11. H. J. Kimble, A. T. Rosenberger and P. D. Drummond, Ref. 5, p. 1; H. J. Kimble, D. E. Grant, A. T. Rosenberger and P. D. Drummond, in Laser Physics, (Lecture Notes in Physics, Vol. 182), eds., J. D. Harvey and D. F. Walls (Springer-Verlag, New York, 1983); A. T. Rosenberger, L. A. Orozco and H. J. Kimble, Phys. Rev. A28, 2569 (1983)
12. Technical Digest, Topical Meeting on Instabilities and Dynamics of Lasers and Nonlinear Optical Systems, OSA, University of Rochester, NY, June 1985, paper WA2; Technical Digest, Topical Meeting on Optical Bistability (OB3), OSA, Tucson, AZ, December 1985, paper W2
13. K. Ikeda, Opt. Commun. 30, 257 (1979); K. Ikeda, H. Daido and O. Akimoto, Phys. Rev. Lett. 45, 709 (1980)
14. H. M. Gibbs, F. A. Hopf, D. K. Kaplan and R. L. Shoemaker, Phys. Rev. Lett. 46, 474 (1981)
15. M. J. Feigenbaum, J. Stat. Phys. 19, 25 (1978)
16. G. Broggi, L. A. Lugiato and A. Colombo, Phys. Rev. A32, 2803 (1985)
17. C. M. Bowden and C. C. Sung, Phys. Rev. A19, 2392 (1979)
18. C. M. Bowden, in Ref. 4, p. 405-429
19. L. N. Cooper, Phys. Rev. 104, 1189 (1956); J. Bardeen, L. M. Cooper and J. R. Schrieffer, ibid, 106, 162 (1957); 108, 1175 (1957)
20. L. P. Gorkov, Sov. Phys. JETP 34, 505 (1958)
21. C. Flytzanis and C. S. Tang, Phys. Rev. Lett. 45, 441 (1980)
22. J. A. Goldstone and E. Garmire, Phys. Rev. Lett. 53, 910 (1984)
23. B. Ritchie and C. M. Bowden, Phys. Rev. A32, 2293 (1985)
24. L. M. Narducci, S. S. Mitra, R. A. Shatas and C. A. Coulter, Phys. Rev. A16, 247 (1977)
25. J-M Yuan, E. Liu and M. Tung, J. Chem. Phys. 79, 5034 (1983); E. Liu and J-M Yuan, Phys. Rev. A29, 2257 (1984)
26. F. A. Hopf, C. M. Bowden and W. H. Louisell, Phys. Rev. A29, 2591 (1984)
27. J. Van Kronendonk and J. E. Sipe, in Progress in Optics XV, edited by E. Wolf (North-Holland, Amsterdam, 1977), p. 245
28. F. A. Hopf and C. M. Bowden, Phys. Rev. A32, 268 (1985)
29. J. Hajto and I. Janossy, Philos. Mag. B47, 346 (1983)
30. K. Bohnert, H. Kalt and C. Klingshirn, Appl. Phys. Lett. 43, 1088 (1983)
31. H. Rossmann, F. Henneberger and H. Voigt, Phys. Status Solidi B115, k63 (1983)
32. H. E. Schmidt, H. Haug and S. W. Koch, Appl. Phys. Lett. 44, 787 (1984)
33. D. A. B. Miller, A. C. Gossard and W. Wiegmann, Opt. Lett. 9, 162 (1984)
34. M. Dagenais and W. F. Sharfin, Appl. Phys. Lett. 45, 210 (1984)
35. J. W. Haus, C. C. Sung, C. M. Bowden and J. M. Cook, J. Opt. Soc. Am. B2, 1920 (1985)
36. Y. C. Li, C. C. Sung, J. W. Haus and C. M. Bowden, "Model for Mirrorless Optical Bistability in Multiple Quantum Well GaAs/GaAlAs Due to Induced Absorption", to be published
37. C. M. Bowden, J. W. Haus, C. C. Sung and J. M. Cook, "Longitudinal Spacial Inhomogeneities in Optical Bistability Due to Induced Absorption", to be published
38. Y. Ben-Aryeh and C. M. Bowden, "Mirrorless Optical Bistability Due to Coherent Dipole-Dipole Interactions in a Spacially Distributed Collection of Two-level Systems", to be published
39. P. D. Drummond and D. F. Walls, J. Phys. A13, 725 (1980)
40. R. H. Dicke, Phys. Rev. 93, 99 (1954)
41. K. Hepp and E. H. Lieb, Ann. Phys. (NY) 76, 360 (1973); Y. K. Wang and F. T. Hioe, Phys. Rev. A7, 8311 (1973); R. Gilmore and C. M. Bowden, Phys. Rev. A13, 1898 (1976); R. Gilmore and C. M. Bowden, J. Math. Phys. (NY) 17, 1617 (1976)
42. J. Ben-Aryeh and C. M. Bowden, post deadline paper, Topical Meeting on Optical Bistability (OB3), Tucson, AZ, December 2-4, 1985
43. P. W. Milonni and P. L. Knight, Phys. Rev. A10, 1096 (1974)
44. I. R. Senitzky, IEEE J. Quant. Elec. QE-21, 1477 (1985)
45. H. J. Carmichael and D. F. Walls, J. Phys. B10, L685 (1977)
46. S. S. Hassan and D. F. Walls, J. Phys. A11, L87 (1978)
47. C. C. Sung and C. M. Bowden, Opt. Comm. 45, 273 (1983)
48. W. D. Jackson, Classical Electrodynamics, 2nd Ed, (Wiley, New York, 1962), Ch. 4

Chaos in Simple Quantum Systems

W.C. Schieve and T.Y. Petrosky

Center for Statistical Mechanics, University of Texas at Austin,
Austin, TX 78712, USA

Abstract

Chaos in classical systems will be reviewed. A test function for the existence of topological chaos, the Melnikov function, is shown to determine the existence of homoclinic points which imply classical topological chaos. The Melnikov function is generalized to semiclassical quantum ensembles and it is shown that quantum ensembles suppress the appearance of topological chaos. A kinetic equation is shown to exist governing fully quantum chaos. The conditions for its existence are new necessary conditions in the presence of quantum mechanics when classical topological ideas fail.

I. Introduction

The complex and rich behaviour of simple classical nonlinear systems was first discovered by POINCARÉ [1]. The incredibly intricate structure of the flows (called chaos) near hyperbolic points is now well known and the subject of continuing interest both in Hamiltonian and dissipative systems [2-6]. In quantum optics laser-like systems have long been known to exhibit this essentially classical behaviour [7-10]. Important questions are: What is the effect of quantum mechanics upon this classical chaos? Is there an intrinsic quantum effect (quantum chaos!) [11-13]? We will here discuss both these questions for simple Hamiltonian systems. It is our view that qualitatively similar results will hold for the dissipative systems of quantum optics in which "strange" attracting sets appear.

In classical dynamical systems the topological appearance of the homoclinic points near a hyperbolic fixed point is central to the understanding of chaos. Indeed a remarkable theorem has been proved [2,3]. Near a homoclinic point the motion may be represented by a shift map. The system is nonintegrable! It is Markovian, and thus statistical. One might expect that the motion may be treated by the methods of non-equilibrium statistical mechanics [14,15,16]. In this case there exists a simple topological test function for the presence of homoclinicity due to MELNIKOV [17,5]. The extension of the constant Melnikov function to quantum mechanics as well as its role in a non-equilibrium statistical mechanical description [18,19] is a central theme of these comments.

In Section II we define the Melnikov function and extend its definition to semi-classical quantum mechanics. In this semi-classical approximation we use the moment expansion technique to evaluate the quantum fluctuations. We find the principle lowest order perturbation effect is a constant shift of the Melnikov function which may suppress the appearance of the classical topological chaos due to the non-locality of the wave packet. We apply this to a driven anti-Duffing oscillatory (double-well potential). Simple estimates are made of the observability of the semiclassical topological chaos on the molecular, atomic and sub-atomic level.

In Section III we turn to a statistical mechanic treatment of the full quantum mechanical motion in regions of parameter space of classical chaos. Here we demonstrate the loss of memory in time of the initial condition of the off-diagonal elements of the density matrix to lowest order in perturbation theory. As a result of this we no longer have a quantum pure state but a statistical mixture. It is shown that this memory term is proportional to the time derivative of the generalized Melnikov function and thus vanishes a necessary condition [14,16] for irreversible behaviour as in statistical mechanics and a new necessary condition for chaos in quantum mechanics.

We then derive a kinetic equation analogous and of a similar form to the Pauli master equation first derived by VAN HOVE [20,14,15,16]. This reduces to the classical Fokker-Planck equation previously obtained [19]. In this case the diffusion coefficient is proportional to the "square" of the amplitude of the Melnikov function. In this section we then consider an oscillating saw-tooth model which has many of the properties of usual statistical mechanics systems such as free motion and duration of collision. We then compare the classical and quantum kinetic equations describing the motion in a chaotic region and observe the essential role of quantum interference which enhances the extent of the chaotic motion. It is this effect we would call quantum chaos.

II. The Semi-Classical Quantum Melnikov Function

Let us consider a classical nonlinear system with one degree of freedom driven by a time dependent external field.

$$H(xp,t) = \frac{p^2}{2m} + V_0(x) + gV_1(x,t) = H_0 + gV_1(x,t) \; , \qquad |g| \ll 1 \tag{2.1}$$

V_1 is periodic in t with period $2\pi/\omega$. Such systems are well known to exhibit chaos classically for various parameter ranges [4,5,6]. In this section we assume there is a separatrix which intersects the hyperbolic fixed point. For $g \neq 0$ the separatrix splits into stable and unstable manifolds. The homoclinic point is the point of transverse intersection of the stable and unstable manifolds in a stroboscopic surface of sections of period $2\pi/\omega$.

The Melnikov function, $\Delta(t,t_0)$, is defined by [17]

$$\Delta(t,t_0) = \Delta^+(t,t_0) - \Delta^-(t,t_0) \quad ,$$

$$\Delta^\pm(t,t_0) = \dot{x}_{sx}(t-t_0)p^\pm(t,t_0) - \dot{p}_{sx}(t-t_0)x^\pm(t,t_0) \quad . \tag{2.2}$$

Here (x_{sx},p_{sx}) is the unperturbed solution on the separatrix and $(x^\pm,p^\pm)$ are the perturbed solutions on the stable and unstable manifolds, (-,+). One may show

$$|\Delta(t_0,t_0)|^2 = \{[\dot{x}_{sx}(0)]^2 + [\dot{p}_{sx}(0)]^2\}^{1/2}$$

$$\times \{[x^+(t_0,t_0) - x^-(t_0,t_0)]^2 + [p^+(t_0,t_0) - p^-(t_0,t_0)]^2\}^{1/2} \, . \tag{2.3}$$

Hence, $\Delta(t_0,t_0) = 0$ at a homoclinic crossing and $\Delta(t_0,t_0) \equiv 0$ (for all t_0) when (+) and (-) manifolds coincide. Applying perturbation theory [17,5] we obtain to the lowest order the central formula

$$g\Delta_1 = \int_{-\infty}^{+\infty} dt' \{H_0,V_1\}_{sx} \quad . \tag{2.4}$$

The Poisson bracket is evaluated at the unperturbed solutions. This is the simplicity and power of the Melnikov function in locating the homoclinic crossing,thus testing for chaos locally [3,5]. We note from (2.4) that

$$\dot{\Delta}_1^\pm = 0 \; . \tag{2.5}$$

Let us extend the definition of the Melnikov function to quantum mechanics [18]. We define

$$\Delta_q \equiv \Delta_q^+ - \Delta_q^- \; , \qquad \text{where}$$

$$\Delta_q^\pm = \langle\psi^\pm|\{\frac{1}{2}[\dot{\hat{x}}_0(t-t_0)\hat{p}(t,t_0) + \hat{p}(t,t_0)\dot{\hat{x}}_0(t-t_0)]$$

$$- \frac{1}{2}[\dot{\hat{p}}_0(t-t_0)\hat{x}(t,t_0) + \hat{x}(t,t_0)\dot{\hat{p}}(t-t_0)]\}|\psi^\pm\rangle \; . \tag{2.6}$$

Here $(\hat{x}_0,\hat{p}_0)$ and $(\hat{x},\hat{p})$ are Heisenberg operators governed by H_0 and H respectively. The states $|\psi^{\pm}\rangle$ are such that if we neglect quantum fluctuations the first moments

$$x^{\pm} \equiv \langle\hat{x}(t)\rangle_{\pm} \quad , \quad p^{\pm} \equiv \langle\hat{p}(t)\rangle_{\pm} \tag{2.7}$$

coincide with the classical solutions on the stable and unstable manifolds. Coupled moment equations to all orders may be obtained from the Heisenberg equations for $\hat{x}$, $\hat{p}$. Closed sets of moment equations to second order may be obtained in, e.g., semi-classical truncation by means of a Taylor expansion around the classical values (x,p) [21].

If we further expand to lowest order in g $x^{\pm}$ and the fluctuations etc. we obtain from (2.6)

$$\Delta_q^{\pm}(t) = \varepsilon_0^{\pm}(t) + g\Delta_1^{\pm}(t) + g[\varepsilon_1^{\pm}(t) - \frac{1}{2}\,\xi_0^{\pm}(t)V_1^{(3)}(x_0^{\pm}(t),t)] \quad , \tag{2.8}$$

where $\varepsilon_0^{\pm}(t)$ and $\varepsilon_1^{\pm}(t)$ are the zero and first order energy fluctuations and $\xi_0^{\pm}$ is the mean deviation of the position. If as in the following example, $V_1^{(3)} = 0$, then (2.8) shows $\Delta_q^{\pm}(t)$ is just the quantum energy fluctuation to this order. An equation for $\varepsilon^+(t) - \varepsilon^-(t)$ may be obtained from the Ehrenfest equations and the corresponding fluctuation equations [21] and we may show to this order

$$\varepsilon_0^+(t) - \varepsilon_0^-(t) = \text{constant} \tag{2.9}$$

and thus

$$\Delta_q = (\varepsilon_0^+ - \varepsilon_0^-) + g\Delta_1 \, , \tag{2.10}$$

where we have neglected as higher order terms in products of $g\xi_0^{\pm}$ etc. (2.10) is the principle result of this section. The quantum Melnikov function to this order is just the classical value plus a <u>time-independent</u> correction due to <u>initial quantum energy fluctuations</u>. (2.10) shows that the energy fluctuations may <u>remove the zeros of the classical Melnikov function</u>, $g\Delta_1$, (2.4). Thereby, the homoclinic points and thus the topological nature of the chaos may not be observed.

Numerical estimates of the energy fluctuation may be easily obtained [18]. Choosing the anti-Duffing oscillator model

$$H(x,p,t) = \frac{p^2}{2m} - \frac{(\alpha x)^2}{2} + \frac{(\beta x)^4}{4} - \frac{g}{2}\,(\alpha x)^2\cos(\omega t+\phi) \, . \tag{2.11}$$

The classical Melnikov function is obtained easily from (2.4). It is

$$g\Delta_1(t_0) = -4\pi g\delta_V\left[\frac{\omega}{\gamma}\right]^2 \mathrm{csch}\left[\frac{\pi\omega}{2\gamma}\right]\sin(\omega t_0+\phi) \quad , \tag{2.12}$$

where $\gamma \equiv \alpha/\sqrt{m}$ and $\delta_V \equiv \alpha^4/(4\beta^4)$. Infinitely many homoclinic points are apparent in the oscillations with respect to ϕ and ωt_0 for any $g \neq 0$.

The <u>maximum</u> quantum mechanical energy shift may be estimated from (2.10)

$$|\varepsilon_0^+ - \varepsilon_0^-| < \varepsilon_0 = \frac{(\Delta p_0)^2}{2m} + \frac{5}{2}\,(\alpha\Delta x_0)^2 \quad , \tag{2.13}$$

where $\Delta x_0 \equiv [\xi_0(t)]^{1/2}$, $\Delta p_0 \equiv [\eta_0(t_0)]^{1/2}$. Assuming $\Delta x_0\Delta p_0 = \hbar/2$, and minimizing the energy fluctuation the parameters of the double well may be obtained. They are shown in Table I for electrons and protons for $x_{cl} = 10^{-7}$ and 10^{-13} cm respectively.

Table I. Well parameters and energy fluctuations for the electron and proton. Here $\Delta p_0 = cp_{cl}$ where $p_{cl} = \sqrt{2m}\,(\alpha/\beta)^2$.

	Electron	Proton
$x_{c\ell}$	10^{-7} cm	10^{-13} cm
c	0.1	0.1
α	3.9×10^{2}	9.1×10^{12}
β	7.4×10^{4}	1.1×10^{13}
γ	1.3×10^{16} $\sec^{-1}$	7.0×10^{24} $\sec^{-1}$
ε_0	1.5×10^{-11} erg	8.3×10^{-3} erg
$\varepsilon_0/4\pi v$	6.4×10^{-3}	6.4×10^{-3}

From (2.13) we see that if

$$\frac{\varepsilon_0}{4\pi\delta_v} < g\left[\frac{\omega}{\gamma}\right]^2 \mathrm{csch}\left[\frac{\omega}{2\gamma}\right] \quad , \tag{2.14}$$

then classical chaos may be observed in the presence of non local quantum fluctuations. Table II shows values of the right-hand side of (2.14).

Table II. Frequency dependence of the classical Melnikov function with g = 1.

ω/γ	$(\omega/\gamma)^2 \mathrm{csch}(\pi\omega/2\gamma)$
10^{-2}	6.4×10^{-3}
10^{-1}	6.4×10^{-2}
1	4.3×10^{-1}
2	3.4×10^{-1}
3	1.6×10^{-1}
4	6.0×10^{-2}
5	1.9×10^{-2}
7	1.6×10^{-3}
10	3.0×10^{-5}

Comparison of $\varepsilon_0/4\pi\delta_v$ from Table I with this shows that the Melnikov function may have observed zeros in a narrow frequency range, i.e. if g = 0.01 $1 < \omega < 3$. Observation of the topological nature of the chaos in the presence of quantum fluctuations depends sensitively on the frequency of the driving term.

III. Quantum Kinetics of Motion

In this section we will obtain to lowest order in $gV_1(x,t)$ a quantum kinetic equation for the Markov statistical motion of a simple nonlinear oscillator driven by a harmonic perturbation. We simplify the Hamiltonian, (2.1),

$$H = \frac{p^2}{2m} + gV_1(x)\sin(\omega t+\phi) \quad \text{(periodic 2L in x)} \tag{3.1}$$

and neglect tunnelling effects. The unperturbed states are $|k\rangle$; $k = n\pi/L$, $n = 0,\pm1...$ having energy $E_k = \hbar^2k^2/2m$ and momentum $p = \hbar k$.

The purpose here is to obtain a kinetic equation for the probability that the system is in state $|k\rangle$, $\langle k|\rho(t)|k\rangle$, $\rho(t)$ being the density operator governed by the von Neumann equation

$$i\dot{\rho}(t) = (L^0 + gL^1)\rho(t) \quad ,$$

$$L^0 = \frac{1}{\hbar}\left[\frac{p^2}{2m},\ \right] \quad ,$$

$$L^1 = \frac{1}{\hbar}[V_1(x,t), \;] \quad . \tag{3.2}$$

In the $|k\rangle$ representation,

$$\langle k|L^0|k'\rangle = \frac{h}{2m}(k^2 - (k')^2) \quad .$$

The techniques of diagramatic perturbation theory [14] or projection operators [15] may be used to obtain such an equation, the Generalized Master equation. To lowest order in g the equation is Markovian. The first important question is, does the contribution of $\langle k|\rho|k'\rangle$ for $k \neq k'$ zero as $t \to \infty$? This is a necessary condition for the existence of a kinetic equation [14]. The relevant term to order g in the Generalized Master equation is [14,15,16]

$$D_{kk}(t) = [L^1 \exp(-\frac{iL^0}{\hbar}t)(1-P)\rho(0)]_{kk}$$

$$= \frac{1}{\hbar}[V_1(x,t)\exp(-\frac{iH_0}{\hbar}t)(1-P)\rho(0)\exp(\frac{iH_0}{\hbar}t)$$

$$- H^0\exp(-\frac{iH_0}{\hbar}t)(1-P)\rho(0)\exp(\frac{iH_0}{\hbar}t)V_1(x,t)]_{kk} \quad , \tag{3.3}$$

where the projection operator, P, has the properties $P^2 = P$, $(P\rho)_{kk'} = \rho_{kk}\delta_{kk'}$. We are interested in the average of a diagonal observable with $D_{kk}(t)$. For the energy, $\bar{E}$, this is

$$\dot{\bar{E}}(t) = \sum_k E_k D_{kk}(t) \quad . \tag{3.4}$$

It has the physical meaning of the time derivative of the energy change during a collision for a system in the initial state $\rho(0)$. This may be written utilizing the trace and introducing the Heisenberg operator

$$V(x_0(t),t) = \exp(\frac{iH_0}{\hbar}t)V(x,t)\exp(-\frac{iH_0}{\hbar}t)$$

as

$$\dot{\bar{E}}(t) = \mathrm{Tr}\dot{\hat{\Delta}}(t)\rho(0) \quad , \tag{3.5}$$

where

$$\dot{\hat{\Delta}}(t) = \frac{1}{\hbar}[H_0, V_1(x_0(t),t)] \quad . \tag{3.6}$$

Now we may, using time reversal symmetry, show that

$$\bar{E}(\infty) - \bar{E}(-\infty) = \frac{1}{\hbar}\int_{-\infty}^{\infty} dt\,\mathrm{Tr}[H_0, V_1(x_0(t),t)]\rho(0) \quad . \tag{3.7}$$

By comparison of (2.4) and (3.7) we see a new generalization of the quantum Melnikov function as the energy transfer during a single collision in the presence of ensembles. (3.5), 3.6), (3.7) are a generalization of this to quantum mechanics.

If $\dot{\bar{E}} = 0$

and $\bar{E}(\infty) - \bar{E}(-\infty) \neq 0$ identically ,

this implies $D_{kk}(t) \to 0$ as $t \to \infty$,

which further implies the system loses the dependence in the kinetic equation upon $\langle k|\rho(0)|k'\rangle$, $k \neq k'$.

Equation (3.7) may be further simplified, working out the commutator we have

$$\bar{E}(\infty) - \bar{E}(-\infty) = \frac{g\pi}{4\hbar m} \int_{-\infty}^{+\infty} p^2 \sum_{n=-1}^{+1} \sum_k n \exp(in\phi) V_{-k}(\eta^k - \eta^{-k}) \left\{\frac{1}{kp - n\omega - i\varepsilon} - \text{c.c.}\right\} \rho_k(p) \quad . \tag{3.8}$$

Here V_k is the Fourier transform of the spacial part of the potential. We have introduced the notation $\rho_k(p) \equiv \langle p| - \hbar k/2|\rho|p + \hbar k/2\rangle$ [14,23]. Here η^k is a finite displacement operator having properties

$$\eta^k \rho_k(p) = \rho_k(p + \hbar k/2)$$

$$(\eta^k - \eta^{-k}) = 2\sinh\left(\frac{\hbar k}{2}\frac{\partial}{\partial p}\right) \quad . \tag{3.9}$$

Expanding and partially integrating we obtain

$$\bar{E}(\infty) - \bar{E}(-\infty) = -i\pi g L \sum_n n \exp(in\phi) \int_{-\infty}^{+\infty} dk \int_{-\infty}^{+\infty} dp\, p^2 \frac{\partial}{\partial p} V_{-k}\, \delta(kp - n\omega)\, \rho_k(p) \quad . \tag{3.10}$$

This has exactly the same form as the classical result. The difference lies in $\rho_k(p)$ which is the Fourier transform of a Wigner function in the $|p\rangle$ representation. If one chooses initially $\rho_k(p) = \exp(ikx_0)\delta(p-p_0)$ one obtains the classical trajectory result. Otherwise quantum effects enter through the non-classical part of the initial valve $\rho_k(p)$. Note that classical ensembles will have the same result. This can also be seen in Section II, (2.10).

Let us consider the quantum kinetic equation. Utilizing time-dependent perturbation theory to order g^2 we may in an analogous way to familiar non-equilibrium statistical mechanics obtain a Pauli-like equation [14,15,16,23]

$$i\frac{\partial \rho_0(p,t)}{\partial t} = \psi(+i\varepsilon)\rho_0(p,t) \quad , \quad \text{where} \tag{3.11}$$

$\rho_0(p,t)$ is the momentum distribution function and

$$\psi(+i\varepsilon) = -i\left(\frac{g\pi}{2\hbar}\right)^2 \sum_{n=-1}^{+1} \sum_k |V_k|^2 (\eta^k - \eta^{-k}) \frac{1}{kp + n\omega - i\varepsilon} (\eta^{-k} - \eta^k) \quad . \tag{3.12}$$

The limit $\varepsilon \to 0^+$ is implied here and in the following equations.

Since we are dealing with a finite system we do not overall let $L \to \infty$ and take the thermodynamic limit $L \to \infty$, $\mathcal{N} \to \infty$, $\mathcal{N}/L$ = constant [14,15,16]. ($\mathcal{N}$ is the number of particles). For purposes of evaluating the summation over k in (3.12) we will approximate it, letting

$$\frac{\pi}{L}\sum_k \to \int dk \ .$$

This leaves one uncompensated factor of $1/L$ reflecting the fact that (3.10) depends on system size. In statistical mechanics the many-body nature of the collision introduces a factor of $\mathcal{N}$ which compensates the L in the thermodynamic limit, and then (3.11) is size independent in this sense. (3.11) cannot be viewed as arising from the singular limit of VAN HOVE [20], $g \to 0$, $t \to \infty$, g^2t finite (plus thermodynamic limit). Rather from this present point of view the singular limit is $g \to 0$, $t \to \infty$, $L \to \infty$ such that $g^2t/L = (g^2 \times \text{number of collisions})$ is finite.

We assert that this quantum-mechanical equation is valid for finite simple non-linear systems in the regime of chaos. Chaos will be present if (3.4) vanishes thus not contributing additively to (3.10) and the dissipative condition of PRIGOGINE [14] holds, $\psi(+i\varepsilon) \neq 0$. This and the conditions that $\bar{E} = 0$ is zero and $E(\infty) - E(-\infty)$, not identically zero are our new dynamical criteria for the existence of chaos in the presence of quantum ensembles when the topological picture can no longer be taken as in Section II. It happens to be the same as the condition for the existence of kinetic equations in statistical mechanics. Such a criteria would also be true in the presence of classical ensembles [22,25].

PETROSKY [19] has previously obtained a classical Fokker-Planck equation for the chaotic motion of a nonlinear driven oscillator. This gave the diffusion coefficient of CHIRIKOV [4]. This classical Fokker-Planck equation may be readily obtained from (3.11) [14] by introducing the representation of $\eta^k - \eta^{-k}$, (3.9), and letting $\hbar \to 0$. We thus have a Fokker-Planck operator

$$\psi_{cl}(+i\varepsilon) = -i\left(\frac{g\pi}{2}\right)^2 \sum_{n=-1}^{+1} \sum_k |V_k|^2 k \frac{\partial}{\partial p} \frac{1}{kp+n\omega-i\varepsilon} k \frac{\partial}{\partial p} , \tag{3.13}$$

which leads to a diffusion coefficient

$$D_{cl} = -2i\left(\frac{g\pi}{2}\right)^2 \sum_{n=-1}^{+1} \sum_k |V_k|^2 \frac{k^2}{kp+n\omega-i\varepsilon} . \tag{3.14}$$

Let us now consider a simple model of (3.1), the periodic saw-tooth potential

$$V(x) = \begin{cases} x & \text{for} \quad 0 < x < \ell \\ (2\ell - x) & \text{for} \quad \ell < x \leq 2\ell \qquad \text{(period 2L)} \\ 0 & \text{for} \quad 2\ell < x \leq 2L \end{cases} \tag{3.15}$$

This may be simply realized as a charge moving in a circuit containing a capacitor and driven by an AC signal. It is easily shown that the spatial Fourier transform is simply

$$V_k = -\frac{1}{2L}\left(\frac{1-\exp(-ik\ell)}{k}\right)^2 .$$

The classical properties of this model have been investigated. It may be reduced to a mapping exactly which in a limit is the standard map [24]. Thus, for all $g \neq 0$ the motion is chaotic at least locally. Details of the classical model will be published elsewhere [25]. The interesting physical aspect of this model is that it has a finite range potential and thus a region of free propagation. The stable and unstable manifolds are formed from the free propagation in the + and - directions.

We may evaluate $D_{kk}(t)$ or $\bar{E}(\infty) - \bar{E}(-\infty)$, (3.10) in detail and verify that the initial state is a finite localized pulse outside the region of interaction letting $\rho k = \exp(ikx_0)\delta(p-p_0)$ we obtain the classical result

$$\bar{E}(\infty) - \bar{E}(-\infty) = 2g \sin^2\left[\frac{m\omega\ell}{2p_0}\right] \frac{1}{\frac{(m\omega)^2}{p_0}} \sin(\omega t_0 + \phi) .$$

Consider (3.12). Recognizing in the limit $\varepsilon \to 0$ that the denominator of (3.12) is $\delta(kp+n\omega)$ and performing the η^k operations, it becomes

$$\psi(+i\varepsilon) = \left(\frac{g\pi}{\hbar}\right)^2 \frac{2\pi}{L} \sum_{n=-1}^{+1} \int_{-\infty}^{+\infty} dk\, |V_k|^2 \delta(E(p+\hbar k)-E(p)+n\hbar\omega)[1-\eta^{2\hbar k}] \quad , \tag{3.16}$$

which has the exact form of a Pauli equation. The contributions to the k integrations come from the roots of the argument of the δ function

$$\hbar^2 \frac{k^2}{2m} + \hbar \frac{kp}{m} + n\hbar\omega = 0 \quad , \qquad n = \pm 1 \quad . \tag{3.17}$$

Call these roots $k_{i,n}$ where $i = 1,2$, $n = +1,-1$

$$\psi(+i\varepsilon) = \frac{2\pi}{L} (g\pi)^2 \sum_{n=-1}^{+1} \sum_{k_{i,n}} \frac{1}{\sqrt{p^2-2n\hbar\omega m}} \frac{\sin^4 k_{i,n}}{k_{i,n}^4} [1 - \eta^{2\hbar k_{i,n}}] \; . \tag{3.18}$$

By the same procedure classically the diffusion coefficient (3.14) of the Fokker-Planck equation (3.11) is

$$D_{cl} = \frac{2\pi}{L} (g\pi)^2 \sum_{k_n} \frac{1}{p} \frac{\sin^4 k_n \ell}{k_n{}^2} , \tag{3.19}$$

where k_n, $n = \pm 1$ are roots of the <u>resonance condition</u>

$$k_n + n\omega m = 0 \quad . \tag{3.20}$$

The difference between (3.18) and (3.19) demonstrates the interesting principal quantum effect of chaos. First, as expected because of the finite displacements, $\eta^{2\hbar k_{i,n}}$, the quantum dynamics is a discrete random walk in $|k\rangle$ space rather than continuous diffusion process in momentum space, p. Because of Pawula's theorem [26] we know there is no valid truncation of the expansion (3.11) to $\hbar^j$ for finite j. The diffusion process is a classical effect in the case $\hbar = 0$. Both the classical and quantum cases have singularities where perturbation fails when the particle (wave) obtains small kinetic energy in the interaction. We have in (3.18) the restriction that $p^2 \gg 2m\hbar\omega$. We should comment that $\psi(+i\varepsilon)$ is essentially the "square" (without the phase ϕ) of the $D_{kk}(t)$. If $\psi_0(+i0) = 0$ then $D_{kk}(t)$ is also. Thus $\psi_0(+i0)$ is intimately related to the Melnikov function as is $D_{kk}(t)$ [22].

The interesting classical result is $D_{cl} = 0$ if

$$k_n \ell = 2\pi M \qquad\qquad M = 0, \pm 1, \pm 2 \ldots$$

or, from (3.20),

$$\pi M = \frac{\omega \ell m}{2p} \quad . \tag{3.21}$$

There is no interaction and no classical chaos at these infinite values of p for any ω. Physically this is a <u>dynamical</u> anti-resonance effect. The potential well accelerates and decelerates the particle in a compensative fashion so the net momentum is unchanged by the interaction. For arbitrary cut-off potential which may be Taylor expanded one may show that the high momentum behavior is just that of (3.21). Consequently there are regions of no chaos. The Melnikov function is identically zero, independent of the phase ϕ.

Quantum mechanically the situation is different. Evaluating the <u>four</u> terms in (3.18) we have

$$\pi(M_3 - M_1) = k\ell \quad ,$$

$$\pi(M_4 + M_2) = k\ell \quad ,$$

$$\pi(M_3 + M_1) = k\ell[1 - \frac{2m\omega}{k^2\hbar}]^{1/2} \quad ,$$

$$\pi(M_4 + M_2) = k\ell[1 + \frac{2m\omega}{k^2\hbar}]^{1/2}, \qquad (M_j = 0,\pm1,\pm2,\ldots) \quad , \tag{3.22}$$

where $M_i = 0,\pm1,\pm2\ldots$, $i = 1,2,3,4$. There are no solutions for arbitrary M_i. However, there may be relationships between the M_i's. There are values of M_i for which for a specified ω (depending on these M_i) there is a p which satisfies (3.22). We have constructed solutions but they appear sparse in number. Most important is they are for special ω. In this subtle way quantum mechanics is more chaotic than the classical result.

The additional conditions in (3.22) are due to quantum interference within the interaction range, i.e. wave function superposition.

Quantum interference makes impossible this classical anti-resonance effect. In regions of momentum space and ranges of ω there is chaotic behavior where there was none classically. In this region the quantum dynamics is governed by the quantum kinetic equation, (3.11). This effect is what we would term quantum chaos. Interference may amplify the classical chaos.

IV. Summary

We have shown that quantum ensembles in the semi-classical approximation may suppress classical topological chaos. We have derived a quantum kinetic equation, (3.11), which will exist if the asymptotic energy transfer, (3.7, 3.8), is constant and non-zero. This is the new dynamical criteria for chaos in the presence of quantum ensembles when the semi-classical topological ideas fail. (3.7) shows the close connection to a quantum generalization of the Melnikov equation.

V. Acknowledgments

We wish to thank Prof. I. Prigogine for his comments and Mr. E. Lewis for many discussions pertaining to the last paragraphs of this paper.

VI. References

1 H. Poincaré: New Methods of Celestial Mechanics, Vol. 3, (Gauthier-Villiers, Paris 1899), p.391; NASA Technical Translation F451, May 1967
2 V. I. Arnold and A. Avez: Ergodic Problems of Classical Mechanics (Benjamin, New York 1968)
3 J. Moser: Stable and Random Motions in Dynamical Systems (Princeton University Press, Princeton, N.J., 1973)
4 R. H. G. Helleman: in Fundamental Problems in Statistical Mechanics V, ed. by E. G. D. Cohen (North-Holland, Amsterdam 1980)
5 J. Guckenheimer and P. Holmes: Nonlinear Oscillations, Dynamical Systems and Bifurcations of Vector Fields (Springer-Verlag, New York 1983)
6 J. Ford: in Lectures in Statistical Physics, Lecture Notes in Physics 28, ed. by W. C. Schieve and J. S. Turner (Springer-Verlag, New York 1974)
7 J. C. England, R. R. Snapp and W. C. Schieve: "Fluctuations, Instabilities and Chaos in Laser-Driven Non Linear Ring Cavity," in Progress in Optics XXI, ed. by E. Wolf (North Holland, New York 1984)
8 New Zealand Symposium: Laser Physics, ed. by J. D. Harvey and D. Walls, Lecture Notes in Physics, No. 183 (Springer-Verlag, New York 1983)
9 J. R. Ackerhalt, P. W. Milloni and M. L. Shih: Physics Reports 128, 205 (1985)
10 Instabilities in Quantum Optics, Rochester 1986 Proceedings to be published
11 Stochastic Behavior in Classical and Quantum Hamiltonian Systems, Vol. 93 Lecture Notes in Physics, ed. by G. Casati and J. Ford (Springer-Verlag, New York 1979)

12 M. J. Berry: Semiclassical Mechanics of Regular and Irregular Motion, Les Houches School 1981 (North-Holland, Amsterdam 1981)
13 Chaotic Behavior in Quantum Systems, ed. by G. Casati (Plenum, New York 1985)
14 I. Prigogine: Non-Equilibrium Statistical Mechanics (Interscience 1962); see also P. Resebois: in Physics of Many-Particle Systems, ed. by E. Meeron (Gordon and Breach, New York 1966)
15 R. Zwanzig: Lectures in Theoretical Physics 3 (Boulder 1960), p.106; see also Physica 30, 139 (1973)
16 W. C. Schieve: Lecture Notes in Physics 28, eds. W. C. Schieve and J. Turner (Springer-Verlag, New York 1974)
17 V. K. Melnikov: Trans. Moscow Math. Soc. 12, 1 (1963)
18 T. Y. Petrosky and W. C. Schieve: Phys. Rev. A 31, 3907 (1985)
19 T. Y. Petrosky: Phys. Rev. A 29, 2078 (1984)
20 L. van Hove: Physica 21, 517 (1955); 23, 441 (1957)
21 A. Messiah: Quantum Mechanics (North-Holland, Amsterdam 1961) and A. Andrews: J. Phys. A 14, 1123 (1981)
22 T. Petrosky: Phys. Rev. A 32, 3716 (1985)
23 P. Resebois: Physica 27, 541 (1961)
24 B. V. Chirikov: Physics Reports 52, 263 (1979)
25 T. Petrosky and W. C. Schieve: in preparation
26 R. F. Pawula: Phys. Rev. 162, 186 (1967)

Part III

Fundamental Problems in Quantum Mechanics

Are There Quantum Jumps?*

A. Schenzle **, *R.G. DeVoe, and R.G. Brewer*

IBM Almaden Research Center, 650 Harry Road, San Jose, CA 95120, USA

Abstract: A quantum statistical theory is developed for a single atom, three-state quantum system where a strong (1-3) and a weak (2-3) transition are driven resonantly by an incoherent field. From a second order correlation function of the fluorescence intensity, it is concluded that a quantum jump can occur, in the sense that the single atom can undergo large fluctuations between a state of full emission in the strong transition to a state of no emission when the electron is shelved in the metastable state (2).

Quantum mechanics is an inherently statistical theory that describes the behavior of microscopic systems in a probabilistic way. For example, one cannot predict the result of a single event or a single measurement such as the instant in time that a single excited atom decays to a lower state or when a detector will observe the emitted photon. However, it is possible to predict the average time an atom spends in its excited state or the average time it takes to observe the photon after excitation. In the case of a single atom, these averages are realized experimentally in a series of repeated measurements where each time an atom is prepared in an identical way. For a large ensemble of independent atoms, a single measurement also yields an average of the individual atomic emission signals which follows the well-known exponential decay law.

Now that a single atom or ion can be observed spectroscopically, either in a faint atomic beam or in an electromagnetic trap [1,2], a fundamental test of an atom's quantum statistical behavior is at hand. An important example is the observation of anti-bunching [3] in resonance fluorescence – an effect that is based on the properties [4] of a single atom and that is erased by the presence of other excited atoms.

Another one-atom effect "shelving" has been proposed by Dehmelt [5] and is the subject of this article [6]. It involves, for example, the three-level quantum system of Fig. (1a) where a strong ($1 \leftrightarrow 3$) and a weak ($2 \leftrightarrow 3$) transition are coupled, each being driven by a resonant field. Many photons will be scattered in resonance fluorescence by the strong transition but occasionally this scattering will be suppressed or perhaps even extinguished when the weak transition shelves the atom in the metastable state (2). Thus, the strong transition registers the presence of the weak transition, and because the weak transition linewidth may be exceptionally narrow, this scheme has been proposed for an ultimate laser frequency standard [5].

Consider now the case of a large ensemble of atoms where one subgroup undergoes spontaneous emission in the strongly driven $1 \leftrightarrow 3$ transition, the events occurring randomly in time, while a second subgroup doesn't contribute to the $1 \leftrightarrow 3$ scattering because they are shelved. An individual atom may fluoresce randomly in time and be shelved for other periods, but the time-average of the number in each subgroup is a constant. Therefore, for an ensemble the fluorescence intensity of the strong transition would merely be slightly reduced due to the shelving effect [6].

---

*A preliminary version of this article appeared in Phys. Rev. **A33**, 2127 (1986).

**On leave from the Department of Physics, University Essen, Gesamthochschule, Essen, West Germany.

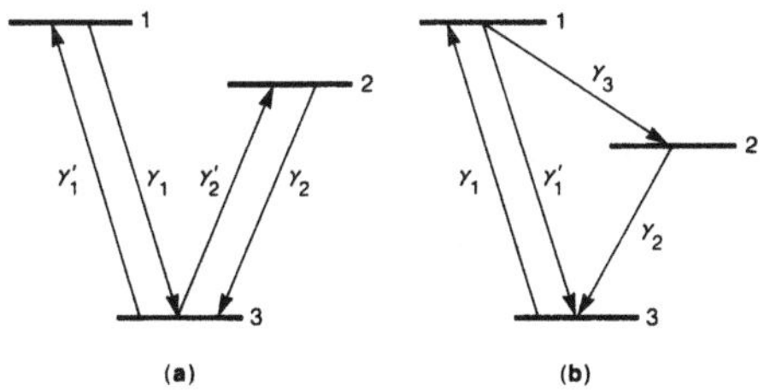

Fig. 1. Energy-level diagrams for the case where (a) the transitions 1-3 and 2-3 are driven by two resonant fields and (b) only the 1-3 transition is excited.

We now ask "How is the reduction of the fluorescence intensity realized for the case of a single atom?" Does the single atom emit light continuously but at a reduced rate, or will the fluorescence intensity undergo large fluctuations between a state of full emission, while the $1 \leftrightarrow 3$ transition is driven, to a state of no emission when the electron is shelved? If the second possibility is correct, then the single atom would display an individual quantum jump, not by the emission of a single photon but by the turning on and off of the strong fluorescence of the $1 \leftrightarrow 3$ transition. While a forbidden transition may have a feeble rate of emission, (say 1 photon/second) which would be difficult to detect, the presence or absence of fluorescence from the strong transition could be of macroscopic order, say 10^8 photons/second and easily detected.

The potential of the shelving principle to indicate individual quantum events through a macroscopic signal has been discussed recently by Cook and Kimble [7]. Indeed, their discussion adopts the intuitive view that quantum jumps occur whereas *in this paper we predict their existence.* The answer to whether quantum jumps occur or not can only come from a detailed quantum statistical calculation. We acknowledge, however, that for a single atom such a theory will never describe any single event, a discontinuous quantum jump, or an individual time trajectory of the fluorescence intensity. On the other hand, a calculation of the average fluorescence intensity clearly would be insufficient in deciding whether the intensity is continuous in time or widely fluctuating between periods of emission and darkness. Detailed statistical information about the dynamics of the emission process is required and that follows from the hierarchy of correlation functions of the emitted field. As a cautionary note, we mention that the more intuitive classical statistics can be misleading since it can differ substantially from a quantum statistical approach.

We describe the dynamics of the atom, corresponding to either Figs. 1(a) or (b), by a set of fermion operators

$$a_i^+ \text{ and } a_i \ , \qquad i = 1,2,3$$

for the levels 1, 2 and 3. The amplitude of the two fields emitted $E_i^{\pm}(i = 1,2)$ is proportional to the polarization[8] of the medium

$$\begin{aligned} E_i^- &= \sqrt{\Gamma_i}\ a_i^+ a_3 \\ & \qquad\qquad i = 1,2 \\ E_i^+ &= \sqrt{\Gamma_i}\ a_3^+ a_i \end{aligned} \tag{1}$$

where all unimportant factors have been absorbed in the renormalization. The retardation due to the propagation from the source to the detector is not displayed explicitly either. We assume that detectors can spectrally resolve either $E_1^{\pm}$ or $E_2^{\pm}$ and that quantum beats,[9] characteristic of spontaneous emission from multi-level systems, need not be considered.

Some insight into the characteristics of the emitted field is provided by the first order correlation function [10]

$$G_i(t) = \langle E_i^-(t) E_i^+(t) \rangle = \Gamma_i \rho_{ii}(t) \ , \tag{2}$$

which is proportional to the average fluorescence intensity since $\rho_{ii}(t)$ is the density matrix of the three-level system. For long times, the stationary value

$$\lim_{t\to\infty} G_i(t) = G_i^{SS} = \gamma_i \rho_{ii}^{SS} , \qquad i = 1,2 \tag{3}$$

is approached. The existence of the metastable state $|2\rangle$ merely reduces the average fluorescence intensity [6] and hence G_1^{SS}, but there is no indication whether the reduction occurs in a continuous or discontinuous fashion. Insight into the dynamics and statistical nature of the three-level atomic emission is revealed in the stationary second order correlation function [10]

$$G_{ij}^{SS}(t) = \langle E_j^-(0)E_i^-(t)E_i^+(t)E_j^+(0)\rangle , \qquad t > 0 . \tag{4}$$

With the insertion of (1) into (4), we obtain

$$G_{ij}^{SS}(t) = \gamma_i\gamma_j \langle a_j^+(0)a_3(0)a_i^+(t)a_i(t)a_3^+(0)a_j(0)\rangle . \tag{5}$$

where we have made use of the fermion commutation relation $a_{33}^+a_{33} + a_{33}a_{33}^+ = 1$. Under the assumption that spontaneous emission establishes a Markov process, the above correlation function can be evaluated using the quantum regression theorem.[11] For this purpose, we use the time evolution operator

$$a_j^+(t)a_i(t) = \sum_{\ell,n} K_{\ell,n}^{ij}(t)a_n^+(0)a_\ell(0) + \text{Noise} . \tag{6}$$

It obeys the same equations of motion as the density matrix

$$\rho_{ij}(t) = \sum_{\ell,n} K_{\ell n}^{ij}(t)\rho_{\ell n}(0) , \tag{7}$$

except for the noise terms. Inserting this result into the photon correlation function (5), we find

$$G_{ij}^{SS}(t) = \Gamma_i\Gamma_j K_{33}^{ii}(t)\rho_{jj}^{SS} . \tag{8}$$

From Eq. (7), the following properties are evident:

$$K_{\ell n}^{ij}(t=0) = \delta_{i\ell}\delta_{jn} \text{ and } K_{33}^{ij}(t=\infty) = \rho_{ij}^{SS} \tag{9}$$

where $\rho_{33}(0)=1$. Thus, the corresponding correlation functions are

$$G_{ij}(t=0) = 0 , \tag{10a}$$

$$G_{ij}(t=\infty) = \Gamma_i\Gamma_j\rho_{ii}^{SS}\rho_{jj}^{SS} \tag{10b}$$

where (10a) reflects the well-known property of anti-bunching [3,4] in resonance fluorescence. For intermediate times, the two-photon correlation function $G_{ij}(t)$ can be interpreted as the probability of observing a photon at time $t=0$ corresponding to the $j \to 3$ transition and then subsequently at time t a second photon for the $i \to 3$ transition. Since the time origin of G_{ij} has been chosen arbitrarily, the probability of observing the first photon at $t=0$ is just $G_j^{SS} \equiv \Gamma_j\rho_{jj}^{SS}$. Therefore, the conditional probability of observing a second photon at time t, in a time window $1/\Gamma_1$, is proportional to

$$P_{ij}(t) = G_{ij}^{SS}(t)/(\Gamma_i\Gamma_j\rho_{jj}^{SS}) . \tag{11}$$

The only quantity remaining to be evaluated is the coefficient $K_{\ell,n}^{ij}(t)$ which follows from the solution of the three-level density matrix. The level structure of Figs. 1(a) and (b) require the solution of either 8 or 4 coupled linear differential equations, and these are amenable only to numerical integration. To obtain analytic results and thus insight into the properties of spontaneous emission from a three-level system, we simplify the dynamics by assuming incoherent driving fields as suggested in Fig. 1. The resulting rate equations are solved easily.but since the solutions are still involved, an additional simplification is introduced in Fig. 1(a) by driving both transitions strongly, but not at an infinite rate, so that $\gamma_1' \cong \gamma_1$ and $\gamma_2 \cong \gamma_2'$ and in Fig. 1(b) by saturating the $1 \leftrightarrow 3$ transition so that $\gamma_1' \cong \gamma_1$ and assuming that the relaxation rates in and out of the metastable state are equal

($\gamma_2 = \gamma_3$). These restrictions are by no means essential for obtaining a solution, but they reduce the complexity of the result and the number of independent parameters. Here, $\gamma_1' = 4\ \alpha^2/\Gamma_1$, $\gamma_1 = \Gamma_1 + \gamma_1'$, $\gamma_2' = 4\ \beta^2/\Gamma_2$, and $\gamma_2 = \Gamma_2 + \gamma_2'$. Observing the inequality

$$\gamma_1 >> \gamma_2 ,$$

which is physically relevant for this problem, we obtain for either Figs. 1(a) or (b) the simple expressions

$$K_{33}^{11}(t) = \frac{1}{3}\left[1 + \frac{1}{2}\left(e^{-\frac{3}{2}\gamma_2 t} - 3e^{-2\gamma_1 t}\right)\right] + O\left(\frac{\gamma_2}{\gamma_1}\right) , \tag{12a}$$

$$K_{33}^{22}(t) = \frac{1}{3}\left(1 - e^{-\frac{3}{2}\gamma_2 t}\right) + O\left(\frac{\gamma_2}{\gamma_1}\right) . \tag{12b}$$

The conditional probability of observing a photon E_1 at time t after having seen a photon E_1 at t=0 follows from (8), (11) and (12a) as

$$P_{11} = \frac{1}{3}\left[1 + \frac{1}{2}\left(e^{-\frac{3}{2}\gamma_2 t} - 3e^{-2\gamma_1 t}\right)\right] , \tag{13}$$

and is plotted in Fig. 2 as curve 2. Also, the average fluorescence intensity of the strong transition is $G_1^{SS} = 1/3\ \Gamma_1$ since $\rho_{11} = 1/3$ in a fully saturated three-level system.

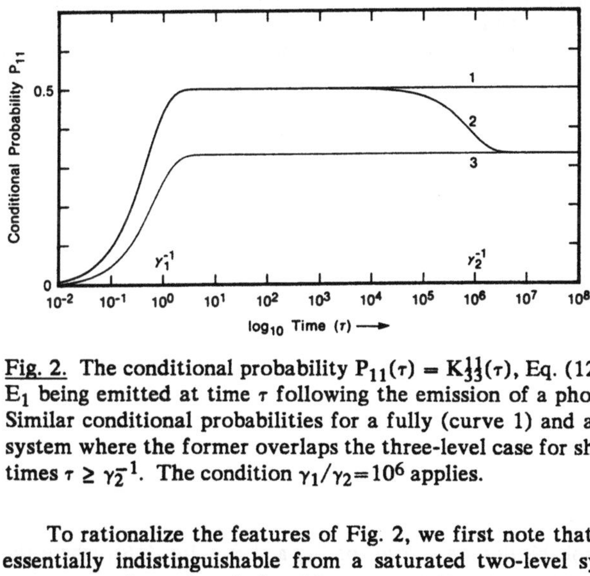

Fig. 2. The conditional probability $P_{11}(\tau) = K_{33}^{11}(\tau)$, Eq. (12a), of a three-level system for a photon E_1 being emitted at time τ following the emission of a photon E_1 at an earlier time τ=0 (curve 2). Similar conditional probabilities for a fully (curve 1) and a partially (curve 3) saturated two-level system where the former overlaps the three-level case for short times $\tau \leq \gamma_1^{-1}$ and the latter for long times $\tau \geq \gamma_2^{-1}$. The condition $\gamma_1/\gamma_2 = 10^6$ applies.

To rationalize the features of Fig. 2, we first note that for short times $t << \gamma_2^{-1}$, Eq. (13) is essentially indistinguishable from a saturated two-level system (curve 1), which with the same parameters obeys the relationships

$$P_{11}^{(a)} = \frac{1}{2}\left(1 - e^{-2\gamma_1 t}\right) , \quad G_1^{SS} = \frac{1}{2}\Gamma_1 . \tag{14}$$

Over this period, the three-level system behaves as if the metastable state (2) didn't exist, and the conditional probability for the two and three level systems are the same. For the shortest times $t << \gamma_1^{-1}$, the conditional probability drops to zero, indicating that there is no second photon available immediately after absorption of the first photon. This is the anti-bunching phenomenon [3,4].

For times $t \sim \gamma_2^{-1}$ and longer, the three-level conditional probability coincides with that of a partially saturated two-level system (curve 3) which obeys

$$P_{11}^{(b)} = \frac{1}{3}\left(1 - e^{-\frac{3}{2}\gamma_1 t}\right) , \quad G_1^{SS} = \frac{1}{3}\,\Gamma_1 . \tag{15}$$

We see that while the two-level solutions (a) and (b) represent continuous fluorescence signals with average intensities $1/2\ \Gamma_1$ and $1/3\ \Gamma_1$ respectively, the three-level result (13) bridges the two cases. Furthermore, at intermediate times $\gamma_1^{-1} < t < \gamma_2^{-1}$ the instantaneous intensity lies significantly above the limiting average value at long times $t >> \gamma_2^{-1}$. Thus, we conclude that there must exist significant periods of darkness or extremely weak emission so that the time-average intensity reduces to the correct asymptotic value $1/3\ \Gamma_1$.

Additional confirmation of these ideas appears in the cross-correlation conditional probabilities

$$G_{12}^{SS}(t)/(\Gamma_1\Gamma_2\rho_{22}^{SS}) = K_{33}^{11}(t) = P_{12}(t) = P_{11}(t) , \tag{16a}$$

$$G_{21}^{SS}(t)/(\Gamma_1\Gamma_2\rho_{11}^{SS}) = K_{33}^{22}(t) = P_{21}(t) = P_{22}(t) . \tag{16b}$$

These relations correspond to the emission of photon 2 ($2 \rightarrow 3$ transition) followed by photon 1 ($1 \rightarrow 3$ transition) or in the inverse order. Since K_{33}^{11} and K_{33}^{22} given by (12) are different, the conditional probability depends on the order of emission. This feature is summarized in Fig. (3) and shows for P_{21} that on the average when photon 1 is emitted, a long time γ_2^{-1} will elapse before photon 2 is emitted. Looking backward in time, the emission of photon 2 signifies that a dark interval, γ_2^{-1} preceded it since on the average a photon 1 was not emitted. Conversely, P_{12} implies that after photon 2 is emitted, only a brief period γ_1^{-1} is required on the average before photon 1 appears. The slow 2-3 transition allows the fast 3-1 transition to follow almost immediately. Then, according to P_{11} a sequence of rapid bursts of photon 1 light follows, to be interrupted again by a period of darkness. A period of photon 1 light lasting $2\gamma_2^{-1}$ with an intensity of $(1/2)\gamma_1$ and a period of darkness lasting γ_2^{-1} would then yield the correct time-averaged value $G_1^{ss} = (1/3)\Gamma_2$. It should be noted that the time asymmetry in P_{12} and P_{21} is a pure quantum effect whereas in classical statistics the behavior is independent of the order of emission.

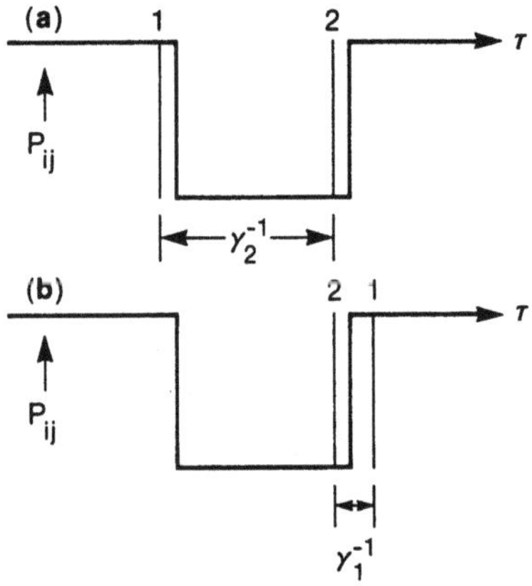

Fig. 3. Cross-correlation conditional probabilities are shown for P_{12} and P_{21}, Eq. (16), where the asymmetry in the delay times of photons 1 and 2 is evident.

One might question whether the higher order correlation functions, which have been neglected, are required. Since we are dealing with a Markov process, an assumption that is vital to the application of the regression theorem, nothing new is learned by considering higher order correlation functions. This is because the higher order function decomposes into a product of second order correlations.

It thus appears that the intuitive picture of jumps in a three-level quantum system as initially envisioned by Dehmelt agrees with the quantum statistical theory developed here. Clearly, an experimental test would now be timely, but the assumption of an incoherent driving field should be replaced by a coherent source, and this case will be treated in a forthcoming article.

We acknowledge having benefited from conversations on this subject with F. T. Arecchi, K. Jungmann, and J. H. Eberly. We also appreciate receiving from J. Javanianen a preprint of his related paper [12]. This work was partially supported by the Office of Naval Research.

References

1. W. Neuhauser, M. Hohenstatt, P. Toschek and H.G. Dehmelt: Phys. Rev. Lett. 41, 233 (1978).
2. D.J. Wineland: Science 226, 395 (1984).
3. M. Dagenais and L. Mandel: Phys. Rev. A18, 2217 (1978).
4. H.J. Carmichael and D.F. Walls: J. Phys. B89, 1199 (1976).
5. H.G. Dehmelt: IEEE Trans. Instrum. Meas., Vol. IM-31, No. 2, 83 (1982); Bull. Amer. Phys. Soc. 20, 60 (1974).
6. See also F.T. Arecchi, A. Schenzle, R.G. DeVoe, K. Jungmann and R.G. Brewer: Phys. Rev. A, March 1986.
7. R.J. Cook and H.J. Kimble: Phys. Rev. Lett. 54, 1023 (1985).
8. G.S. Agarwal: Springer Tracts in Modern Physics, 70, (Berlin, Springer, 1974), p. 39; J.R. Ackerhalt and J.H. Eberly, Phys. Rev. D10, 3350 (1974).
9. A. Schenzle and R.G. Brewer: Laser Spectroscopy, Second International Conference (New York, Springer, 1975), edited by S. Haroche, J.C. Pebay-Peyroula, T.W. Hansch and S.E. Harris, p. 420.
10. R.J. Glauber: Quantum Optics and Electronics, (Gordon and Breach, New York, 1965), edited by C. DeWitt, A. Blandin and C. Cohen-Tannoudji, p. 65.
11. M. Lax: Phys. Rev. 157, 213 (1967).
12. J. Javanianen: Phys. Rev. A, to be published.

Modified Optical Bloch Equations for Solids

P.R. Berman[1] *and R.G. Brewer*[2]

[1]Physics Department, New York University, 4 Washington Place, New York, NY 10003, USA
[2]IBM Almaden Research Center, 650 Harry Road, San Jose, CA 95120, USA

In a recent experiment of DeVoe and Brewer [1], it was concluded that the optical Bloch equations are incapable of describing the saturation phenomena observed. Optical free induction decay (FID) measurements of the impurity ion crystal Pr^{3+}:LaF_3 were conducted where the Pr^{3+} ions are coherently prepared by a laser field under steady-state conditions and then freely precess when the driving field is removed. At low optical fields, the observed Pr^{3+} optical linewidth is dominated by magnetic fluctuations arising from pairs of fluorine nuclear flip-flops. At high optical fields, this nuclear broadening mechanism is quenched and the Bloch equations are seriously violated. On physical grounds, this failure is due to a time-averaging of the magnetic interaction as the optical nutation frequency increases [2]. The phenomenological dipole dephasing time T_2 of the Bloch equations is therefore not a true constant but lengthens with increasing field strength.

Several theories [2]-[8] have been proposed to explain the DeVoe-Brewer results. Most of these theories are an extension of a method developed by Redfield [9] for modifying the Bloch equations to include the effects of local, magnetically induced fluctuations of the transition frequency of the radiating spins.

In this report, we present theoretical results for the optical FID decay rate obtained by assuming specific models for the local field fluctuations imposed on the ions' optical resonance frequency, and compare the results with "diffusion-type" theories [2]-[5] of frequency fluctuations. Whereas all the theories explain the overall qualitative structure of the experimental results, especially in strong fields, the predictions in the low-field regime differ sufficiently to allow for the possibility of an experimental test of one theory over another. The low-field data of DeVoe and Brewer are not yet precise enough to provide a critical test of any of the theories, but future experiments may be more conclusive.

Without going into the details of the calculation, a good qualitative understanding of the FID decay rate can still be obtained using simple physical considerations. The FID signal results when an ensemble of ions such as Pr^{3+} is initially prepared by an external optical field and then freely radiates when the field is suddenly removed. A large-scale inhomogeneous width Δ^* characterizes the sample, leading to a first-order FID signal that decays rapidly in a time $(\Delta^*)^{-1}$ – an effect that we do not consider here. The contribution to the FID signal which *does* concern us arises from a nonlinear interaction with the external field. As in the case of an inhomogeneously broadened vapor, the external field "burns" a hole in the ground state population and creates a bump in the excited state population, corresponding to that subgroup of ions with transition frequencies that resonantly interact with the external field. The FID decay rate is determined by the power-broadened linewidth of the hole resulting from the preparative phase *plus* the free-precession decay when the field is turned off.

In the absence of frequency fluctuations for a two-state system, the FID decay rate given by the Bloch equations is

$$\gamma_F = \left[(1/T_2^o)^2 + \chi^2(T_1^o/T_2^o)\right]^{1/2} + 1/T_2^o\,, \qquad (1)$$

where T_1^o and T_2^o are relaxation times in the absence of any frequency fluctuations and χ is the Rabi frequency of the external field. The bracketed term in (1) is the power-broadened hole width and the second term is the contribution of the free-decay period.

Frequency fluctuations modify the FID signal by affecting both the width of the hole and the free-decay process. We adopt a simple model, often applicable to solids, in which local field fluctuations produce frequency shifts $\delta\varepsilon(t)$ in the ion resonance frequency. Only Markoffian processes are considered; *i.e.*, the frequency following a fluctuation depends at most only on the frequency before the fluctuation. As such, the frequency jump processes can be characterized by the three parameters:

Γ = rate at which fluctuations occur ("jump rate"),
$\delta\varepsilon$ = rms frequency shift per fluctuation,
$\varepsilon_0 = \sqrt{2}$ frequency associated with the frequency fluctuations distribution.

The quantity ε_0, in effect, is a measure of the maximum frequency displacement produced by the local field fluctuations. It is assumed that Γ, $\delta\varepsilon$ and ε_0 are much smaller than the large inhomogeneous width Δ^*.

In this work, we restrict the discussion to three types of Markoffian processes: (1) a Gaussian-Markoffian or diffusion-type model; (2) a "Difference" model in which the probability of finding a frequency ε following a fluctuation when the frequency was ε' before the fluctuation is a function of $(\varepsilon-\varepsilon')$ only; and (3) a "Strong" model in which the frequency distribution following a fluctuation is the equilibrium distribution characterized by width ε_0. In the diffusion model, Γ, $\delta\varepsilon$ and ε_0 appear only in the combinations [10]

$\beta = \Gamma\delta\varepsilon/\varepsilon_0$ = effective jump rate,
$q = \beta\varepsilon_0^2/2$ = diffusion coefficient.

We now proceed to give the FID decay rates calculated for each model. Only the weak field regime is considered since, for strong fields, all models predict the same decay rate $\gamma_F \sim \chi(T_1^o/T_2^o)^{1/2}$ owing to the fact that, for large χ, frequency fluctuations become unimportant on the coherence time scale associated with the Rabi oscillation.[1] As has been noted [1], this result disagrees with the conventional Bloch prediction $\gamma_F \sim \chi(T_1/T_2)^{1/2}$, where the low field values T_1 and T_2 include the effects of frequency fluctuations.

1. Diffusion Model

Two limits, $\varepsilon_0 << \beta$ and $\varepsilon_0 >> \beta$, can be distinguished. If $\varepsilon_0 << \beta$, one finds[2-5] an FID decay rate

$$\gamma_F \sim 2/T_2; \qquad (1/T_2) = (1/T_2^o) + \varepsilon_0^2/2\beta\ , \tag{2}$$

which implies that the T_2 associated with the weak-field regime can be much smaller than T_2^o if $\varepsilon_0^2/2\beta >> 1/T_2^o$. When $\varepsilon_0 >> \beta$, the problem is best solved by the "Difference" model (discussion below) and, for $\varepsilon_0(\beta T_1^o)^{1/2} >> 1/T_2^o$, we find a decay rate of order

$$\gamma_F \sim 2/T_2; \qquad (1/T_2) = \frac{1}{2}\,(\beta T_1^o/2)^{1/2}\,\varepsilon_0\ . \tag{3}$$

2. Difference Model

The difference model [10] does not obey detailed balancing and cannot be used in the limit $\varepsilon_0 << \Gamma$ since frequency fluctuations *do* restore equilibrium in this limit. Thus, use of the difference model, which is characterized by a "kernel" $W(\varepsilon' \to \varepsilon)$ [$W(\varepsilon' \to \varepsilon)$ is the probability density per unit time for fluctuations to change the frequency from ε' to ε] given by

[1]The zero field intercept extrapolated from the strong field regime differs for the various models and could be used in a consistency check of the theories. Moreover, values of γ_F in the intermediate field regime can also be used to distinguish the theories.

$$W(\varepsilon'\to\varepsilon) = \Gamma(2\pi\delta\varepsilon^2)^{-1/2} \exp[-(\varepsilon - \varepsilon')^2/2\delta\varepsilon^2] , \tag{4}$$

is limited to the case $\varepsilon_0 >> \Gamma$. Within this restriction, one can distinguish two subcases, $\delta\varepsilon < (1/T_2^o) + \Gamma$, $\delta\varepsilon > (1/T_2^o) + \Gamma$.

If $\delta\varepsilon < (1/T_2^o) + \Gamma$, one recovers the diffusion limit. The broadening of the hole is of order $\sqrt{n}\,\delta\varepsilon$, where $n = \Gamma T_1^o$ is approximately the number of frequency jumps occurring in time T_1^o. Considering only the interesting case when this width is considerably larger than the natural width $1/T_2^o$, *i.e.*, when

$$(\Gamma T_1^o)^{1/2} \delta\varepsilon T_2^o = (\beta T_1^o)^{1/2} \varepsilon_0 T_2^o >> 1 , \tag{5}$$

we find an FID signal which decays as

$$\text{FID signal} \propto \frac{\pi}{2} - \tan^{-1}\left[(\beta T_1^o/2)^{1/2} \varepsilon_0 t\right] . \tag{6}$$

The decay rate (3) agrees qualitatively with the numerical work of Javanainen [5] using a diffusion model.

If $\delta\varepsilon > (1/T_2^o) + \Gamma$, fluctuations take the ion frequency outside the hole burned by the field and simply increase the decay rate, one finds

$$\gamma_F \sim 2/T_2; \qquad 1/T_2 = 1/T_2^o + \Gamma , \tag{7}$$

3. Strong Model

In the strong model, the kernel is given by

$$W(\varepsilon'\to\varepsilon) = \Gamma\left(\pi\varepsilon_0^2\right)^{-1/2} \exp\left(-\varepsilon^2/\varepsilon_0^2\right) , \tag{8}$$

and is independent of the initial frequency ε'. If $\varepsilon_0 << \Gamma$, we find [8] a decay rate

$$\gamma_F \sim 2/T_2; \qquad 1/T_2 = 1/T_2^o + \varepsilon_0^2/2\gamma , \tag{9}$$

which is similar to (2). For large fluctuation rates, it is unimportant whether a few strong or many weak fluctuations redistribute the frequency. If $\varepsilon_0 >> \Gamma$, $1/T_2^o$, χ, fluctuations shift the frequency out of the hole burned by the field and simply increase the decay rate as

$$\gamma_F \sim 2/T_2; \qquad 1/T_2 = 1/T_2^o + \Gamma . \tag{10}$$

The physical mechanism is the same as that leading to (7).

All the weak field results for the FID decay rate give values of $\gamma_F \equiv 2/T_2 \neq 2/T_2^o$. Since the form of T_2 is model dependent, the different models lead to different predictions for the weak-field regime. However, with the exception of (6) which indicates a nonexponential decay, one cannot distinguish the various theories in the weak-field regime unless one has an independent means for experimentally varying the parameters $\delta\varepsilon$, Γ and ε_0. This appears feasible in molecular vapors by varying the perturber-to-active-atom mass ratio or the perturber pressure, but may prove difficult in solids. If an independent variation of the parameters is not possible, one must compare the various theories over the range from weak to strong external fields to obtain a best fit to the data, a procedure that may prove useful when new experimental data become available.

This work is partially support by the U.S. Office of Naval Research.

References

1. R.G. DeVoe and R.G. Brewer: Phys. Rev. Lett. 50, 1269 (1983); 52, 1354 (1984)
2. A. Schenzle, M. Mitsunaga, R.G. DeVoe and R.G. Brewer: Phys. Rev. A30, 325 (1984); R.G. Brewer and R.G. DeVoe: Phys. Rev. Lett. 52, 1354 (1984)

3. M. Yamanoi and J.H. Eberly: Phys. Rev. Lett. 52, 1353 (1984); J. Opt. Soc. Am. B1, 751 (1984)
4. E. Hanamura: J. Phys. Soc. Jpn. 52, 2258 (1983); 52, 3265 (1983); 52, 3678 (1983)
5. J. Javanainen: Opt. Comm. 50, 26 (1984)
6. K. Wodkiewicz, B.W. Shore and J.H. Eberly: Phys. Rev. A30, 2390 (1984); K. Wodkiewicz and J.H. Eberly: Phys. Rev. A32, 992 (1985).
7. P.A. Apanasevich, S.Ya. Kilin, A.P. Nizovtsev and N.S. Onishchenko: Opt. Comm. 52, 279 (1984)
8. P.R. Berman and R.G. Brewer: Phys. Rev. A32, 2784 (1985).
9. A.G. Redfield: Phys. Rev. 98, 1787 (1955).
10. P.R. Berman: Phys. Rev. A9, 2170 (1974); P.R. Berman, J.M. Levy and R.G. Brewer: Phys. Rev. A11, 1668 (1975).

Ground State Photons Dressing Atoms and Molecules

F. Persico[+] *and G. Compagno*

Gruppo Nazionale CNR and Centro Interuniversitario MPI di Struttura della Materia, Istituto di Fisica dell'Universita', Via Archirafi 36, I-90123 Palermo, Italy

R. Passante

Istituto per le Applicazioni Interdisciplinari della Fisica (IAIF), CNR Palermo, Italy

1. Introduction

The problem of the detailed form of the cloud of virtual particles dressing the sources of a field [1] has aroused a persistent interest in several branches of physics until very recently. From a qualitative point of view the origin of the cloud can be understood as follows. The source is subjected to quantum fluctuations because of the interaction with the field. In the course of a fluctuation, one or more "virtual" quanta of the field may appear, depending on the form of the source-field coupling. The unperturbed energy E of the source-field system is not necessarily conserved during such a fluctuation, and states of unperturbed energy $E' \neq E$ may be attained; the magnitude of the energy unbalance $\Delta E = |E-E'|$, however, is constrained by the Heisenberg uncertainty principle

$$\Delta E \sim \hbar/\tau \tag{1}$$

where τ is the duration of the fluctuation, at the end of which the energy must balance again and the extra-quanta of the field are to be reabsorbed. Because of their quantum nature, these processes can take place also if the system is in the lowest possible energy state; in this lecture we concentrate on the latter ground-state situations.

Since fluctuations are taking place continuously, the source can be described as surrounded by a steady-state cloud of virtual quanta continuously emitted into, and reabsorbed from, the field. In contrast with a real quantum, which is emitted by an energy-conserving process and which can leave the source in view of its infinite lifetime in (1), a virtual quantum can only reach out at finite distance $r \sim v\tau \sim \hbar v/\Delta E$ from the source, where v is a typical velocity of the field. Consequently one should expect the linear dimensions of the virtual cloud surrounding the source roughly to coincide with r, for virtual transitions characterized by an energy unbalance ΔE.

(+) Also at IAIF

From a quantitative point of view, the total source-field system can be described by a Hamiltonian

$$H = H_0 + H' \qquad (2)$$

where H_0 contains the free source and field Hamiltonians, H' containing the coupling. The dressed ("physical") ground state $|\tilde{0}\rangle$ of H, apart from suitable normalization constants, can be expressed in the Schroedinger picture as

$$|\tilde{0}\rangle = |0\rangle + |0'\rangle \qquad (3)$$

where $|0\rangle$ is the ground state of H_0 ("bare" ground state), and $|0'\rangle$ is a correction due to H' which is a mixture of eigenstates of H_0 with various numbers of field quanta. In order to obtain a quantitative description of the cloud, one introduces an appropriate operator $\Psi(\mathbf{r})$, whose quantum average $\langle\tilde{0}|\Psi(\mathbf{r})|\tilde{0}\rangle$ on the dressed ground state is deemed to yield a reliable idea of the distribution of the virtual quanta as a function of the distance $\mathbf{r}$ from the source.

Well-known examples of such a procedure are

i) the case of a Klein-Gordon field of mass m coupled to a static source [2], where $\Psi(\mathbf{r})$ is taken to be the field amplitude operator $\phi(\mathbf{r})$, yielding a Yukawa-like shape for the meson cloud far from the source, decreasing at a rate $mc/\hbar$ (c= velocity of light);

ii) the optical polaron case [3], in the regime of small and intermediate coupling [4], where $\Psi(\mathbf{r})$ can be taken as $\nabla^2 H'$, the lattice charge density induced by the electron of mass m. In this case the distribution of virtual phonons follows an exponential law far from the electron, with a decrease rate $(2m\omega/\hbar)^{1/2}$, ω being the frequency of the optical phonons [5];

iii) the bare nucleon model with spin and isospin coupled to a 3-component pion field $\phi_\alpha(\mathbf{r})$ (α=1,2,3), where $\Psi(\mathbf{r})$ is taken to coincide with the pion charge density operator $Q_\pi = -e/\hbar(\phi_2\dot{\phi}_1 - \phi_1\dot{\phi}_2)$ [2]. This model yields a meson distribution behaving like the square of a Yukawa function with the same decrease rate as in i), and it has been recently improved by taking into account the three-quark bare nucleon structure, which introduces a further Δ bare nucleon state in addition to the traditional ones p and n; also virtual transitions to this new state contribute to Q_π [6].

The results of the above treatments are in agreement with qualitative considerations similar to those developed at the beginning of this section. It should be remarked, however, that no definite prescription has been given concerning the choice of $\Psi(\mathbf{r})$. In fact one might equally well have chosen the number density operator

$$N(\mathbf{r}) = -i/\hbar[\dot{\phi}_-(\mathbf{r})\phi_+(\mathbf{r}) - \phi_-(\mathbf{r})\dot{\phi}_+(\mathbf{r})] \qquad (4)$$

for Ψ in i), where $\phi_\pm$ are the positive and negative frequency components of the scalar field, or directly the lattice distortion H' in ii). Although it is unlikely that this would change the main qualitative features of the distribution obtained in the examples above, the same does not seem to be true in the analogous QED problem, where the shape of the virtual cloud seems to depend nontrivially on the nature of the operator $\Psi(\mathbf{r})$ which is chosen to describe it. It is tempting to relate this feature of QED to the difficulties connected with the localization of the photon [7].

2. Virtual Electromagnetic Fields

In the previous section we have not discussed the case of the e.m. field coupled to electric charges or currents. The reason is that treatments of this problem have been mainly concerned in the past with the effects of the coupling on the source, such as Lamb shift [8], and only recently interest in effects on the field has become more explicit [9] [10] [11]. The aim of this communication is to review some of the results hitherto obtained in the problem of the virtual fields. In particular, we shall discuss the following problems:

i) Single hydrogen atom. This problem can also be discussed using multipolar coupling and Heisenberg picture [10]. Here we shall summarize results obtained in the minimal coupling and Schroedinger representation [11]. The Hamiltonian for this case is (2) with

$$H_0 = H_A + H_F \; ; \; H_F = \int[\dot{\mathbf{A}}^2/c^2 + (\nabla\times\mathbf{A})^2]d^3r/8\pi \; ;$$
$$H' = V_1 + V_2 \; ; \quad V_1 = (-e/mc)\mathbf{A}(\mathbf{r})\cdot\mathbf{p}; \quad V_2 = (e^2/2mc^2)\mathbf{A}^2(\mathbf{r}) \qquad (5)$$

where H_A is the bare, spinless atomic hydrogen Hamiltonian, H_F is that of the free transverse field expressed in terms of the transverse vector potential $\mathbf{A}$ in the Coulomb gauge, and $\mathbf{p}$ is the electron momentum.

ii) Two hydrogen atoms [12]. Again minimal coupling and Schroedinger representation are used, and the Hamiltonian is an obvious generalization of (5) with

$$H_A = \sum_i H_A^{(i)} \; ; \; V_1 = (-e/mc)\sum_i \mathbf{A}(\mathbf{r}_i)\cdot\mathbf{p}_i \; ; \; V_2 = (e^2/2mc^2)\sum_i \mathbf{A}^2(\mathbf{r}_i) \tag{6}$$

where i=1,2 and the two spinless atoms are assumed to be at large distances, such that no overlap between electronic charges takes place. Moreover the atom-field interaction is taken in the dipole approximation.

iii) Single molecule [13]. Here the energy spectrum of the bare Hamiltonian is taken arbitrarily and the multipolar coupling scheme is used. Only the region of space far from the molecule is considered, such that one can use for the molecule-field coupling the variant obtained by an appropriate unitary transformation by CRAIG and POWER [14]. The Hamiltonian is given by (2) with

$$H_0 = H_M + H_F \; ; \; H_M = \sum_n E_n |n\rangle\langle n| \; ; \; H_F = \int[\mathbf{D}^2(\mathbf{r}) + \mathbf{B}^2(\mathbf{r})]d^3r/8\pi$$
$$H' = V_2 = -\, \alpha_{ij}\, D_i(\mathbf{R})D_j(\mathbf{R})/2 \tag{7}$$

where E_n are eigenvalues of the bare molecule, $\mathbf{D}(\mathbf{r})$ is the transverse displacement vector field, $\mathbf{B}(\mathbf{r})$ is the magnetic field, α_{ij} is the ground-state static symmetric polarizability of the molecule and $\mathbf{R}$ its position.

We remark that V_1 is $O(e)$ and V_2 is $O(e^2)$ in all Hamiltonians (5)-(7).

The question now arises about the quantity $\Psi(\mathbf{r})$ that is best suited to represent the shape of the dressing of the source by the e.m. field. Here we do not discuss the forms used in previous approaches [15], and we concentrate on the forms which make direct use of physically measurable quantities, such as electric or magnetic fields. Thus we define the total energy density of the e.m. fields and other related quantities as

$$\mathcal{H}_f(\mathbf{r}) = \mathcal{H}_\perp(\mathbf{r}) + \mathbf{E}_\parallel^2(\mathbf{r})/8\pi + \mathbf{E}_\perp(\mathbf{r})\cdot\mathbf{E}_\parallel(\mathbf{r})/4\pi = \mathcal{H}_{el} + \mathcal{H}_{mag} \; ;$$
$$\mathcal{H}_{el} = [\mathbf{E}_\perp^2(\mathbf{r}) + \mathbf{E}_\parallel^2(\mathbf{r}) + 2\mathbf{E}_\perp(\mathbf{r})\cdot\mathbf{E}_\parallel(\mathbf{r})]/8\pi \; ; \; \mathcal{H}_{mag} = \mathbf{B}^2(\mathbf{r})/8\pi \tag{8}$$

where $\mathbf{E}_\perp(\mathbf{r}) = -\dot{\mathbf{A}}(\mathbf{r})/c$ and $\mathbf{E}_\parallel(\mathbf{r}) = -\nabla\Phi(\mathbf{r})$ are the transverse and longitudinal part of the electric field, $\mathbf{B}(\mathbf{r})=\nabla\times\mathbf{A}(\mathbf{r})$, and

$$\mathcal{H}_\perp = [\mathbf{E}_\perp^2(\mathbf{r}) + \mathbf{B}^2(\mathbf{r})]/8\pi =$$
$$= W(\mathbf{r}) + ZPT + [\mathbf{E}_\perp^{+}(\mathbf{r})\cdot\mathbf{E}_\perp^{+}(\mathbf{r}) + \mathbf{B}^{+}(\mathbf{r})\cdot\mathbf{B}^{+}(\mathbf{r}) + h.c.]/8\pi \tag{9}$$

In (9) $\mathbf{E}_\perp^+$, $\mathbf{B}^+$, $\mathbf{E}_\perp^-$ and $\mathbf{B}^-$ are positive and negative frequency components of the fields, ZPT are zero-point terms which we shall neglect because their contributions do not depend on the position of the sources, and

$$W(\mathbf{r}) = [\mathbf{E}_\perp^-(\mathbf{r})\cdot\mathbf{E}_\perp^+(\mathbf{r}) + \mathbf{B}(\mathbf{r})^-\cdot\mathbf{B}^+(\mathbf{r})]/4\pi \qquad (10)$$

is the so-called coarse-grained energy density (CGED) of the transverse field [11], such that $\int_{V_1} W(\mathbf{r})d^3r$ can be interpreted as the energy carried by the decorrelated transverse photons within the volume V_1. $W(\mathbf{r})$ in real fields, such as in a laser beam, is obviously directly measurable by a photodetector; when referred to a virtual field, however, the CGED loses this privileged position. Thus in the following we shall consider it as an appropriate candidate for $\Psi(\mathbf{r})$ together with the various operators appearing in (8)-(10), in an attempt to gain a picture of the virtual e.m. cloud as complete as possible.

3. QED Theory of the Virtual Cloud

Two approaches have been used: one is in the Heisenberg representation [10] [13], in which the equations of motion for $\Psi(\mathbf{r},t)$ are solved to the desired order in the source-field coupling constant appearing in H', and subsequently the quantity $\lim_{t\to\infty} \langle 0|\Psi(\mathbf{r},t)|0\rangle$ is evaluated. The second approach is in the Schroedinger representation [11] [13], and it is the same in principle as that outlined in the Introduction, leading to the evaluation of $\langle\tilde{0}|\Psi(\mathbf{r})|\tilde{0}\rangle$. Both approaches yield the same results for the cases in which they have been tested [13]. Here we concentrate on the second.

Use of Rayleigh-Schroedinger perturbation theory, appropriately corrected for normalization, to deal with H' in Hamiltonians (5), (6) and (7) leads to the following contributions to $|0'\rangle = \sum_{i=1}^{\infty} |g^i\rangle$ up to $O(e^3)$:

$$|g^1\rangle = z_0V_1|0\rangle \; ; \; |g^2\rangle = \left(z_0V_2 + z_0V_1z_0V_1 - \langle V_1z_0{}^2V_1\rangle_0/2\right)|0\rangle \; ;$$

$$|g^3\rangle = \left(z_0{}^2\langle V_2\rangle_0V_1 + z_0V_1z_0V_2 + z_0V_2z_0V_1 - z_0{}^2\langle V_1z_0V_1\rangle_0V_1 + z_0V_1z_0V_1z_0V_1 - \right.$$
$$\left. - z_0\langle V_1z_0{}^2V_1\rangle_0V_1/2 - \langle V_1z_0{}^2V_2\rangle_0 - \langle V_1z_0{}^2V_1z_0V_1\rangle_0\right)|0\rangle \qquad (11)$$

where $z_0=(1-P_0)(E_0-H_0)^{-1}$, P_0 is the projection operator onto the bare ground state of H_0 and $\langle Q\rangle_0 = \langle 0|Q|0\rangle$. We remark that $|g^i\rangle$ is of $O(e^i)$, where e is the electron charge; consequently if $\Psi(\mathbf{r})$ is any of the charge-independent operators appearing in (8)-(10), the matrix element of the kind $\langle g^i|\Psi(\mathbf{r})|g^k\rangle$ is $O(e^{i+k})$, and the $O(e^n)$ contribution to the quantum average of $\Psi(\mathbf{r})$ on the ground state is

$$\langle\Psi(\mathbf{r})\rangle_n = \sum_{\ell=0}^{n}\langle g^\ell|\Psi(\mathbf{r})|g^{n-\ell}\rangle \quad ; \quad |g^\circ\rangle = |0\rangle \tag{12}$$

4.Results

i) Single hydrogen atom. Here $|0\rangle$ describes the atom in the bare 1s state and the empty field. The choice $\Psi(\mathbf{r}) = W(\mathbf{r})$ yields from (12)

$$\langle W(\mathbf{r})\rangle_0 = \langle W(\mathbf{r})\rangle_1 = 0$$

$$\langle W(\mathbf{r})\rangle = \langle g^1|W(\mathbf{r})|g^1\rangle = \left(\langle g^1|\mathbf{E}_\perp^-\cdot\mathbf{E}_\perp^+|g^1\rangle + \langle g^1|\mathbf{B}^-\cdot\mathbf{B}^+|g^1\rangle\right)/4\pi \tag{13}$$

in view of $\langle 0|W(\mathbf{r}) = W(\mathbf{r})|0\rangle = 0$. Thus two-photon states coming from the $\mathbf{A}^2$ term in (5) do not play any role in determining the form of the CGED cloud at $O(e^2)$. The matrix elements in (13) can be evaluated exactly, with the following result

$$\langle g^1|\mathbf{E}_\perp^-\cdot\mathbf{E}_\perp^+|g^1\rangle = K\sum_{N\ell}\ell(\ell+1)\{\ell|\int_0^\infty\omega(\omega_{No}+\omega)^{-1}j_{\ell+1}(\omega r/c)f(N,\ell,\omega)d\omega|^2 +$$
$$+ (\ell+1)|\int_0^\infty\omega(\omega_{No}+\omega)^{-1}j_{\ell-1}(\omega r/c)f(N,\ell,\omega)d\omega|^2\} \; ; \tag{14}$$

$$\langle g^1|\mathbf{B}^-\cdot\mathbf{B}^+|g^1\rangle = K\sum_{N\ell}\ell(\ell+1)(2\ell+1)|\int_0^\infty\omega(\omega_{No}+\omega)^{-1}j_\ell(\omega r/c)f(N,\ell,\omega)d\omega|^2 \tag{15}$$

where K is a constant proportional to e^2, N and ℓ are the principal and azimuthal quantum numbers of the atomic electron, $\omega_{No}=(E_N-E_o)/\hbar$, j_ℓ is a spherical Bessel function and $f(N,\ell,\omega)$ is a form factor. Thus the CGED arises at $O(e^2)$ from the incoherent superposition of contributions of the one-photon virtual transitions to the excited atomic states, each state belonging to a subshell N,l. The lifetime of such a virtual state being $\tau_{N\ell}\sim(\omega_{No}+\omega)^{-1}$, the qualitative argument presented in the Introduction gives a range for the corresponding virtual photons $r_{N\ell}\sim c(\omega_{No}+\omega)^{-1}$, and a limit $c/\omega_{No}=\lambda\!\!\!^{-}_{No}$ for long-wavelength photons. This limit depends on N, and in this sense the CGED may be thought as structured in concentric shells of virtual photons, each shell of radius $\sim\lambda\!\!\!^{-}_{No}$ arising from the contributions $\langle W(\mathbf{r})\rangle_{N\ell}$ of the virtual transitions leading to the excited atomic subshells N,ℓ. In fact (14) and (15) can be solved analytically, and both of them for each N,ℓ contribution behave like $r^{-2\ell-2}$ for $r\lesssim\lambda\!\!\!^{-}_{No}$ and like $r_{-2\ell-4}$ for $r\gtrsim\lambda\!\!\!^{-}_{No}$. In the dipole approximation $f(N,\ell,\omega)=0$ for ℓ different from unity, and for each N

$$\langle W(\mathbf{r})\rangle_{2,N,\ell=1} \propto \begin{cases} r^{-4} & (r\lesssim\lambda\!\!\!^{-}_{No}) \\ r^{-6} & (r\gtrsim\lambda\!\!\!^{-}_{No}) \end{cases} \tag{16}$$

ii) Two hydrogen atoms. Here $|0\rangle$ refers to both atoms in 1s states and to empty field. We wish to discuss the possibility of interference between the two virtual clouds. Again we choose $\psi(\mathbf{r})=W(\mathbf{r})$, and we push the perturbation to include contributions up to $O(e^4)$. From (12)

$$\langle W(\mathbf{r})\rangle_1 = \langle W(\mathbf{r})\rangle_3 = 0 \ ; \quad \langle W(\mathbf{r})\rangle_2 = \langle g^1|W(\mathbf{r})|g^1\rangle \ ;$$

$$\langle W(\mathbf{r})\rangle_4 = \langle g^1|W(\mathbf{r})|g^3\rangle + \langle g^2|W(\mathbf{r})|g^2\rangle + \langle g^3|W(\mathbf{r})|g^1\rangle \qquad (17)$$

From (11) we see that two-photon processes from $\mathbf{A}^2$ come into play at $O(e^4)$ through V_2. The contributions to the various terms of the virtual cloud in (17) can be conveniently analyzed in terms of diagrams [12]. It can be shown that the only contributions to $\langle W(\mathbf{r})\rangle_2$ come from

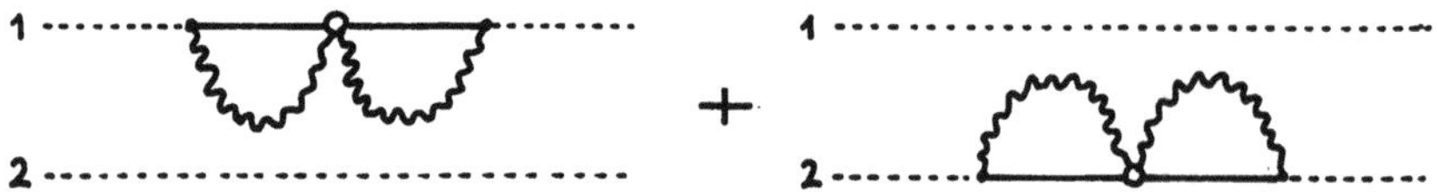

(18)

where the numbers refer to atoms 1 and 2. Although it is not possible here to explain the diagrammatic technique used, it should be obvious that the diagrams in (18) refer to processes where one of the atoms is involved, the other remaining completely idle. This leads to conclude that at $O(e^2)$ the clouds of virtual photons do not interfere, since the dressing of each atom is independent of that of the other. It should be remarked that this is different from the real photon case discussed by SCULLY [16], where interference appears at $O(e^2)$. Interference between virtual photons takes place at $O(e^4)$, where diagrams of the form

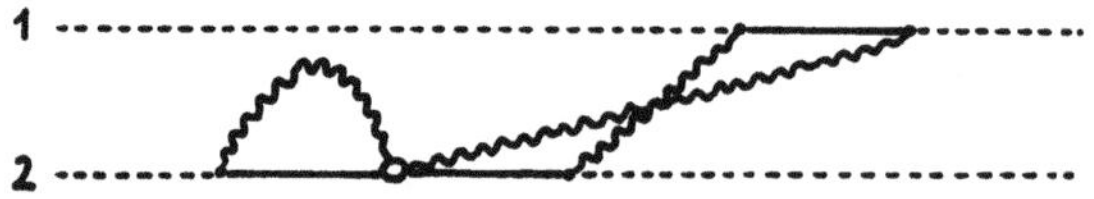

(19)

appear to contribute to $\langle W(\mathbf{r})\rangle_4$. As it is clear from (19), photons are exchanged between the atoms, and the total cloud is not additive. An explicit evaluation of the shape of the cloud at $O(e^4)$ has not yet been performed.

iii) Single molecule. Here the molecule is permanently in the bare ground state and $V_1=0$. We discuss the total magnetic energy density and we choose $\Psi(\mathbf{r})= \mathcal{H}_{mag}$ in (8). Since (7) is derived within the multipolar coupling scheme, the results may be used to test the minimal coupling scheme versus the multipolar one [13]. From (11) and (12) we have

$$|g^1\rangle = 0 \; ; \quad |g^2\rangle = z_0 V_2|0\rangle \; ;$$

$$\langle\mathcal{H}_{mag}\rangle_1 = 0 \; ; \; \langle\mathcal{H}_{mag}\rangle_2 = \langle g^0|\mathcal{H}_{mag}|g^2\rangle + \langle g^2|\mathcal{H}_{mag}|g^0\rangle \quad (20)$$

Carrying the calculations through up to $O(e^2)$, we find

$$\langle\mathcal{H}_{mag}\rangle = \langle\mathcal{H}_{mag}\rangle_2 = (-7/32\pi^2)\hbar c\alpha_{ij}(\delta_{ij}-\hat{r}_i\hat{r}_j)/r^7 \quad (21)$$

consistently with the theory of Van der Waals forces [17]. For an atom $\alpha_{ij} = \alpha\delta_{ij}$ and

$$\langle\mathcal{H}_{mag}\rangle = (-7/16\pi^2)\hbar c\alpha/r^7 \quad (22)$$

which looks different from (16) for $r \gg \lambda_{No}$. This paradox has been resolved recently by evaluating the $O(e^2)$ contributions to $\langle\mathcal{H}_{mag}\rangle$, in the minimal coupling scheme, coming from

$$[\mathbf{B}^+ \cdot \mathbf{B}^+ + \mathbf{B}^- \cdot \mathbf{B}^-]/8\pi \quad (23)$$

in (9). These contributions are also $O(r^{-6})$ and they cancel the $O(r^{-6})$ magnetic part of $\langle W(\mathbf{r})\rangle$, leaving exactly (22) in total.

5. Final Comments

The results indicate that the Schroedinger representation-minimal coupling scheme [11] and the Heisenberg representation-multipolar approach [10] yield the same results in the far zone. This is also confirmed by further results in the near zone [13] [17]. In order to represent the form of the virtual e.m. cloud it is possible to choose among a number of operators. This choice, however, seems to be related to the particular features of the cloud one wishes to evidence. While it is obvious that $\mathcal{H}_{el}$ and $\mathcal{H}_{mag}$ are directly related to the Van der Waals forces in the far zone, it is not clear at the moment which experiment can be devised to yield directly the CGED $W(\mathbf{r})$.

The authors wish to acknowledge fruitful conversations and communication of unpublished results by E.A. Power and T. Thirunamachandran in the framework of a collaboration supported by NATO under Grant RG85/0112. General support by Comitato Regionale Ricerche Nucleari e Struttura della Materia, Palermo, Italy is also acknowledged.

References

1. L. Van Hove, Physica 18, 145 (1952); 21, 901 (1955); 22, 343 (1956)

2. E.M. Henley, W. Thirring, Elementary Quantum Field Theory (McGraw-Hill 1962)

3. H. Frolich, H. Pelzer, S. Zienau, Phil. Mag. 41, 221 (1950)

4. T.D. Lee, F. Low, D. Pines, Phys. Rev 90, 297 (1953)

5. F.M. Peeters, J.T. Devreese, Phys. Status Sol. B115, 285 (1983)

6. S. Theberge, A.W. Thomas, G.A. Miller, Phys. Rev. D22, 2838 (1980); 24, 216 (1981)

7. T.D. Newton, E.P. Wigner, Rev. Mod. Phys. 21, 400 (1949)

8. H.A. Bethe, Phys. Rev. 72, 339 (1947)
T.A. Welton. Phys. Rev. 74, 1157 (1948)
H.E. Moses, Phys. Rev. A8, 1710 (1973)
C.K. Au, G. Feinberg, Phys. Rev. A9, 1794 (1974)
J. Dupont-Roc, C. Fabre, C. Cohen-Tannoudji, J. Phys. B11, 563 (1978)
H. Grotch, Am J. Phys. 49, 48 (1981)

9. L. Davidovich, H.M. Nussenzveig, in Coherence and Quantum Optics IV, Eds. L. Mandel, E. Wolf (Plenum N.Y. 1978) pp. 953-974
J. Dalibard, J. Dupont-Roc, C. Cohen-Tannoudji, J. Physique, 43, 1617 (1982)
G. Compagno, R. Passante, F. Persico, Phys. Lett. A98, 253 (1983)

10. E.A. Power, T. Thirunamachandran, Phys. Rev. A28, 2649 (1983)

11. R. Passante, G. Compagno, F. Persico, Phys. Rev. A31, 2827 (1985)

12. G. Compagno, F. Persico, R. Passante, Phys. Lett. A112, 215 (1985)

13. F. Persico, E.A. Power, Phys. Lett. A

14. D.P. Craig, E.A. Power, Int. J. Quantum Chem. 3, 903 (1969)

15. L. Mandel, Phys. Rev. 144, 1071 (1966)
R.J. Cook, Phys. Rev. A25, 2164 (1982); 26, 2754 (1982)

16. M.O. Scully, K. Druhl, Phys. Rev. A25, 2208 (1982)

17. G. Feinberg, J. Sucher, Phys. Rev. A2, 2395 (1970)

18. R. Passante, E.A. Power, to be published

Overdamping and Quasi Quantum Nondemolition Measurements

F. Haake[1] *and D.F. Walls*[2]

[1]Fachbereich Physik, Universität-Gesamthochschule Essen, D-4300 Essen, Fed. Rep. of Germany

[2]Physics Department, University of Waikato, Hamilton, New Zealand

An exactly soluble model of a linear oscillator undergoing damping or amplification under the influence of a heat bath is discussed from the point of view of measurement theory. In the limit of overdamping the oscillator displacement becomes a quasi quantum nondemolition variable measured by the bath.

A device capable of measuring a quantum observable X must turn an initial superposition of eigenstates of X into a mixture such that the diagonal elements of the density matrix in the X representation are unchanged while the off-diagonal ones are collapsed to zero. It has been well known since the early days of quantum mechanics that a macroscopic apparatus with many degrees of freedom is necessary to achieve a practically irreversible decay of off-diagonal elements [1-9].

Recently, Zurek [10] has proposed measuring devices consisting of a "meter" (with one or a few degrees of freedom) and a heat bath. The Hamiltonian governing the complete measurement then has the structure

$$H = H_S + H_M + H_B + H_{SM} + H_{MB} \tag{1}$$

where H_S, H_M, and H_B are the free Hamiltonians of the system to which the observable X pertains, the meter, and the bath, respectively. The interaction H_{SM} serves to correlate different eigenstates of X with different eigenstates of a suitable "pointer" variable Q of the meter. The interaction H_{MB} is designed such as to destroy coherences between different eigenstates of the pointer variables. Walls and co-workers have constructed many concrete model systems for which the environment-induced collapse of the wave function could be displayed explicitly using master equation techniques [11-14]. It has proven important in these investigations to let the pointer variable Q enter the meter-environment interaction as

$$H_{MB} = Q\Gamma \tag{2}$$

where Γ is an observable pertaining to the bath. Such a structure of H_{MB} distinguishes the eigenstates of Q as the preferred basis with respect to which the reduced density operator of the meter tends to diagonalize.

We shall here be concerned with a slightly simplified class of models where the meter itself is the quantum system subject to measurement, the pointer variable Q being the measured observable. As in most cases investigated previously, we shall take the bath to consist of N harmonic oscillators (with displacements q_n and momenta p_n) and employ the Hamiltonian

$$\begin{aligned} H &= H_M + H_B + H_{MB} \\ H_B &= \tfrac{1}{2}\sum_{n=1}^{N}\left(p_n^2 + \omega_n^2 q_n^2\right) \\ H_{MB} &= Q\sum_n \epsilon_n q_n \end{aligned} \tag{3}$$

where the ϵ_n are suitable coupling constants.

We would face an ideal case of a measurement of Q if Q were a constant of the motion, i.e. if $[Q, H_M] = 0$. In that case the diagonal elements of the reduced density operator of the meter in the Q representation would remain strictly constant, while the off-diagonal ones would relax to zero as discussed in ref.[14]. Such ideal measurements, also called quantum nondemolition measurements, have recently been discussed quite extensively [15,16,13,14].

For a good measurement it is by no means necessary that Q be a QND variable. It suffices that Q be the slowest one among the observables pertaining to the meter. There must be two time scales, a long one characterizing the dynamics of the diagonal elements of the meter density operator and a short one for the off-diagonal elements. We shall here be concerned with such "quasi QND" variables.

In order to illustrate how to construct meters with two distinct time scales we shall let the meter be a linear system with the free Hamiltonian

$$H_M = \tfrac{1}{2}\left(P^2 + \omega_0^2 Q^2\right). \tag{4}$$

We shall mostly be concerned with real ω_0, i.e. the case of the harmonic oscillator. We shall also comment on the oscillator with an inverted parabolic potential corresponding to imaginary ω_0. The Hamiltonian (3,4) has first been employed to model linear damping by Ullersma [17]. More recently, strong-damping and low-temperature properties imparted to the central oscillator by the heat bath have been investigated by Haake and Reibold [18].

The initial situation of relevance is one without correlations between the bath and the meter. The bath can be taken to be in thermal equilibrium but we need not restrict the state of the meter. The combined initial density operator thus reads

$$\rho_{MB}(0) = Z_B^{-1} e^{-\beta H_B} \rho_M(0). \tag{5}$$

While most applications of models like the above are related to weak damping, we have set ourselves the task of providing the central oscillator with two vastly different time scales. We are thus facing a strong coupling between the meter and the bath, and must refrain from using any perturbative procedure in constructing the reduced density matrix $\rho_M(t)$ of the meter. Fortunately, the Hamiltonian (3,4) allows for explicit diagonalization and therefore for a completely rigorous calculation of $\rho_M(t)$.

For our discussion of $\rho_M(t)$ we shall need the mean displacement [17,18]

$$\langle Q(t)\rangle = \dot{A}(t)\langle Q(0)\rangle + A(t)\langle P(0)\rangle, \tag{6}$$

the mean momentum $\langle P(t)\rangle = \langle \dot{Q}(0)\rangle$, and the variances

$$\begin{aligned}
\sigma_{QQ}(t) &= \dot{A}^2\sigma_{QQ}(0) + 2\dot{A}A\sigma_{QP}(0) + A^2\sigma_{PP}(0) + X(t)\\
\sigma_{PP}(t) &= \ddot{A}^2\sigma_{PP}(0) + 2\ddot{A}\dot{A}\sigma_{QP}(0) + \dot{A}^2\sigma_{PP}(0) + Y(t)\\
\sigma_{QP}(t) &= \tfrac{1}{2}\langle Q(t)P(t) + P(t)Q(t)\rangle - \langle Q(t)\rangle\langle P(t)\rangle = \tfrac{1}{2}\dot{\sigma}_{QQ}(t).
\end{aligned} \tag{7}$$

It is, of course, due to the harmonicity of H and the Gaussian nature of the equilibrium ensemble exp $(-\beta H_B)$ of the bath that all of these quantities as well as all higher-order moments of Q and P and even the density matrix ρ_M itself can be expressed in terms of only three functions of time, A(t), X(t) and Y(t).

While the amplitude A(t) is temperature independent, the two quantities X(t) and Y(t) describe the build-up of thermal equilibrium in the central oscillator. Once A(t) is known, the thermal amplitude X(t) can be obtained from [18]

$$X(t) = \sum_{n=1}^{N} \frac{\epsilon_n^2}{\omega_n^2} \left|\int_0^t dt' e^{i\omega_n t} A(t')\right|^2 \hbar\omega_n \left(n_{th}(\omega,T) + \tfrac{1}{2}\right). \tag{8}$$

By replacing A(t) on the r.h.s. in (8) with $\dot{A}(t)$ we get the other thermal amplitude, Y(t). The temperature dependence enters X(t) and Y(t) through $n_{th}(\omega_n,T)$, the mean number of quanta in an harmonic oscillator with frequency ω_n at the temperature T.

Let us now turn to the reduced density matrix of the meter. By denoting as $|x \pm \frac{1}{2}y\rangle$ the eigenstate of Q with eigenvalue $x \pm \frac{1}{2}y$ and by assuming, for the sake of simplicity, the meter to be in a coherent state initially, we have the beautifully transparent expression [19]

$$\begin{aligned} &\langle x - \tfrac{1}{2}y|\rho_M(t)|x + \tfrac{1}{2}y\rangle = \\ &= (2\pi\sigma_{QQ})^{1/2} e^{-(x - \langle Q\rangle)^2/2\sigma_{QQ}} \\ &\qquad e^{-y^2(\sigma_{QQ}\sigma_{PP} - \sigma_{QP}^2)/2\hbar^2\sigma_{QQ}} \\ &\qquad e^{-iy[\langle P\rangle + (x - \langle Q\rangle)\frac{\sigma_{QP}}{\sigma_{QQ}}]/\hbar}. \end{aligned} \tag{9}$$

The phase factor in (9) obviously describes an oscillatory behaviour with respect to y. More important for us are the two real Gaussians in (9). Since we are considering matrix elements with respect to eigenstates of Q it is quite intuitive that the x dependent Gaussian is centered at the mean displacement ⟨Q⟩ and has the variance σ_{QQ} as its squared width. The parameter y, on the other hand, which measures the "distance" of the off-diagonal matrix element from the diagonal has a Gaussian distribution centered at zero. The squared width of that distribution,

$$\sigma_{yy} = \sigma_{QQ}\,[(\sigma_{QQ}\sigma_{PP} - \sigma_{QP}^2)/\hbar^2]^{-1} \tag{10}$$

differs from the one in X by a dimensionless "collapse factor" $(\sigma_{QQ}\sigma_{PP} - \sigma_{QP}^2)\,\hbar^2$.

In order to find the explicit time dependence of the density matrix we must first evaluate the amplitude A(t). As explained in [17,18] a convenient starting point is the following representation of A by a contour integral,

$$A(t) = \frac{1}{2\pi i}\oint dz\, \frac{\sin zt}{g(z)} \tag{11}$$

$$g(z) = z^2 - \omega_0^2 - \sum_n \frac{\epsilon_n^2}{z^2 - \omega_n^2} \tag{12}$$

where the closed path of integration encircles all N + 1 pairs of zeros $\pm z_\nu$ of the function g(z) which are the eigenfrequencies of the Hamiltonian H. A simple graphical analysis of g(z) reveals all of these eigenvalues to be real, provided the coupling constants are not so large as to violate

$$\omega_0^2 > \sum_n \epsilon_n^2/\omega_n^2. \tag{13}$$

For the bath to deserve its name the frequencies ω_n must look, from the time scales characteristic of the meter, like a continuum. Correspondingly, the dense sequence of poles $\pm z_\nu$ in the integrand of (11) must look like a cut. By introducing the spectral density

$$\Delta\omega\gamma(\omega) = \sum_{\substack{n \text{ with } \omega_n \\ \text{in } \Delta\omega \text{ around } \omega}} \epsilon_n^2 \tag{14}$$

we may replace sums over bath oscillators like in (12) or (8) by integrals,

$$g(z) = z^2 - \omega_0^2 - \int_0^\infty d\omega\gamma(\omega)/(z^2-\omega^2) \tag{15}$$

The complex contour integral in (1) can then be given the form of a Fourier integral

$$A(t) = \int_{-\infty}^{+\infty} d\omega \, \frac{\gamma(|\omega|)\sin\omega t}{2g_+(\omega)g_-(\omega)\omega} \tag{16}$$

where $g_\pm(\omega) = g(\omega \pm i0)$ are the limits of $g(z)$ on the upper and lower lip of the cut along the real axis.

Adopting Ullersma's spectral function

$$\gamma(\omega) = \frac{2}{\pi} \frac{\kappa\alpha^2\omega^2}{\alpha^2+\omega^2} \tag{17}$$

we can evaluate $A(t)$ as well as the thermal amplitudes $X(t)$ and $Y(t)$ in closed form. Note that (17) is the spectral density of a field in three spatial dimensions, cut off by a Lorenzian factor at some frequency α. We are interested in the limit where $1/\alpha$ is a time much shorter than any of the time scales on which the meter can be observed. With the choice (17) the Fourier integral in (16) can again be evaluated by complex integration. The integrand turns out to have three poles in the upper half of the complex ω plane. One, $i\lambda$, is purely imaginary and of order α. The other two, much smaller in modulus, are either complex or, for the coupling constant κ sufficiently large, both imaginary, $\pm i\Gamma_\pm$. Only the latter case is of interest to us now. The three damping rates obey

$$\begin{aligned} &\lambda + \Gamma_+ + \Gamma_- = \alpha \\ &\Gamma_+\Gamma_- + \lambda(\Gamma_+ + \Gamma_-) = \omega_0^2 \\ &\lambda\Gamma_+\Gamma_- = \alpha(\omega_0^2 - \alpha\kappa). \end{aligned} \tag{18}$$

We refer the reader to [17,18] for the exact form of the resulting amplitude $A(t)$. For our present discussion it suffices to consider the limiting version obtained to leading order in $1/\alpha$ for $t >> 1/\alpha$,

$$A(t) = \frac{e^{-\Gamma_- t} + e^{-\Gamma_+ t}}{\Gamma_+ - \Gamma_-} . \tag{19}$$

Obviously, this amplitude and thus the corresponding expectation values $\langle Q(t)\rangle$ and $\langle P(t)\rangle$ also solve the equation of motion of an overdamped oscillation,

$$\langle Q\rangle + (\Gamma_+ + \Gamma_-)\langle Q\rangle + \Gamma_+\Gamma_-\langle Q\rangle = 0. \tag{20}$$

From the last of eqs. (18) we infer that the renormalized frequency $\sqrt{\Gamma_+\Gamma_-}$ of the meter is smaller than the unperturbed frequency ω_0. The bath is thus revealed to soften the restoring force for the meter displacement Q. We shall now limit our consideration further to situations where that softening has gone so far as to make $\omega_0^2 - \alpha\kappa$ comparable to or even smaller yet than ω_0^4/α^2. This case indeed deserves special attention since the three roots γ, Γ_+ Γ_- are then all very different in magnitude,

$$0 < \Gamma_- << \Gamma_+ << \lambda \approx \alpha. \tag{21}$$

We are therefore facing a heavily overdamped meter.

As is obvious from (19) and (20) and as is in fact well known, the momentum of a heavily overdamped oscillator tends to follow the displacement adiabatically, $\langle P(t)\rangle = \Gamma_-\langle Q(t)\rangle$, for times $t > 1/\Gamma_+$ after the imposition of initial data. To within corrections of relative order Γ_-/Γ_+ we can expect both the variance $\sigma_{pp}(t)$ and the squared width $\sigma_{yy}(t)$ to equilibrate through the fast transient $\exp(-2\Gamma_+ t)$ but the variance of Q, $\sigma_{QQ}(t)$, to contain the slow transient $\exp(-2\Gamma_- t)$. This is indeed what turns out.

Even though vigorous expressions, valid for all values of α, w_0, κ and all temperatures are available [18],we shall write down the variances in the limit.

$$\Gamma_- << \Gamma_+ << \alpha << kT/h, \qquad t >> 1/\Gamma_+ \tag{22}$$

which is the most interesting one with respect to measurement theory. We obtain, with a mass M restored,

$$\begin{aligned}
\sigma_{QQ}(t) &= \frac{kT}{M\Gamma_+\Gamma_-}\left(1 - e^{-2\Gamma_- t}\right) \\
\sigma_{PP}(t) &= MkT \\
\sigma_{QP}(t) &= \frac{kT}{\Gamma_+}\, e^{-2\Gamma_- t} \\
\sigma_{yy}(t) &= \hbar^2/MkT.
\end{aligned} \tag{23}$$

It is noteworthy that these results are in complete agreement with the ones obtained by Savage and Walls [12] using a master equation. The agreement is not completely trivial,since the traditional derivation of the master equation for the damped harmonic oscillator involves a perturbative approximation of lowest order with respect to the coupling to the heat bath [20,21]. Even though the traditional derivation thus does not cover strong coupling effects such as overdamping,the master equation itself does [18], as long as the temperature is, in frequency units, much larger than the damping rates $\Gamma_\pm$, as required in (22).

At any rate the results (23) imply that our model displays the expected collapse of the wavefunction. The off-diagonal elements of the density matrix (9) with $|y|$ larger than the thermal de Broglie wavelength of a particle with mass M assume exponentially small values on the fast time scale $1/\Gamma_+$ [22]. For times of that order the diagonal elements $\langle x|\rho_M(t)|x\rangle$ have not yet departed appreciably from their initial values since the variance σ_{QQ} begins to grow large much later only, for $t > 1/\Gamma_-$.

We would like to conclude with a remark on an interesting variant of our model. We have up to now restricted the discussion to a positive definite Hamiltonian H by imposing the inequality (14) which reads in the continuum limit with Ullersma's spectral strength

$$\omega_0^2 > \int_0^\infty d\omega\, \frac{\gamma(\omega)}{\omega^2} = \kappa\alpha. \tag{24}$$

We have seen the most interesting behaviour of heavy overdamping to arise when this inequality is just marginally respected, $\kappa\alpha \lesssim \omega_0^2$.

If we increase the coupling constant κ just a bit further, so as to have $\omega_0^2 \lesssim \alpha\kappa$, the central oscillator becomes softened beyond stability. One of the N+1 zeros z_ν^2 of g(z) then turns negative and gives rise to an imaginary eigenfrequency $\pm i|z_0| = \pm i\Gamma_-$, such as one would have for a particle exposed to an inverted parabolic potential. In other words, we are now facing a model of an amplifier [24-26].

The analysis sketched above needs only a slight modification for the amplifier case. In evaluating the complex contour integral (11) we must account for the isolated poles $\pm i\Gamma_-$ separately from the 2N real ones, since only the latter merge to a cut along the real axis of the complex z plane. By adding the contributions of the isolated poles and the cut we obtain as the analogue of (19)

$$A(t) = \frac{e^{+\Gamma_- t} - e^{-\Gamma_+ t}}{\Gamma_+ + \Gamma_-} \tag{25}$$

just as if the rate Γ_- had changed sign with respect to the damping case. In fact, Γ_+, $-\Gamma_-$, and λ still solve the cubic equation (18). If we now assume the critical coupling just barely exceeded, $0 \lesssim \kappa\alpha - \omega_0^2 < \omega_0^4/\alpha^2$ the growing exponential still lives on a much longer time scale than the decaying one, i.e. we still have (21). By simply letting $\Gamma_- \to -\Gamma_-$ in (23) we then obtain

$$\begin{aligned} \sigma_{QQ}(t) &= \frac{kT}{M\Gamma_+\Gamma_-}\left(e^{2\Gamma_- t} - 1\right) \\ \sigma_{yy}(t) &= \hbar^2/MkT. \end{aligned} \tag{26}$$

Like the heavily overdamped oscillator the linear amplifier with two vastly different eigenvalues (the one associated with amplification much smaller than the one associated with damping) can serve as a meter in a measurement device.

We would like to thank Craig Savage for interesting discussions. We gratefully acknowledge financial support by the Deutsche Forschungsgemeinschaft to Fritz Haake. Last not least, Fritz Haake would like to express his gratitude to the Physics Department of the University of Waikato for its hospitality during a very pleasant and fruitful stay.

References

1) J. von Neumann: "Mathematische Grundlagen der Quantenmechanik", Springer, Berlin (1932); "Mathematical Foundations of Quantum Mechanics", Princeton University Press, Princeton, New Jersey (1955).
2) C.F. von Waizsaecker: Naturwissenschaften 42, 521, 545 (1955); Z. fuer Naturforschung 13a, 245 (1958)
3) G. Ludwig: Z. Physik 135, 483 (1953); 150, 346 (1958); 152, 98 (1958)
4) E.P. Wigner: Am. J. Phys. 31, 6 (1963)
5) A. Daneri, A. Loinger, and A.M. Prosperi: Nuclear Physics 33, 297 (1962); Nuovo Cimento 44B, 119 (1965)
6) W. Weidlich: Z. Physik 205, 199 (1967)
7) F. Haake and W. Weidlich: Z. Physik 213, 451 (1968)
8) H.Z. Zeh: "Foundations of Quantum Mechanics", proceedings of the International School of Physics "Enrico Ferm" ed. by B. D'Espagnat, Academic, New York (1971)
9) E.P. Wigner: "Quantum Optics, Experimental Gravity, and Measurement Theory", ed. by P. Meystre and M.O. Scully, NATO Advanced Study Institute Series B, Vol. 94, Plenum, New York.

10) W.H. Zurek: Phys. Rev. D24, 1516 (1981); D26, 1862 (1982)
11) D.F. Walls and G.J. Milburn: Phys. Rev. A31, 2403 (1985)
12) C.M. Savage and D.F. Walls: Phys. Rev. A32, 2316 (1985); A32, 3487 (1985)
13) D.F. Walls, M.J. Collett, and G.J. Milburn: Phys.Rev. D32, 3208 (1985)
14) R.F. O'Connell, C.M. Savage, and D.F. Walls: to be published
15) V.B. Braginski, Y.I. Voroutsov, and K.S. Thorne: Science 209, 547 (1980)
16) C.M. Caves, K.S. Thorne, K.W.P. Drever, V.D. Sandberg, and M. Zimmermann: Rev. Mod. Phys. 52, 341 (1980)
17) P. Ullersma: Physica 32, 27 (1966); 32, 56 (1966); 32, 74 (1966); 32, 90 (1966).
18) F. Haake and R. Reibold:, Phys.Rev. A32, 2462 (1985); Acta Phys. Austriaca 56, 37 (1984).
19) A special case of (9) was discussed in (1)
20) G.S. Agarwal: 178, 2025 (1969); A4, 739 (1971)
21) F. Haake: Statistical Treatment of Open Systems, Vol. 66 of Springer Tracts in Modern Physics (Springer, Berlin, 1973)
22) A closer inspection of the exact solution for $\sigma_{yy}(t)$ reveals that the collapse of the off-diagonal elements at high temperatures is thermally enhanced to a time span of roughly $h \sqrt{\Gamma_+ \Gamma_-}/kT\Gamma_+$. The relaxation of σ_{QQ}, on the other hand, does not enjoy such enhancement. The effect is similar to the thermally enhanced damping of macroscopic quantum coherence discussed by Caldeiro and Leggett in ref. [23].
23) A. O. Caldeira and A.J. Leggett: Phys.Rev. A31, 1059 (1985)
24) R.J. Glauber: "Group Theoretical Methods in Theoretical Physics", Vol. 1 International Conference, Zvenigorod, USSR, 1982 (Marwood Academic Publishers, 1985)
25) D.F. Walls: J.Phys. A6, 496 (1973)
26) M.J. Collett: to be published

First Order Coherence of Vacuum Fluctuations: Optical Pumping Experiments in Presence of Electromagnetic Boundaries

F. De Martini and G. Innocenti

Dipartimento di Fisica dell'Universita', I-00185 Roma, Italy

1. Introduction

The existence of 0-point fluctuations of any field,and in particular of the electromagnetic one,can be taken as a semantic boundary between the realms of classical and quantum physics (1,2).It seems therefore important to investigate,by simple and noncontroversial experiments,the properties of physical systems which are believed to be substantially affected by its presence.In electrodynamics the famous Casimir experiments belong to this class.In the present work the spontaneous emission (SE) of optical radiation from excited atoms is taken as test process of the underlying quantum fluctuations associated with radiation states with zero or,maximum,one photon per mode.Of course the causal correspondence between SE (or Lamb shift (LS)) effects and field fluctuations has been questioned by several authors in the last years (3-7).In these works radiation-reaction has been assumed as an alternative model allowed by the commutation properties of atomic- and field-operators in the Heisenberg theory of quantized atom-radiation interaction.However we believe that the problem has in fact been recently and definitely solved by theorists of the Ecole Normale Superieure in Paris,in favor of the correspondence Spontaneous Emission-Vacuum Field on the grounds of the operator hermiticity that should necessarily correspond to the separate obvervability of the time rates of change of any atomic operator interacting respectively with vacuum and radiation-reaction fields (8).This leads to favor,among possible operator-ordering procedures in the nondifferential part of the Heisenberg equation,the completely symmetric one.The soundness of this interpretation can be satisfactorily taken as the basis of our model,which could be also recast pictorially in the following way: Spontaneously emitted (or scattered) photons on a field mode k can be considered as "stimulated" by the 0-point field existing in that mode. A similar conclusion can be extended to other nonresonant interaction effects like LS etc.Apart from close correspondence of this model to the "real" microscopic processes it is intended to describe, it proves to be, in our case, a very useful representation of the phenomena and therefore it leads, in a surprising way, to novel ways of interpretation and to devise new kinds of significant experiments. Several such experiments will be presented in the following text. The quantum theory of the effects we investigate will be then outlined.

2. Periodic Optical Pumping Spontaneous Emission.

Let us explain the physics of our experiments by focusing our attention on two situations related to the geometry of the electromagnetic boundary conditions affecting the field in the zone of space where the excited (and then emitting) atoms are placed.

(a): Atoms in front of one infinite conducting plane.

Let us explain this experiment on very simple grounds. Suppose that the vacuum field could be described by a sort of stochastic Maxwell field, a viewpoint taken in Stochastic Electrodynamics (3). An "observer" that could measure the energy density associated with a particular mode k of this field, after suitably filtering through a passband Δf centered on f, should detect in the vicinity of a plane mirror a field interference extending over a distance of the order of the field coherence length i.e. $l_c = c/(\Delta f)$. Of course, in the case of vacuum fluctuations the absence of photons to be annihilated (or nonadiabatically scattered) excludes the observability in the usual sense. However, according to our model, the presence of the field can still be probed, and the average intensity measured, by an indirect, "stimulated" process as Spontaneous Emission (SE). Therefore an assembly of excited atoms with spontaneous linewidth Δf centered on f can be used as "active filters" allowing a (linear) measurement of the "stimulating" field intensity belonging to a mode k by mere observation of the SE emitted on the same mode. In our experiment excited Cr atoms belonging to a thin ruby crystal slab are used to probe the field distribution close to a plane, 100% reflecting Ag layer deposited in the same slab. Suppose, for sake of simplicity, that the field mode $k = \hat{k}\,(2\pi n)/\lambda$ under investigation is orthogonal to the mirror (Fig. 1). Following our model the 0-point energy density belonging to k should be zero at the boundary (or close to zero, for a conducting metal) and proportional to $\sin^2(k{*}z)$ along the z-direction, normal to the mirror, owing to the field self-interference due to mirror reflection (Wiener effect) (9). In other words, and more generally, "classical" interference arising from mirror

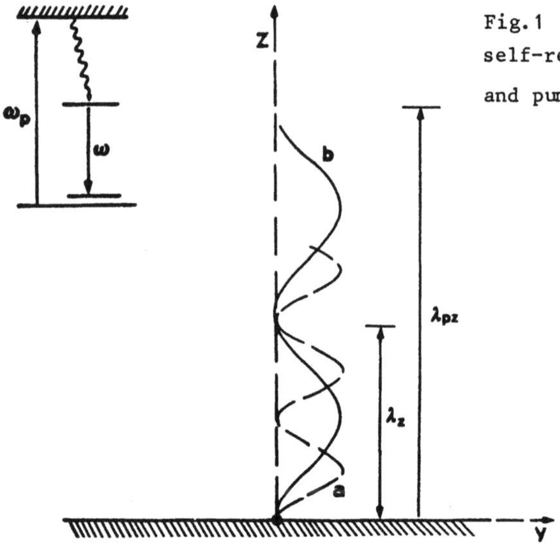

Fig.1 Wiener-Fringe patterns due to mirror self-reflections of 0-point field (a) and pump field (b).

reflection of vacuum fields creates in the region close to the boundary and in the direction of the boundary allowed k-modes standing wave patterns of 0-point Wiener fringes (OWF) (9). Then the SE emitted on a particular k mode by excited Cr atoms placed close to the maxima of the OWF fringes belonging to the same k-mode,and detected (through an interference filter) at the dominant fluorescence wavelength of ruby, $\lambda = (2\pi n/k) = 0{,}69\ \mu m$, through two pinholes placed along the direction of k (making a angle ϑ with the z-axis),is strongly "stimulated" (4 times more than in open space),while atoms placed close to the corresponding OWP interference zeros tend not to emit at all on the mode k. This affects the respective atom emission intensity and, slightly for atoms not too close to the mirror, the SE lifetime (10). In the present case we want to investigate mainly the former dominant effect that can extend over a distance $l = l_c$ from mirror which is virtually infinite for extremely sharp fluorescence SE lines.

In order to detect this effect let us apply to this case the novel technique of Periodic Optical Pumping,which literally consists of imposing on the atomic medium a regular periodic excitation pattern along a certain axis,say -z,with an externally controlled periodicity.This is a technique of general application that,in this case,can be easily realized using a highly collimated laser beam as pump source and by creating,by simple interference,a pattern of plane fringes orthogonal to -z at the pumping wavelength, and with fringe periodicity λ_{pz} in the zone where the atoms are placed.

In our experiment with ruby we have used as pump the 4880 Å line of an Ar laser with beam angular divergence $\sim 8.10^{-4}$. The pump beam, with proper polarization, was made to interfere with itself by reflection over the Ag mirror deposited on the crystal slab. If ϑ_p is the angle made by this beam, with the z-axis normal to the slab, the interference pumping period along z is $\lambda_{pz} = (\lambda_p/(2\cos\vartheta_p)) = (\pi n_p/k_{pz})$ where n, n_p are the refractive indexes of ruby at λ and λ_p for the respective field polarizations (11). As the energy density associated with the pumping interference pattern is proportional to the local density of excited Cr atoms in the slab, we are able, by scanning over the excitation angle ϑ_p on the sample, to accumulate pumping on the atoms that, by virtue of the local intensity of OWF, are more "stimulated" to emit by SE in the mode k under investigation. Then, if our physical model for SE is correct, a suitably filtered detector placed behind two observational pinholes along k should detect a sharp increase of SE intensity in correspondence of the k-space resonant condition: $k_z = k_{pz}$, or $(\lambda/n) = (\lambda_p/n_p \cos\vartheta_p)$. In addition, for a given ϑ_p, this resonant condition leads to a sharply peaked angular distribution of the SE intensity, a strikingly peculiar effect that, to our knowledge, has never been observed before in connection with chaotic or noise fields emitted by arrays of classical or quantum antennas, in the context of radio, microwave or optical radiation (12). In our experiment the combined action of two independent interference effects is able to create a sharply defined radiation pattern which is not determined by mere diffraction.

In our experiment, an optically polished L= 8 μm thick, 1 cm diameter ruby crystal disk (0,05 wt % Cr_2O_3 doping) is glued with a transparent adhesive on a side of a SF10 (flint) glass 90 prism. The orientation of the crystal is such so the emitting dipoles at the R_1 (6943 A) and R_2 (6929 A) SE lines are oriented along the x-axis of the system (Fig. 2). The coherence length of the SE radiation, detected via an interference filter by a RCA C31025C (Quantacon), high quantum efficiency photomultiplier, is l_c = 30 μm. A computer interfaced Tektronix 7D20 fast digitizer provided signal processing, display and data collection. The prism, together with a 1° aperture, SE collimating telescope connected to a goniometer for measurement of the output angle ϑ, and to an optical fiber, was mounted at the center of a larger goniometer by which the pumping angle ϑ_p on the active slab could be measured. The experimental x-axis was parallel to the common axis of the goniometers. The intensity of the (λ_p=4880 A, divergence $\sim 8.10^{-4}$) Ar laser pump beam was taken sufficiently low in order to avoid Stimulated-Emission (STE) processes in the

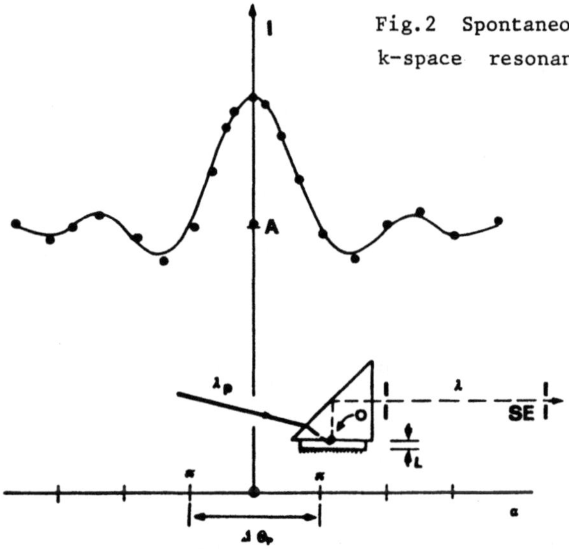

Fig.2 Spontaneous emission intensity as function of k-space resonance parameter (α).

sample.The polarization of the pump beam was taken parallel to the x-axis.In Fig.2 the SE output intensity at ϑ =0,is plotted as function of ϑ_p .A sharp resonant peak,with a typical sinc(x)=(sinx/x) shape,is found to overlap a virtually flat SE distribution.A similar peak,with 3° aperture outside the prism,shows up as function of ϑ in angular distribution experiments with fixed ϑ_p.As it will be shown later in the theory,assuming with no loss of generality that the detector is excited by a single SE radiation mode k,solving the coupled atom-field Heisenberg equations for the field operator ($\hat{a}_k^{+} \hat{a}_k$),and then inserting in Tr($\rho \cdot \hat{a}_k^{+} \hat{a}_k$) expressing photodetection (13),the maximum intensity of the detected SE is found to be proportional to:

$$I(\vartheta,\vartheta)=(2/L)*\int_0^L \sin^2(k_z z)*\sin(k_{pz}^2 z)\,dz \qquad (1)$$

where L is the sample thickness.Integrating,we get close to resonant condition: $I(\alpha)=((2+\mathrm{sinc}(2\alpha k_z L))/4)$,where $\alpha=((k_z - k_{pz})/k_z)$.$I(\alpha)$ is also plotted in Fig.2.

(b) Atoms in a microscopic electromagnetic cavity.

Adding a second plane boundary (mirror) parallel to the first one and at a distance d of order λ along the normal to the surfaces (z-axis),i.e. making a "microscopic" Fabry-Perot (FP) interferometer,leads, for atoms placed in the cavity,to still more pronounced effects on all physical processes involving coupling with radiation i.e. Spontaneous Emission and Lamb Shifts (13).
Owing to multiple field interference a strong selection of radiation modes is determined,together with a large enhancement of mode density along the allowed k-directions (9).

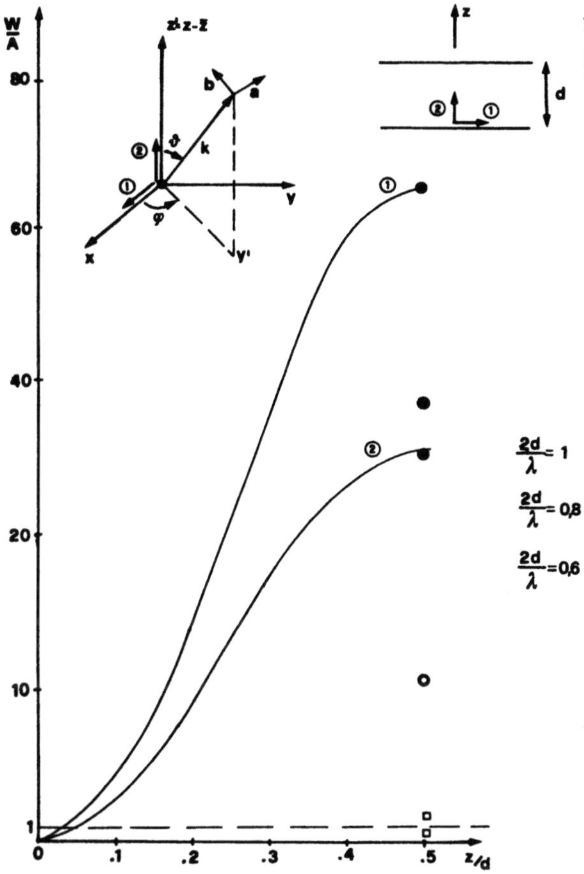

Fig.3 Spontaneous emission decay probability in a microscopic Fabry-Perot cavity

This results, in the context of our experiment, in large changes of decay probability $W(z)=1/\tau$ for excited atoms placed at different z-positions in the cavity. This is shown for the critical case: $d = \lambda/2$ in Fig.3, where the free-space value of W =(Einstein's A-coefficient)$=1/\tau_0$ is also indicated.

This peculiar effect has been investigated making use of our Wiener-Fringe technique in a Fabry-Perot optical cavity driven by a piezoelectric (pzt) transducer (Fig.4). A 10^{-4} ethanol solution of Rubrene dye ($\tau_0=(1/A)=15$ nsec.) flowing between the mirrors was the active medium. The SE radiation was filtered by a $\Delta\lambda$ =30 A interference filter centered on λ =6328 A. Periodic Optical Pumping Spontaneous Emission was made available in our experiment by the fact that the FP providing a high-Q cavity effect for SE wavelength λ had only one mirror 100% reflecting the pumping wavelength λ_p (mirror B in Fig.4) the other mirror being antireflection-coated for the same wavelength. The well-collimated pump beam at λ_p=0,53μ was provided by SHG of a Nd-Yag laser pulses of 5 nsec duration. As in the

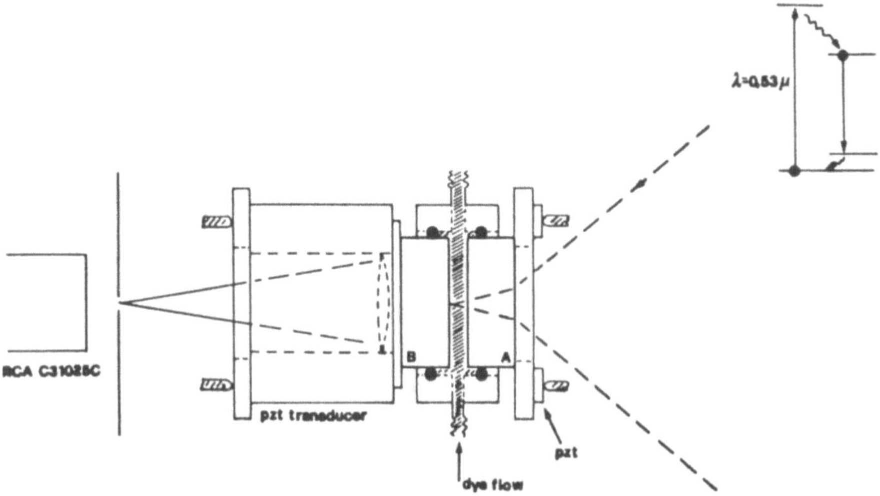

Fig.4 Microscopic Fabry-Perot cavity (Section)

previous case (a),changes of the pumping incidence angle ϑ_p on the FP pumping-entrance mirror leads to selective excitation of atoms placed at different z-positions in the FP,making available a direct investigation of the effect shown in Fig.3.By setting $d<(\lambda/2)$ we reproduced,in a different context,a Casimir-like effect:no modes are now available for atom-radiation coupling and then W falls quickly to zero for any z and the fluorescence time T gets increasingly longer (decay eventually occurs over channels different from electric-dipole coupling).
In our experiment we have verified by the Periodic Pumping technique the behavior of the function W(d) shown in Fig.3.For $(2d/\lambda)<1$ we have obtained an increase of fluorescence lifetime for rubrene of from 3 to 5 times the corresponding "natural" lifetime.Fig.5 shows this effect for the same medium.A detailed account of these results will be reported in a following paper.

3. Quantum theory of spontaneous emission with electromagnetic boundaries.

The vacuum-field picture we have adopted to anticipate and to interpret the results reported in the previous chapters has a rigorous counterpart in the theoretical description provided by Quantum Electrodynamics (13,14).
In the present chapter we outline the theory following two paths: (a) we calculate the SE decay times for a single atom by a perturbative approach, (b) we outline the relevant results of the complete Heisenberg equation formulation.

(a) Single atom spontaneous decay times.

Assume Fermi "golden rule" for evaluating the probability W of photon emission over the radiation mode-k (13).

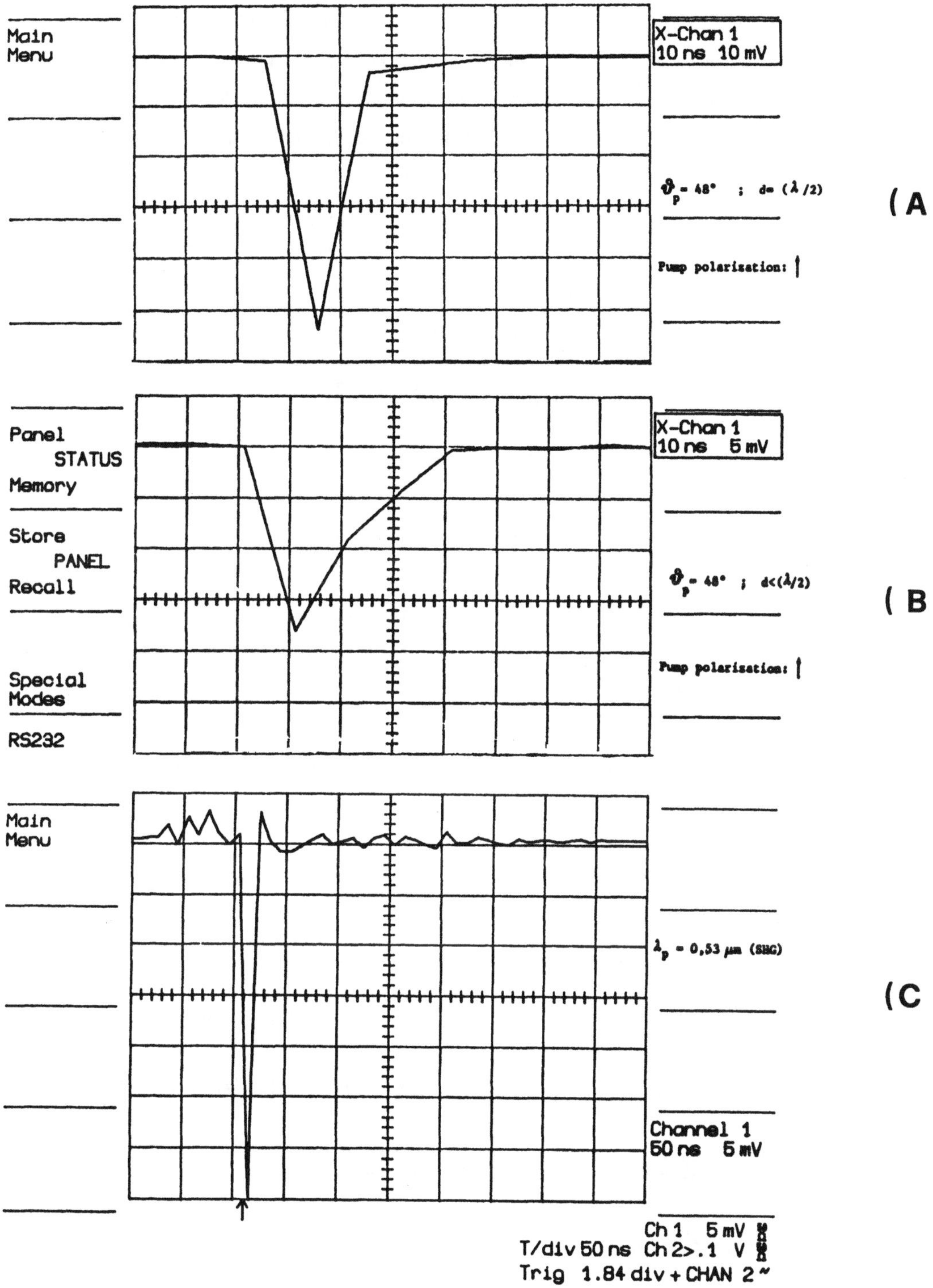

Fig.5 Spontaneous emission decay with Fabry-Perot interferometer. (A) with $(2d/\lambda)=1$ (B) with $(2d/\lambda)<1$. Plot (C) shows the exciting laser pulse (Plotter HP 7470A).

Refer to the inset of Fig.3 for determination of polar coordinates, keeping in mind that the atom is at the origin of these coordinates and is placed in space at z=z. The mirror (B) of the Fabry Perot shown in Fig.4 coincides with plane (x-y) i.e. z=0 of our geometry and d is the distance between the mirrors. Labels a) and b) correspond to two E-field polarizations respectively orthogonal and parallel to the incidence plane (y'-z'). Labels (1) and (2) correspond to orientations of the electric-dipole vector respectively parallel and orthogonal to the FP mirrors.

The general expression of the vector-potential quantum operator for the propagation mode-k may be written in the form:

$$\vec{A}_k = \sqrt{(h/(V\omega_k))*Y} \; *(\hat{a}^{+}_k * \vec{f}_{ki} *\exp(i\ \omega_k t) \ + \text{h.c.}\ ). \qquad (2)$$

where $Y = 1/((1-R)^2 *(1+F \sin^2(\delta)))$ is the Airy function; $\delta = k\ d \cos\vartheta$; $k = (2\pi n/\lambda)$; $R^2 = R_1 * R_2$; R_1 and R_2 the reflectivities of the FP mirrors at the SE wavelength ; $F = 4R/(1-R)^2 = ((2/\pi)*(\text{"FP finesse"}))^2$. (ref.(9)).

Furthermore $\vec{f}_{ki}$ is a quantity expressing field-polarization and dependent on the z atomic coordinate $\bar{z}$. It also expresses phase-matching of the two sets of multiple-reflecting field waves to which the atom is coupled in the cavity. For our purposes we can write:

$$\vec{f}_{ki} = 2i\,\vec{\varepsilon} * \sin\ (k\ z\ \cos\vartheta) * \exp(-i\vec{k}\cdot(\hat{x} + \hat{y})) \qquad (3)$$

By evaluating the expression of W in the usual way and by summing-up these quantities over the allowed radiation modes, we obtain the general expressions of the overall probability of single-atom decay by Spontaneous Emission for $\vec{\mu}$: electric-dipole coupling and for $\vec{\mu}$ respectively parallel and orthogonal to the mirrors:

$$W(z) = 1/T = A*3/(2(1-R)^2)*(I_1 + I_2) \qquad \text{case 1: } (\vec{\mu} \text{ parallel to mirrors})$$

$$W(z) = 1/T = A*3/(1-R)^2 \quad *(I_1 - I_2) \qquad \text{case 2: } (\vec{\mu} \text{ orthogonal to mirrors}) \qquad (4)$$

where the free-space emission probability is: $A = 1/T = (2|\mu|)^2/(3\hbar)*(k)^3$

and:

$$I_1 = \int_0^{\pi/2} (\sin^2(\xi\ \delta)*\sin(\vartheta)/(1+F \sin^2(\delta)))\cdot d\vartheta \qquad \xi = (z/d)$$

$$I_2 = \int_0^{\pi/2} (\sin^2(\xi\ \delta)*\sin(\vartheta)*\cos 2(\vartheta)/(1+F \sin^2(\delta)))\cdot d\vartheta$$

We assume $k = (2\pi/\lambda)$ where λ = SE wavelength, henceforth. The results of computer evaluation of W(z) for cases 1) and 2), "finesse"=155, d=(λ/2) are plotted in Fig.3

where the values of $W(d/2)$ for two values of $(2d/\lambda)<1$ are also shown. We see that the large enhancement of the vacuum-field due to the high Q of the cavity determines a large enhancement of $W(z)$ respect to the free-space value. However $W(z)$ drops dramatically for decreasing values of $(2d/\lambda)<1$ owing to mode elimination.

The theory is easily extended to the single mirror case by simply assuming $R_2=0$ i.e. by eliminating the corresponding mirror. In this case we get: Y=1, F=0, "finesse"=0 and, with correspondence to the cases above,

$$W(z)= A*(1-(3/2)*(\mathrm{sinc}(2kz)/(1-(2kz)^{-2})+\cos(2kz)/(2kz)^2) \quad \text{case 1}$$

$$(4')$$

$$W(z)= (A/3)*(1-3(\mathrm{sinc}(2kz)-\cos(2kz))/(2kz)^2) \quad \text{case 2}$$

Equations (4') are similar to the ones obtained by Philpott by another method (6).

(b) Heisenberg equations.

Write the time evolution of an operator $\hat{O}$ in Heisenberg representation $d\hat{O}/dt=(1/h)*[H,\hat{O}]$ (13). The Hamiltonian $H= H_R + H_A + H_{RA}$ describes radiation (R), 2-level (single) atom (A) and R-A coupling (13). Assume for simplicity and with no loss of generality that our quantum detector is excited by only one radiation mode-k. Our aim is to solve the Heisenberg equation for the operator $\hat{O} = (\hat{a}^{\dagger}_k\hat{a}_k)$ which is directly involved in the dynamics of photodetection. Note that, in our case the expression of the field Vector-Potential operator includes the "0-point Wiener Fringe" effect by the terms Y and f appearing in expression (2).

The solution of the Heisenberg equation for $(\hat{a}^{\dagger}_k\hat{a}_k)$ involves operator-ordering considerations and a previous solution of a hierarchy of similar equations for: operators $O = \hat{a}^{\dagger}_k, \hat{s}^{\dagger}_i, (\hat{s}^{\dagger}_i s_i)$, where $s^{\dagger}_i, s_i$ are the "fictitious spin" 2-level atom (i) displacement operators (8,13). The operator $(\hat{a}^{\dagger}_k\hat{a}_k)$ should then be averaged over the vacuum k-state and over the assembly of atomic states that, in our experiment, should express the preparation induced by our Periodic Pumping, i.e. "Wiener-Fringe" technique. Then, as we have already noted, the relevant density operator for complete atom-radiation system to be used in the ensemble average expressing photodetection is proportional to $f=\sin^2(k_{pz}z)$, apart for a phase factor in the argument. This leads to expression $I(\vartheta,\vartheta_p)$ already given by equation (1) showing a novel k-space resonance effect.

1. C.Itzykson, J.B.Zuber: Quantum Field Theory (Mc-Graw Hill, New York 1980) Ch 11. For Casimir effect: Ch 3. H.Casimir, D.Polder; Phys.Rev. 73,360 (1948).
2. J.A.Wheeler, W.Zurek: Quantum Theory and Measurement (Princeton University, 1983)

3. A.O.Barut (Editor): Foundations of Radiation Theory and Quantum Electrodynamics (Plenum,New York,1980).Cfr.papers by P.W.Milonni,J.H.Eberly,K.Wodkiewich. A survey of Stochastic Electrodynamics is given by T.H.Boyer.
4. J.R.Ackerahlt,P.L.Knight,J.H.Eberly: Phys.Rev.Letts.30,456 (1973) and Phys.Rev. D 10,3350 (1974).
5. P.W.Milonni:Optics Comm. 9,119 (1973) (with P.L.Knight); Phys.Rep.31,955 (1976)
6. M.R.Philpott:Chem.Phys.Lett.19,435 (1973);see also H.Morawitz,Phys.Rev.187,1792 (1969).In Philpott paper a sign mistake in expression (13) corresponding to our (4'),case 2,leads to an incorrect statement in the discussion of the result of the same equation.
7. I.R.Senitzky:Phys.Rev.Lett.31,955 (1973).
8. J.Dalibard,J.Dupont-Roc and C.Cohen Tannoudij:J.Phys.43 1617 (1982).For QED with Lippman fringes see E.Fermi,Rev.Mod.Physics 4,87 (1932) and .S.Agarwal Phys.Rev.A 12,1987 (1975).
9. M.Born and E.Wolf:Principles of Optics (Macmillan,New York,1980) For interference and Fabry-Perot interferometry : Ch.VII. O.Wiener:Ann.d.Physik,40,203 (1890).A Wiener technique was used to study transfer of electron excitation in solids by P.Liao,H.Weber and B.Tofield,Solid State Comm.16,881 (1975).
10 SE lifetimes T by molecular monolayers close to single metal surfaces have been investigated by the Goettingen group:cfr. for a comprehensive covering of the subject and bibliography the review paper by K.H.Drexhage in Progress in Optics,North-Holland,Amsterdam,1974.Vol.XII.That work is connected with the present one.However cavity effects,intensity effects on radiation pattern, collective effects from a thick medium as well as the Wiener pumping technique we are reporting here have not been previously studied or anticipated.Owing to the large number of radiation modes which determine T with very limited mode discrimination close to a single conducting surface,this one affects T only over a small distance of order λ.In order to interpret the Goettingen results,which seem not to have been reproduced elsewhere (likely,because of the difficulty of the monolayer technique),a large amount of interesting theoretical work,mostly based on classical arguments,has been produced in the past.Among the heuristic models proposed:effect of boundaries on antenna impedance and dipole-image interaction.This last model has also been treated using field quantization as a kind of superradiant effect.However,apart from the "image" concept that seems to be irrelevant,by definition,in reality

consideration,this model,when translated into quantum theory,corresponds to the atomic "self-reaction" process which is ruled out by the rigorous quantum mechanical considerations of Ref.8.So far,the vacuum field interpretation has not been explicitly proposed to explain the Goettingen results.
Our work on cavity effects is connected with the following publications:D.Kleppner,Phys.Rev.Lett. 47,223,1981; 55,2137,1985 (with R.G.Hulet, E.S.Hilfer); P.Goy,J.L.Raymond,M.Gross and S.Haroche,Phys.Rev.Lett.50,1903, 1983; D.Meschede,H.Walther and G.Muller,Phys.Rev.Lett. 54,551,1985. Note that,dealing with optical spectra ($h\nu \gg kT$) allows us to work at room-T.

11 W.Koechner; Solid State Engineering (Springer-Verlag,Berlin,1976).Pumping bands emission lines and refractive indexes of ruby are given in Ch.2.

12 F.E.Terman;Electronic and Radio Engineering (McGraw-Hill, New York,1955).Ch 23.

13 R.J.Glauber,Phys.Rev.131,2766 (1963); R.Loudon; The Quantum Theory of Light (Oxford,1983).Ch.5.

14 The conceptual indirectness of QED has been pointed out by many leading scientists in the past.See for instance:R.P.Feynmann;The Quantum Theory of Fields,Interscience,New York,1961;V.F.Weisskopf,Physics of Twentieth Century M.I.T.Press,Cambridge,1972 and our Ref.4.Nevertheless we believe that in the particular context of the processes we investigate in the present paper,the explanation based on vacuum-fluctuations has a direct and simple correspondence with its description given by QED.This doesn't seem the case for other "heuristic" interpretations we refer to in (10).To our knowledge no interpretation other than vacuum-fluctuations has been proposed so far for the Casimir effect.This one has been experimentally tested many times (1). This seems to be sufficient for concluding in favor of a real existence of the vacuum field whatever definition of "reality" we decide to adopt.

Rydberg States in Laser Fields: Wave Packets and Near-Threshold Phenomena

P. Zoller, G. Alber, and H. Ritsch

Institute for Theoretical Physics, University of Innsbruck, A-6020 Innsbruck, Austria

1. Introduction

The description of the behavior and motion of Rydberg electrons excited by laser radiation is an essential element in our understanding of multiphoton processes in atoms and molecules [1,2]. Here we consider two aspects of this problem: (i) We discuss generation of Rydberg wave packets by short laser pulses; and their detection in a two-photon experiment with time-delayed pulses [3,4], (ii) We develop a theory of excitation of (autoionizing) Rydberg states which from the outset includes the laser-induced mixing between **all** bound and continuum states of a Rydberg series [5].

When a short laser pulse of duration τ excites electrons from a low-lying atomic level to Rydberg states, wave packets are formed because (i) photon absorption from the initial state to the Rydberg series is localized in space in a volume of the size of the initial state, which is small compared with the spatial extent of Rydberg states; and (ii) provided the laser pulse is so short that its spectral width simultaneously excites many Rydberg levels around some mean energy $E_{\bar{n}} < 0$:

$$\hbar/\tau \gg \frac{1}{2\pi}(E_{\bar{n}+1} - E_{\bar{n}}) \simeq \frac{1}{2\pi}\frac{dE}{dn}\bigg|_{\bar{n}} = \hbar/T_{E_{\bar{n}}} . \tag{1}$$

Here E_n are Rydberg energies for the principal quantum number n and $T_{E_{\bar{n}}}$ is the classical orbit time corresponding to a particle of energy $E_{\bar{n}}$ moving on a Kepler orbit. According to Equ. (1) wave packets are excited when the pulse duration is short compared with the classical orbit time. In this sense Rydberg wave packets can be interpreted as quantum beats between n-states. In Sec. 2 we give a brief summary of a recently developed semiclassical formalism to describe generation and detection of Rydberg wave packets [3]. In particular, we present a semiclassical analysis of a two-photon process with two time-delayed laser pulses: a first short pulse excites a wave packet whose motion is probed by a second pulse. This provides us with a time and space-sensitive measurement since absorption or induced emission of a probe laser photon can only occur near the inner turning point of the classical Kepler orbit where acceleration of the electron is largest [6]. Thus the two-photon transition probability will show peaks, whenever the time delay between the laser pulses in a multiple of a classical orbit time (as long as the wave packet has a well-defined structure).

Our foregoing discussion has emphasized that absorption (emission) of laser photons occurs in a **finite** interaction volume, much smaller than the size of Rydberg orbitals and almost independent of energy. This physical picture is reflected in the fact that the oscillator strength distribution from Rydberg states to other bound or free orbits is a smooth function of energy across the Rydberg threshold. We can make this a basis of a theory of near-threshold behavior of Rydberg states in laser fields. The idea is to solve the problem of interaction of laser radiation with the Rydberg electron above threshold and then to extrapolate this information to the bound Rydberg series [5] using the properties of Coulomb functions [7]. The essential feature of this theory is that it considers the interaction with a Rydberg series as a whole (as opposed to treating the interaction with each Rydberg level separately). Thus from the outset laser-induced mixing between all bound and continuum states of one series is included. Sec. 3 briefly summarizes results obtained for strong field excitation of autoionizing resonances [5].

2. Generation and Detection of Rydberg Wave Packets by Short Laser Pulses

We consider an atom with a single valence electron which is excited by a short pulse at time t_a from an initial state $|i\rangle$ (energy E_i) to Rydberg states. The electric field has an amplitude $\mathcal{E}_a(t)$, mean frequency ω_a and pulse duration τ_a. In perturbation theory we find for the electron wave function at a time $t > t_a + \tau_a$ after the pulse

$$|\psi(t)\rangle = |i\rangle e^{-iE_i t/\hbar} + |\psi_i^{(a)}(t)\rangle \qquad \text{with} \tag{2}$$

$$|\psi_i^{(a)}(t)\rangle = \sum\!\!\!\!\!\!\!\int_{n,l,m_l} |nlm_l\rangle\, e^{-iE_{nl}t/\hbar}\, \frac{i}{\hbar}\, \mu^{(a)}_{nlm_l,i}\, \tilde{\mathcal{E}}(\delta^{(a)}_{in})\, e^{\delta^{(a)}_{in} t_a} \;. \tag{3}$$

Here $|nlm_l\rangle$ are eigenstates of the atomic Hamiltonian H_A for the energy E_{nl}; l and m_l are angular momentum quantum numbers. The dipole operator is denoted by μ, $\delta^{(a)}_{in} = (E_{nl} - E_i - h\omega_a)/\hbar$ are detunings. The spectral density of the pulse is defined by

$$\tilde{\mathcal{E}}(\Delta\omega) = \int_{-\infty}^{+\infty} dt\; \mathcal{E}_a(t)\; e^{i\Delta\omega(t-t_a)}. \tag{4}$$

If the laser pulse $\mathcal{E}_a(t)$ is short in the sense of Equ. (1), $|\psi_a^{(i)}(t)\rangle$ describes a Rydberg wave packet; at least as far as the radial motion is concerned: due to dipole selection rules only few angular momenta contribute in the sum (3), so there is no further localization in the angular variables. Since the spectral density $\tilde{\mathcal{E}}_a$ is concentrated in an energy band $-\hbar/\tau_a \lesssim E - E_i + \hbar\omega_a \lesssim \hbar/\tau_a$ with the maximum centered around $E_{\bar{n}} = E_i + \hbar\omega_a$, only Rydberg states in this energy interval will contribute in Equ. (3).

We now turn to a discussion of the motion of the wave packet (3). This serves as a guideline for our interpretation of two-photon processes with time-delayed pulses. We expect that the radial center of the wave packet (to the extent it is well defined) follows a classical Kepler orbit. The propagator which governs the time evolution in Equ. (3) for $t \geq t'$ is

$$G^+(\vec{x},t|\vec{x}',t') = e^{-iH_A(t-t')} = -\frac{1}{2\pi i}\int_{-\infty}^{+\infty} dE\, e^{-iE(t-t')} \times \tag{5}$$
$$\times \sum_{lm} \frac{g_{El}(r,r')}{rr'} Y_{lm}(\hat{x})\, Y^*_{lm}(\hat{x}')$$

where $g_{El}(r,r')$ is a radial Green's function of the atom with outgoing wave boundary conditions. The key to a semiclassical analysis and physical interpretation is to rewrite this Green's function in the form [8]

$$g_{El}(r,r') = \begin{cases} g^{(s)}_{El}(r,r') & (E > 0) \qquad (6a) \\ g^{(s)}_{El}(r,r') - 2\pi i \dfrac{S(E,l,r)S(E,l,r')}{e^{-2\pi i(\nu+\alpha)} - e^{-2\pi\beta}} & (E < 0,\ \beta \to 0)\ . \qquad (6b) \end{cases}$$

The Green's function for an energy E above threshold ($E > 0$) has been denoted by $g^{(s)}_{El}(r,r')$ where the superscript indicates that it is a smooth Green's function, i.e. free of singularities in E. The explicit mathematical form is not relevant for us at this point. Of course, $g^{(s)}_{El}(r,r')$ is not a physical Green's function below threshold since it diverges for $r_2 < r \to \infty$ with r_2 the outer turning point of the Kepler orbit for the energy $E < 0$. This asymptotic divergence is compensated by the second term in Equ. (6b) where $S(E,l,r)$ is the regular solution of the radial Schrödinger equation (normalized on the energy scale). $S(E,l,r)$ diverges exponentially for $E < 0$ except at the bound state energies $E_n = -Ry/\nu_n^2$ with $\nu_n = n - \alpha$ (n = integer, α = quantum defect) [7]. As expected the second term in Equ. (6b) has poles at these bound state energies. It is the resonant part of the Green's function.

Let us assume that the laser pulse excites a superposition of bound Rydberg states. The propagation of $|\psi_i^{(a)}(t)\rangle$ is then described by

$$g_{El}(r,r') = g^{(s)}_{El}(r,r') - 2\pi i \sum_{m=1}^{\infty} e^{2\pi i(\nu+\alpha)m} S(E,l,r)S(E,l,r') \tag{7}$$

where we have expanded the resonance denominator in Equ. (6b). Substituting Equ. (7) into (5) and (3), and evaluating the resulting integral over the energy (5) in a stationary phase approximation leads to the following interpretation:

(i) For times $t - t_a < T_{E_{\bar{n}}}/2$ only the first term in Equ. (7) contributes. It describes a radial wave packet moving on a Kepler orbit

$$\int_{r_1}^{r(t)} \frac{\mu}{p(r')} dr' = t - t_a \qquad (t - t_a > 0) \qquad (8)$$

with $p(r)$ the radial momentum of the electron, μ the electron mass and r_1 the inner turning point. We emphasize that - although we have a bound wave packet - its motion is governed by a continuum propagator $g_{El}^{(s)}(r,r')$. This is not surprising because for $t - t_a < T_{E_{\bar{n}}}/2$ does not "know" that its radial motion is bounded, as it did not have time to "see" the potential barrier near the outer turning point r_2. In a two-photon experiment with short pulses (no time delay) and Rydberg levels as intermediate states this corresponds to the absence of intermediate state resonances (**direct** process).

(ii) For times $t - t_a > T_{E_{\bar{n}}}/2$ the wave packet is reflected from the outer potential barrier. This periodic motion on a Kepler orbit is described by the second term in Equ. (7)

$$\pm \int_{r_1}^{r(t)} \frac{\mu}{p(r')} dr' = t - t_a - m \cdot T_{E_{\bar{n}}} \qquad (m = 1,2,\ldots) \qquad (9)$$

with

$$T_E = \frac{\pi\hbar}{Ry} (-E/Ry)^{-3/2} \equiv \frac{\pi\hbar}{Ry} \nu^3 \ . \qquad (10)$$

According to Equ. (9) the m = 1,2,... terms in Equ. (7) correspond to the first, second return of the wave packet to its inner turning point. It is possible to observe this periodic motion in a two-photon process with time-delayed pulses: a first pulse at t_a excites the wave packet which is probed at a later time t_b by a second short pulse. As we remarked earlier the transition probability $P_{f\leftarrow i}$ will show peaks at $\Delta t = t_b - t_a = mT_{E_{\bar{n}}}$ $(m = 1,2,\ldots)$ since absorption of the probe photon occurs near the inner turning point of the Kepler orbit. We emphasize the close connection between a resonant two-photon process and the observation of wave packets in a time-delayed pulse experiment (**indirect or two-step** process).

In Fig. 1 the radial function of r (in units of the Bohr radius) is shown at time $t - t_a$ = = 1/9, 2/9 and 3/9 × $T_{E_{\bar{n}}}$ for a Gaussian laser pulse (duration 8 psec) exciting hydrogen p states around $\bar{n}$ = 85 ($T_{E_{\bar{n}}}$ = 94 psec). Fig. 2 is a plot of a two-photon Raman transition probability as a function of the time delay $t - t_a$ between the two pulses measured in units of a classical orbit time ($\bar{n}$ = 85; $\tau_a = \tau_b$ = 5 psec). The dashed line is the semiclassical (stationary phase) approximation which compares well with a calculation based on an explicit summation over n-states (solid line). A sweeping of the laser frequency during the pulse can lead to an initial contraction of the wave packet (Fig. 3). Finally, Fig. 4 shows

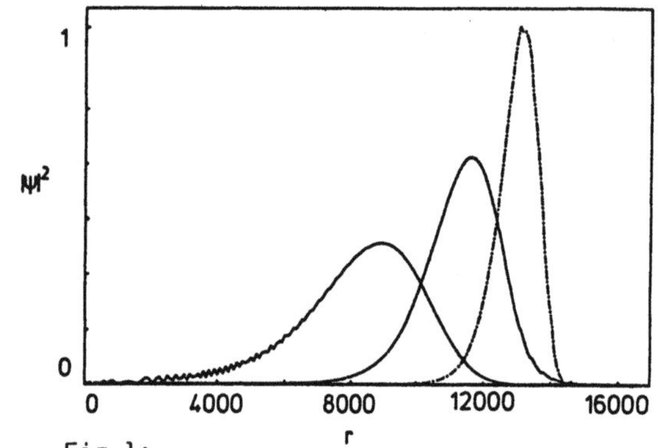

Fig.1:
Radial motion of the wave packet

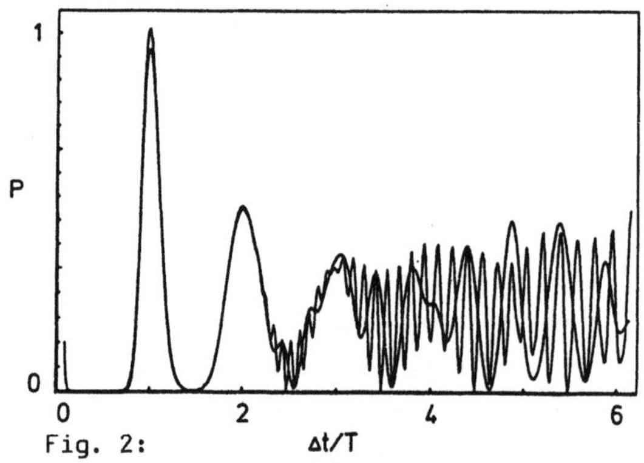

Fig. 2:
Raman transition probability as a function of the time delay between the two pulses in units of the classical orbit time (dashed line = semiclassical approximation)

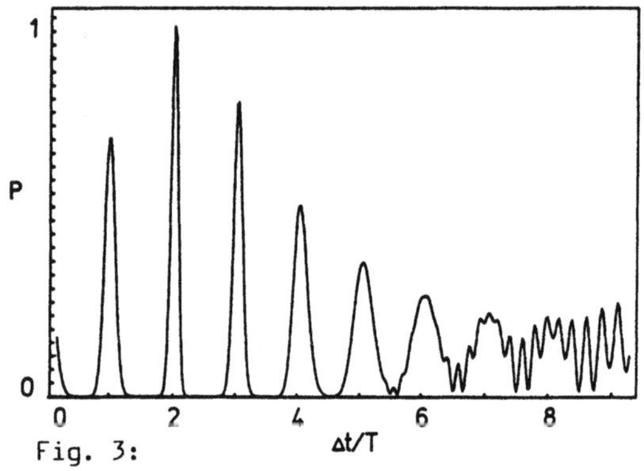

Fig. 3:
Same as Fig. 2 but with a change of the laser frequency during the pulse

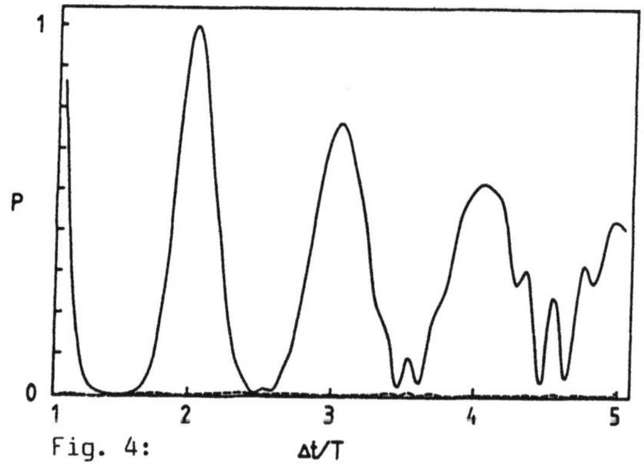

Fig. 4:
Constructive (solid line) and destructive interference (dashed line) between wave packets in a Raman experiment.

interference between two wave packets: the first one is excited at t_a, the second one by a short pulse at time $t_{a'} = t_a + T_{E_{\bar{n}}}$ ($\bar{n} = 85$). It is possible to probe this interference by a third time-delayed pulse at $t_b = t_{a'} + T_{E_{\bar{n}}}$; depending on the relative phase between the first and second pulse the two wave packets interfere constructively (solid line) or destructively (dashed line).

3. Strong Field Excitation of an Autoionizing Rydberg Series Near Threshold

We consider excitation of an autoionizing Rydberg [9] series in a configuration where a first laser excites electron from the groundstate $|g\rangle$ with energy E_g to one of the excited states $|e\rangle$ (energy E_e) by N-photon absorption which is coupled by a stronglaser of frequency ω_2 to an autoionizing Rydberg series. For weak excitation by the first the ionization is

$$R = \frac{1}{2}\,\Omega_{eg}^{2} \quad \mathrm{Im}\left\{\frac{-1}{E_g+N\omega_1-E_e-\Sigma}\right\} \quad \text{with} \tag{11}$$

$$\Sigma = \langle e|D\;\frac{1}{E_g+N\omega_1+\omega_2-H_A+i\epsilon}\;D|e\rangle \qquad \epsilon\to 0 \tag{12}$$

the resonant part of the complex selfenergy of $|e\rangle$ H_A is the atomic Hamiltonian, D the dipole operator describing interaction with the strong field. For $E_I < E_e+\omega_2$, i.e. in the unstructured continuum (Fig. 5), Σ is a slowly varying function energy; in the autoionizing region $0 < E_1+{}_2 < E_I$. Σ contains the infinite series of Rydberg resonances. To extract this rapid energy dependence we relate the "above" and "below" threshold self-energy. We obtain

$$\Sigma = \begin{cases} \delta\omega_{bg} - i\frac{1}{2}\gamma\;\dfrac{1+q^2r^2}{1+r^2} & \text{(flat continuum)} \\[2ex] \delta\omega_{bg} - i\frac{1}{2}\gamma\;\left(1+\dfrac{(q-i)^2}{x+i}\right) & \text{(autoionization)}\,. \end{cases} \tag{13}$$

Here $\delta\omega_{bg}$ is a back ground shift in the flat continuum above the second threshold; γ is the ionization width state $|e\rangle$ to the first continuum. The detuning variable is defined as

$$x = \tan\pi\,(\nu_2 + \alpha)/r^2 \tag{14}$$

with the effective quantum number ν_2 given by

$$E_g + N\,\hbar\omega_1 + \hbar\omega_2 = E_I - Ry/\nu_2^2\;. \tag{15}$$

α is the unperturbed quantum defect of series 2 and r is a measure of the configuration interaction between the two channels; the Fano parameter q is a measure of the interference between direct ionization of $|e\rangle$ to the first continuum and excitation to the autoionizing resonance followed by decay to the continuum. It is of central importance that the parameters γ, q, α and $\delta\omega_{bg}$ are essentially constant along the Rydberg series and across the threshold.

We emphasize the following features:

(i) Σ and, therefore, R contain n-mixing by the second laser of the infinite number of Rydberg states in terms of a small set of parameters which vary slowly with energy.

(ii) The two-level result (for a single autoionizing resonance) is obtained by expanding the detuning variable x around the Rydberg state n,

$$x \to (E_g + N\,\hbar\omega_1 + \hbar\omega_2 - E_I + Ry/\nu_2^{(n)2}) \quad / \,\frac{1}{2}\,\Gamma \tag{16}$$

with $\Gamma = 4r^2/\nu_2^{(n)3}$ the autoionization width. This result is valid when Γ and the Rabi frequency for $|e\rangle \rightarrow |n\rangle$ (which scales proportional to $\sim\nu_2^{(n)-3/2}$) are both much smaller than the level separation of the Rydberg series. It is remarkable that the generalization from two-level theory to infinite Rydberg series can be obtained by the simple substitution rule (16).

(iii) Σ averaged over the resonance is continuous across threshold.

In Figs. 6a and b the ionization R is plotted as a function of the detuning of the first (weak) laser. At sufficiently low intensities and for well isolated autoionizing resonances the two-level approximation for excitation from $|e\rangle$ to the autoionizing state is valid. In this case a doublet splitting (AC-Starksplitting) of the ionization signal as a function of detuning of the first laser is observed

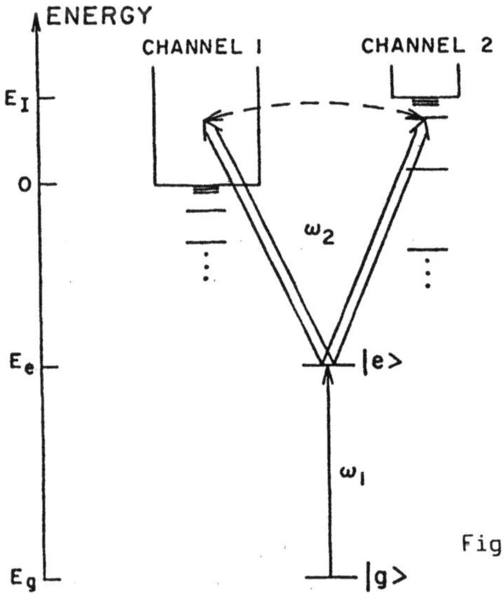

Fig. 5: Atomic configuration

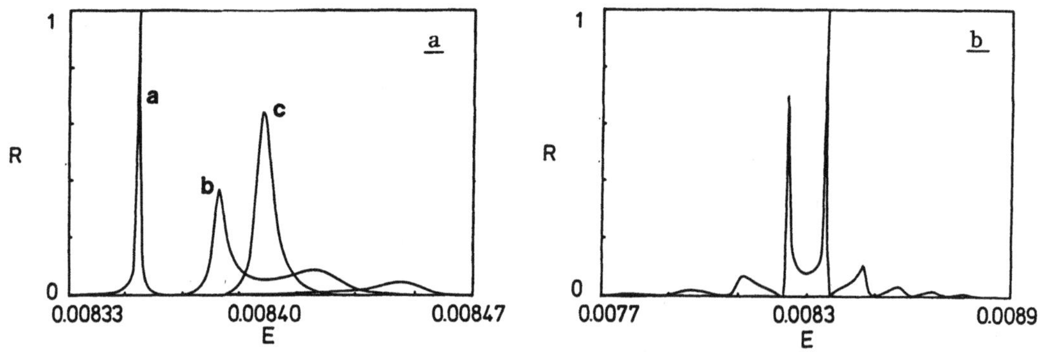

Fig. 6: Ionization rate R as a function of the detuning of the first (weak) laser (AC-Starksplitting of autoionizing resonances). E denotes the final state energy in Rydbergs. (a) The two-level approximation is valid. (b) AC-Starksplitting of overlapping autoionizing resonances.

(Fig. 6a). The asymmetry of the spectrum is a consequence of the interference between direct ionization to the continuum and the indirect (resonant) transition via the autoionizing resonance [9]. In Fig. 6b the ionization rate is plotted at high intensities and for overlapping autoionizing resonances. Laser-induced mixing of many n-states of the Rydberg series is observed.

REFERENCES

1 J.A.C. Gallas, G. Leuchs, H. Walther and H. Figger: Adv. At. Molec. Phys. 20, 413 (1985)

2 See the various contributions in "Rydberg states of atoms and molecules", ed. by R.F. Stebbings and F.B. Dunning (Cambridge University Press, 1983)

3 G. Alber, H. Ritsch and P. Zoller: submitted for publication

4 For discussion of spontaneous emission of Rydberg atoms excited by short pulses see D. Parker and C.R. Stroud, Jr., preprint

5 P. Zoller: to be published; Proceedings of the 6th Summer School on Coherent Quantum Optics; Ustron, Poland (18-26 Sep. 1985), to be published; see also A. Giusti-Suzor and H. Lefebvre-Brion: Phys. Rev. A 30, 3057 (1984)

6 S.P. Goreslavskii, N.B. Delone and V.P. Krainov: Sov. Phys. JETP 55, 1032 (1983); I. Ya. Bersons, Sov. Phys. JETP 56, 731 (1982)

7 M.J. Seaton: Rep. Prog. Phys. 46, 167 (1983)

8 C.H. Greene, U. Fano and G. Strinati: Phys. Rev. A 19, 1485 (1979); G.H. Greene, A.R.P. Raund, U. Fano, Phys. Rev. A 26, 2441 (1982)

9 For a review see P. Lambropoulos and P. Zoller in "Multiphoton Ionization of Atoms", ed. S.L. Chin and P. Lambropoulos (Academic Press, 1984)

Squeezing in the Degenerate Parametric Oscillator Using the Positive P-Representation

H.J. Carmichael[1] *and M. Wolinsky*[2]

[1]Department of Physics, University of Arkansas, Fayetteville, AR 72701, USA

[2]Department of Physics, University of Texas at Austin, Austin, TX 78712, USA

1. Introduction

The theoretical description of quantum-dynamical processes which generate squeezed states meets with some interesting questions. Squeezed light has quantum fluctuations in one field quadrature reduced below the quantum limit for a coherent state [1]. If the theoretical description of a process which generates squeezed light is cast in terms of a Fokker-Planck equation, or equivalent stochastic differential equations, negative diffusion is needed to accomplish this reduction. Phase-space representations, such as the Glauber-Sudarshan P-representation [2,3], have been widely used in quantum optics to convert operator equations describing the interaction of optical fields into Fokker-Planck equations [4-6]. Traditional approaches based on these representations fail in the treatment of squeezing precisely because they give Fokker-Planck equations with non-positive-definite diffusion.

Non-positive-definite diffusion is a clear signature of dynamics generating a manifestly quantum-mechanical state; it also appears in the treatment of systems which produce antibunched light--for example, in the theory of optical bistability [7-9]. It is interesting and important to ask how the theoretical description of such nonclassical dynamics is to proceed. There is, of course, no single answer to this question; more than one method can generally be used to solve a given problem. It is sometimes possible to solve Heisenberg equations for field operators directly, whereby non-positive-definite diffusion is not an issue at all. Squeezing in the parametric amplifier without pump depletion provides a case in point [10-15]. Here the Heisenberg equations are linear; their solution is straightforward. The Heisenberg approach is not so useful if the operator equations are nonlinear, however (the parametric amplifier <u>with</u> pump depletion, for example). The addition of dissipation brings further complication, introducing coupling between field and reservoir operators. The Fokker-Planck equation is derived within the Schrödinger picture as the phase-space-equivalent of an operator master equation [4-6]. This approach accommodates dissipation rather readily and has been used extensively for the treatment of quantum dissipative systems. It is appealing for the simple conceptual formulation it gives to a quantum-mechanical problem, setting it in correspondence with a classical stochastic process. The Fokker-Planck approach can be particularly helpful in dealing with the difficulty of nonlinearity. Various exact (for special cases) and approximate methods for solving the Fokker-Planck equation are known [16]. With access to modern computers there is always the option of numerically simulating the corresponding stochastic differential equations, even for rather complicated multidimensional systems [17]. But, as we have seen, to use this approach to describe squeezing we must first resolve the issue of non-positive-definite diffusion.

Drummond and Gardiner have provided one resolution in the form of a new representation which they call the positive P-representation [18]. This is a generalization of the Glauber-Sudarshan representation and leads to a Fokker-Planck equation in very similar form to that obtained by the traditional methods. The only formal difference is that phase-space variables

α^* and α corresponding to field creation and annihilation operators $a^\dagger$ and a now become independent complex variables α_* and α. These variables are only required to be complex conjugate in the mean. The corresponding quantum-stochastic description is now set within twice as many dimensions. It is then possible to guarantee that the Fokker-Planck equation will have positive-definite diffusion in the higher-dimensional space.

In this paper we present a theory of squeezing in the parametric oscillator using the positive P-representation. Our work is intended to illustrate the usefulness of this representation in the treatment of nonclassical dynamical effects in quantum optics, and to point to some issues which remain to be resolved surrounding its use in nonlinear situations. Our results also highlight the possible use of the degenerate parametric oscillator for the generation of squeezed light. Milburn and Walls [19] dampened the initial enthusiasm for this system as a source of squeezing with their calculation showing a maximum reduction of fluctuations by a factor of two at the threshold for parametric oscillation, and a decreasing effect above threshold. Recently, we have shown that the reduced squeezing above threshold is due to pump depletion, and suggest that during an initial transient regime squeezing should continue to increase with pump intensity above threshold [20]. Our calculation is based on a model which neglects pump depletion. In this paper we solve the transient nonlinear problem including pump depletion by numerical simulation of stochastic differential equations derived using the positive P-representation. We find that our suggestion of increased transient squeezing above threshold is correct. Above threshold the initial vacuum state for the subharmonic field is unstable, and decays to one of two stable states with depleted pump and nonzero amplitude for the subharmonic. Our previous calculation holds during the interval before the subharmonic turns on, while we await the decay of the unstable state. The duration of the transient squeezing is sensitive to the threshold photon number in the pump mode, the presence of any weak initial subharmonic field, and the distance the system is taken above threshold.

In section 2 we introduce our model for the degenerate parametric oscillator and its formulation as a stochastic process in eight dimensions using the positive P-representation. Steady-state solutions to the deterministic equations obtained by neglecting the quantum noise terms are obtained in section 3. In section 4 these provide the basis for a linearized treatment of quantum fluctuations. Our numerical results for the full nonlinear problem are presented in section 5. Section 6 provides a brief summary and conclusions.

2. Quantum-Stochastic Model in the Positive P-Representation

We consider two quantized field modes, a and b, with frequencies ω and 2ω, respectively. Both modes are resonant in an optical cavity where they interact via an intracavity crystal with second-order susceptibility $\chi^{(2)}$. At cavity input and output mirrors they see the same reflection and transmission coefficients, R and T, respectively, and a phase change ϕ_T on transmission. Mode b is excited by an injected classical field with complex amplitude E_i. We introduce reservoir interactions to describe the losses at the cavity mirrors. Then the complete Hamiltonian for modes a and b plus reservoir interactions is

$$H = H_S + H_R^A + H_R^B + (a\Gamma_A^\dagger + a^\dagger\Gamma_A) + (b\Gamma_B^\dagger + b^\dagger\Gamma_B) \ , \qquad (2.1)$$

where H_R^A and H_R^B are free Hamiltonians for reservoirs coupling to modes a and b, respectively, $\Gamma_A^\dagger$, Γ_A and $\Gamma_B^\dagger$, Γ_B are reservoir operators which accomplish this coupling, and the system Hamiltonian H_S is given by

$$H_S = \hbar\omega a^\dagger a + \hbar 2\omega b^\dagger b + i\hbar\frac{1}{2}(g a^{\dagger 2} b - g^* a^2 b^\dagger) + i\hbar(\tilde{E}_i e^{-i2\omega t} b - \tilde{E}_i^* e^{i2\omega t} b) . \quad (2.2)$$

In Eq.(2.2) g is the mode coupling constant,

$$g = -\kappa(1 - R)^{-1}[\omega L/2n(\omega)c]\chi^{(2)}(\hbar\omega/2\varepsilon_0 V_Q)^{1/2} , \quad (2.3)$$

and $\tilde{E}_i$ is the driving field amplitude,

$$\tilde{E}_i = -i\kappa(\varepsilon_0 V_Q/\hbar\omega)^{1/2}(1 - R)^{-1}\sqrt{T}\exp(i\phi_T)E_i . \quad (2.4)$$

We have assumed perfect phase matching, and $\kappa = (1 - R)/\tau'$ is the cavity linewidth, where τ' is the cavity round-trip time, V_Q is the mode quantization volume, L is the length of the intracavity crystal and $n(\omega)$ is its refractive index at the frequency of the subharmonic.

The master equation for the reduced density operator ρ describing the coupled cavity modes is derived in the Schrödinger picture by tracing over the reservoir variables [4-6]. If we neglect thermal fluctuations the master equation reads:

$$\dot{\rho} = (i\hbar)^{-1}[H_S,\rho] + \kappa(2a\rho a^\dagger - a^\dagger a\rho - \rho a^\dagger a) + \kappa(2b\rho b^\dagger - b^\dagger b\rho - \rho b^\dagger b) . \quad (2.5)$$

Using the positive P-representation we transform this into an equation of motion for a phase-space distribution function P. The positive P-representation establishes a correspondence between normally-ordered averages of the operators a, $a^\dagger$, b, and $b^\dagger$, and phase-space averages of the independent complex variables α, α_*, β, and β_*, respectively, via the characteristic function [6,8,18]

$$\chi_N(\mu,\mu_*,\nu,\nu_*) = \mathrm{tr}(\rho e^{i\mu_* a^\dagger} e^{i\mu a} e^{i\nu_* b^\dagger} e^{i\nu b}) . \quad (2.6)$$

A distribution function $P(\alpha,\alpha_*,\beta,\beta_*)$ is assumed to exist at all times with

$$\chi_N(\mu,\mu_*,\nu,\nu_*) = \int d^2\alpha \int d^2\alpha_* \int d^2\beta \int d^2\beta_*\, P(\alpha,\alpha_*,\beta,\beta_*) e^{i\mu_*\alpha_*} e^{i\mu\alpha} e^{i\nu_*\beta_*} e^{i\nu\beta} , \quad (2.7)$$

from which it follows that

$$\langle a^{\dagger n} a^m b^{\dagger p} b^q\rangle = \int d^2\alpha \int d^2\alpha_* \int d^2\beta \int d^2\beta_*\, \alpha_*^n \alpha^m \beta_*^p \beta^q P = (\alpha_*^n \alpha^m \beta_*^p \beta^q)_{av} . \quad (2.8)$$

Here $(\ )_{av}$ distinguishes the classical phase-space, or stochastic, average from the quantum average $\langle\ \rangle$. The Glauber-Sudarshan representation takes $\alpha_* = \alpha^*$, $\beta_* = \beta^*$ and $\mu_* = \mu^*$, $\nu_* = \nu^*$ so that the phase-space integrals extend over a two-dimensional rather than a four-dimensional complex space. The generalization introduced by the positive P-representation makes no essential difference, however, in the passage from Eq.(2.5) to the equation of motion for P--standard methods yield [4-6]

$$\frac{\partial P}{\partial \tau} = [-\frac{\partial}{\partial\bar{\alpha}}(-\bar{\alpha} + \bar{\alpha}_*\bar{\beta}) - \frac{\partial}{\partial\bar{\alpha}_*}(-\bar{\alpha}_* + \bar{\alpha}\bar{\beta}_*) - \frac{\partial}{\partial\bar{\beta}}(-\bar{\beta} - \bar{\alpha}^2 + \lambda) - \frac{\partial}{\partial\bar{\beta}_*}(-\bar{\beta}_* - \bar{\alpha}_*^2 + \lambda) + \frac{1}{2}(2n_{thr})^{-1}(\frac{\partial^2}{\partial\bar{\alpha}^2}\bar{\beta} + \frac{\partial^2}{\partial\bar{\alpha}_*^2}\bar{\beta}_*)]P , \quad (2.9)$$

where $\tau = \kappa t$ measures time in cavity lifetimes,

$$\lambda = |g\tilde{E}_i|/\kappa^2 \tag{2.10}$$

is a pump parameter, and

$$n_{thr} = (\kappa/|g|)^2 \tag{2.11}$$

appears in the following section as the threshold photon number in the pump mode b. We have removed the secular oscillation of the fields with the introduction of the scaled variables

$$\bar{\alpha} = e^{i\omega t}\exp[-i(\phi_g + \phi_i)/2](2n_{thr})^{-1/2}\alpha \ , \tag{2.12a}$$

$$\bar{\alpha}_* = e^{-i\omega t}\exp[i(\phi_g + \phi_i)/2](2n_{thr})^{-1/2}\alpha_* \ , \tag{2.12b}$$

$$\bar{\beta} = e^{i2\omega t}\exp(-i\phi_i)n_{thr}^{-1/2}\beta \ , \tag{2.12c}$$

$$\bar{\beta}_* = e^{-i2\omega t}\exp(i\phi_i)n_{thr}^{-1/2}\beta_* \ , \tag{2.12d}$$

where ϕ_g and ϕ_i are the phases of g and $\tilde{E}_i$, respectively.

The distinguishing feature of the positive P-representation comes with the interpretation of Eq.(2.9). First consider this equation with the restriction $\bar{\alpha}_* = \bar{\alpha}$ and $\bar{\beta}_* = \bar{\beta}$, as in the Glauber-Sudarshan representation. If we set $\bar{\alpha} = \bar{\alpha}_x + i\bar{\alpha}_y$ and $\bar{\beta} = \bar{\beta}_x + i\bar{\beta}_y$, and write Eq.(2.9) in terms of the four real variables $\bar{\alpha}_x$, $\bar{\alpha}_y$, $\bar{\beta}_x$, and $\bar{\beta}_y$, the resulting Fokker-Planck equation has diffusion terms

$$(2n_{thr})^{-1}\left(\frac{\partial^2}{\partial\bar{\alpha}^2}\bar{\beta} + \text{c.c.}\right) = (2n_{thr})^{-1}\,\frac{1}{2}\left(\frac{\partial^2}{\partial\bar{\alpha}_x^2}\bar{\beta}_x - \frac{\partial^2}{\partial\bar{\alpha}_y^2}\bar{\beta}_x + 2\frac{\partial^2}{\partial\bar{\alpha}_x\partial\bar{\alpha}_y}\bar{\beta}_y\right) . \tag{2.13}$$

This represents a non-positive-definite diffusion. In the positive P-representation $\bar{\alpha}$ and $\bar{\alpha}_*$ are independent complex variables. In the derivation of Eq.(2.9) the analyticity of the product of exponentials in Eq.(2.7) allows $\partial/\partial\bar{\alpha}$ to be interpreted as either $\partial/\partial\bar{\alpha}^x$ or $-i\partial/\partial\bar{\alpha}^y$, and $\partial/\partial\bar{\alpha}_*$ as either $\partial/\partial\bar{\alpha}_*^x$ or $-i\partial/\partial\bar{\alpha}_*^y$, where we write $\bar{\alpha} = \bar{\alpha}^x + i\bar{\alpha}^y$ and $\bar{\alpha}_* = \bar{\alpha}_*^x + i\bar{\alpha}_*^y$. The choice can be made to obtain a Fokker-Planck equation in eight dimensions with positive-definite diffusion. We refer the reader to Refs. [6] and [21] for the formal details in the general case. For our present application we take the following action: Write $\bar{\beta} = (p + iq)^2$ and $\bar{\beta}_* = (u + iv)^2$, where p, q, u, and v are real variables. Then we interpret the diffusion terms in Eq.(2.9) with

$$\begin{aligned}(2n_{thr})^{-1}&\left(\frac{\partial^2}{\partial\bar{\alpha}^2}\bar{\beta} + \frac{\partial^2}{\partial\bar{\alpha}_*^2}\bar{\beta}_*\right)\\ &= (2n_{thr})^{-1}\left[\frac{\partial^2}{\partial\bar{\alpha}^2}(p^2 - q^2 + 2ipq) + \frac{\partial^2}{\partial\bar{\alpha}_*^2}(u^2 - v^2 + 2iuv)\right]\\ &= (2n_{th})^{-1}\left[\frac{\partial^2}{\partial\bar{\alpha}^{x2}}\,p^2 + \frac{\partial^2}{\partial\bar{\alpha}^{y2}}\,q^2 + 2\frac{\partial^2}{\partial\bar{\alpha}^x\partial\bar{\alpha}^y}\,pq\right.\\ &\qquad\left. + \frac{\partial^2}{\partial\bar{\alpha}_*^{x2}}\,u^2 + \frac{\partial^2}{\partial\bar{\alpha}_*^{y2}}\,v^2 + 2\frac{\partial^2}{\partial\bar{\alpha}_*^x\partial\bar{\alpha}_*^y}\,uv\right] ,\end{aligned} \tag{2.14}$$

where for each of the six terms in the square bracket we chose the appropriate interpretation for $\partial/\partial\bar{\alpha}$ or $\partial/\partial\bar{\alpha}_*$ to give the desired result. The right-hand side of Eq.(2.14) represents a positive-definite diffusion. The drift in Eq.(2.9) is interpreted with

$$- \frac{\partial}{\partial\bar{\alpha}}(-\bar{\alpha} + \bar{\alpha}_*\bar{\beta}) = - \frac{\partial}{\partial\bar{\alpha}^x} \mathrm{Re}(-\bar{\alpha} + \bar{\alpha}_*\bar{\beta})$$

$$- \frac{\partial}{\partial\bar{\alpha}^y} \mathrm{Im}(-\bar{\alpha} + \bar{\alpha}_*\bar{\beta}) ,$$

and similarly for the remaining three terms, to complete the construction of a Fokker-Planck equation in the eight-dimensional real space $(\bar{\alpha}^x,\bar{\alpha}_*^x,\bar{\beta}^x,\bar{\beta}_*^x,\bar{\alpha}^y,\bar{\alpha}_*^y,\bar{\beta}^y,\bar{\beta}_*^y)$. This seems rather complicated, but it is not so if we retain our complex notation. The resulting Fokker-Planck equation is equivalent to the Ito stochastic differential equations

$$d\bar{\alpha} = (-\bar{\alpha} + \bar{\alpha}_*\bar{\beta})d\tau + (2n_{thr})^{-1/2}\bar{\beta}^{1/2}dW_1 , \tag{2.15a}$$

$$d\bar{\alpha}_* = (-\bar{\alpha}_* + \bar{\alpha}\bar{\beta}_*)d\tau + (2n_{thr})^{-1/2}\bar{\beta}_*^{1/2}dW_2 , \tag{2.15b}$$

$$d\bar{\beta} = (-\bar{\beta} - \bar{\alpha}^2 + \lambda)d\tau , \tag{2.15c}$$

$$d\bar{\beta}_* = (-\bar{\beta}_* - \bar{\alpha}_*^2 + \lambda)d\tau , \tag{2.15d}$$

where dW_1 and dW_2 are independent Wiener processes.

Equations (2.15) constitute our quantum-stochastic model for the degenerate parametric oscillator, where quantum averages are to be calculated using Eqs.(2.8) and (2.12). This is the model adopted by Milburn and Walls [19], specialized to equal cavity decay rates for the two fields.

3. Steady States and their Stability

Without their noise terms Eqs.(2.15) reduce to a set of nonlinear ordinary differential equations:

$$\dot{\bar{\alpha}} = -\bar{\alpha} + \bar{\alpha}_*\bar{\beta} , \tag{3.1a}$$

$$\dot{\bar{\alpha}}_* = -\bar{\alpha}_* + \bar{\alpha}\bar{\beta}_* , \tag{3.1b}$$

$$\dot{\bar{\beta}} = -\bar{\beta} - \bar{\alpha}^2 + \lambda, \tag{3.1c}$$

$$\dot{\bar{\beta}}_* = -\bar{\beta}_* - \bar{\alpha}_*^2 + \lambda , \tag{3.1d}$$

where the dot denotes differentiation with respect to τ. These are a complexified version of the usual classical equations for the degenerate parametric oscillator. The latter are obtained with $\bar{\alpha}_* = \bar{\alpha}^*$, $\bar{\beta}_* = \bar{\beta}^*$, and then $\bar{\alpha}$ and $\bar{\beta}$ are dimensionless complex field amplitudes. The complexification introduced by the positive P-representation sets the deterministic equations underlying our quantum-stochastic model in a higher-dimensional space than the usual classical equations. Initial conditions with $\bar{\alpha}_* = \bar{\alpha}^*$, $\bar{\beta}_* = \bar{\beta}^*$ remain within the subspace, preserving this symmetry. However, the quantum noise can take trajectories outside this subspace, and then the dynamics in the full eight dimensions are important. These are structured around the steady-state solutions to Eqs.(3.1) and their stability.

Equations (3.1) have the obvious steady-state solution

$$\bar{\alpha}^s = \bar{\alpha}_*^s = 0 , \tag{3.2a}$$

$$\overline{\beta}^s = \overline{\beta}_*^s = \lambda \; . \tag{3.2b}$$

In seeking additional solutions we must recognize that $\overline{\alpha}^s$ and $\overline{\alpha}_*^s$ need not be complex conjugate. There are then four more solutions for all values of λ:

$$\overline{\alpha}^s = \overline{\alpha}_*^s = \pm(\lambda - 1)^{1/2} \; , \tag{3.3a}$$

$$\overline{\beta}^s = \overline{\beta}_*^s = 1 \; , \tag{3.3b}$$

and

$$\overline{\alpha}^s = -\overline{\alpha}_*^s = \pm(\lambda + 1)^{1/2} \; , \tag{3.4a}$$

$$\overline{\beta}^s = \overline{\beta}_*^s = -1 \; . \tag{3.4b}$$

Under the restriction $\overline{\alpha}_*^s = \overline{\alpha}^{s*}$ only the two solutions given by Eq.(3.3) are acceptable, and these only for $\lambda \geq 1$. Thus, the positive P-representation introduces new steady states. What is their significance?

It would be a surprise, and no doubt a problem for the positive P-representation, to find that any of the states with $\overline{\alpha}_*^s \neq \overline{\alpha}^{s*}$ are stable. Indeed, they are not; the stability of the steady states preserves the picture derived from the usual classical equations; of the destabilization of the state with zero subharmonic amplitude above threshold and the establishment of a new stable state with nonzero amplitude. The eigenvalues governing the linearized dynamics are doubly degenerate [17,22], and given by

$$\Lambda_{1,2} = -1 + 1/2(\overline{\beta}_*^s\overline{\beta}^s)^{1/2} \pm 1/2(\overline{\beta}_*^s\overline{\beta}^s - 8\overline{\alpha}_*^s\overline{\alpha}^s)^{1/2} \; , \tag{3.5a}$$

$$\Lambda_{3,4} = -1 - 1/2(\overline{\beta}_*^s\overline{\beta}^s)^{1/2} \pm 1/2(\overline{\beta}_*^s\overline{\beta}^s - 8\overline{\alpha}_*^s\overline{\alpha}^s)^{1/2} \; . \tag{3.5b}$$

Substituting the solutions given by Eqs.(3.2)-(3.4) into Eqs.(3.5) we find the following: from Eqs.(3.2),

$$\Lambda_{1,2} = -1 + 1/2\lambda(1 \pm 1) \; , \tag{3.6a}$$

$$\Lambda_{3,4} = -1 - 1/2\lambda(1 \pm 1) \; , \tag{3.6b}$$

and the steady state is stable for $\lambda < 1$ and unstable for $\lambda > 1$; from Eqs.(3.3),

$$\Lambda_{1,2} = -1/2 \pm 1/2(9 - 8\lambda)^{1/2} \; , \tag{3.7a}$$

$$\Lambda_{3,4} = -\frac{3}{2} \pm \frac{1}{2}(9 - 8\lambda)^{1/2} \; , \tag{3.7b}$$

and the steady states are unstable for $\lambda < 1$ and stable for $\lambda > 1$; from Eqs.(3.4),

$$\Lambda_{1,2} = -\frac{1}{2} \pm \frac{1}{2}(9 + 8\lambda)^{1/2} \; , \tag{3.8a}$$

$$\Lambda_{3,4} = -\frac{3}{2} \pm \frac{1}{2}(9 + 8\lambda)^{1/2} \; , \tag{3.8b}$$

and the steady states are unstable for all λ. Then the change in stability of the steady state given by Eqs.(3.2) at $\lambda = 1$, accompanied by the introduction of a new stable steady state with $\overline{\alpha}_*^s = \overline{\alpha}^{s*}$ for $\lambda > 1$, corresponds to the usual picture for the onset of parametric oscillation. Steady states with $\overline{\alpha}_*^s \neq \overline{\alpha}^{s*}$ are always unstable. The threshold is defined by

$$\lambda_{thr} = 1 \; , \tag{3.9}$$

and, from Eqs.(2.12c), (2.12d), and (3.2b), the deterministic solution has $\beta_*^s\beta^s = n_{thr}\bar{\beta}_*^s\bar{\beta}^s = n_{thr}$ for the mean threshold photon number in the pump mode b.

The unstable steady states with $\bar{\alpha}_*^s \neq \bar{\alpha}^{s*}$ should not now be overlooked. They surely play a role in organizing the global dynamics of Eqs.(3.1) and are therefore potentially important to an understanding of the stochastic differential equations, Eqs.(2.15). Elsewhere, it has been suggested that they might represent a difficulty with the positive P-representation in the vicinity of bifurcation points and unstable states [17,22]. This suggestion is motivated by anomalous trajectories occurring in the numerical simulation of quantum-stochastic models for optical bistability [17] and the single-mode laser [22]. The reader should consult the references for discussion of this point. No obvious anomalies have been encountered in the results reported in section 5--although it must be stated that we have not yet looked at the threshold region. In fact the question of the significance of the additional steady states remains open and merits a thorough investigation in the future. Our present model should provide a good example in which to address this issue, since there is no truncation of derivatives required in passing from Eq.(2.5) to Eq.(2.9), as there is with models involving two-level atoms, and a rigorous basis exists for the representation of harmonic oscillator states using the positive P-representation [18], which is not so for the representation of atomic states. For the present we will accept our quantum stochastic model and proceed with the solution of Eqs.(2.15) to obtain a description of squeezing in the degenerate parametric oscillator.

4. Linearized Treatment of Fluctuations

The size of the quantum noise terms in Eqs.(2.15) is characterized by the pump threshold photon number n_{thr}. If this number is large, a linearized treatment of quantum fluctuations should be adequate. Below threshold the linearization is made about the stable steady state given in Eqs.(3.2), and above threshold about each of the stable steady states given in Eqs.(3.3). After linearization the results for quantum averages no longer depend on n_{thr}; this parameter can be removed from the equations by transforming from the variables $\bar{\alpha}, \bar{\alpha}_*$, $\bar{\beta}$, and $\bar{\beta}_*$ to the new variables

$$\tilde{\alpha} = e^{i\omega t}\exp[-i(\phi_g + \phi_i)/2]\alpha \ , \tag{4.1a}$$

$$\tilde{\alpha}_* = e^{-i\omega t}\exp[i(\phi_g + \phi_i)/2]\alpha_* \ , \tag{4.1b}$$

$$\tilde{\beta} = e^{i2\omega t}\exp(-i\phi_i)\beta \ , \tag{4.1c}$$

$$\tilde{\beta}_* = e^{-i2\omega t}\exp(i\phi_i)\beta_* \ . \tag{4.1d}$$

This is not so for the nonlinear theory, as we will see in the next section. In the following we will calculate variances of the quadrature operators

$$X_1 = 1/2(\tilde{a} + \tilde{a}^\dagger) \ , \tag{4.2a}$$

$$X_2 = -i\ 1/2(\tilde{a} - \tilde{a}^\dagger) \ , \tag{4.2b}$$

where $\tilde{a}$ and $\tilde{a}^\dagger$ are related to a and $a^\dagger$ as $\tilde{\alpha}$ and $\tilde{\alpha}_*$ are related to α and α_*. From Eqs.(4.1) and Eq.(2.8)

$$\langle\Delta X_1^2\rangle = \frac{1}{4} + \langle:\Delta X_1^2:\rangle$$

$$= \frac{1}{4}[1 + (\Delta\tilde{\alpha}^2)_{av} + (\Delta\tilde{\alpha}_*^2)_{av} + 2(\Delta\tilde{\alpha}_*\Delta\tilde{\alpha})_{av}] \ , \tag{4.3a}$$

$$\langle \Delta X_2^2 \rangle = \frac{1}{4} + \langle :\Delta X_2^2: \rangle$$

$$= \frac{1}{4}[\, 1 - (\Delta\tilde{\alpha}^2)_{av} - (\Delta\tilde{\alpha}_*^2)_{av} + 2(\Delta\tilde{\alpha}_*\Delta\tilde{\alpha})_{av}] \,, \qquad (4.3b)$$

where :: denotes normal ordering and $\Delta\tilde{\alpha} = \tilde{\alpha} - \tilde{\alpha}^s$, $\Delta\tilde{\alpha}_* = \tilde{\alpha}_* - \tilde{\alpha}_*^s$.

Linearizing Eqs.(2.15) below threshold we find that the equations for the subharmonic decouple from those for the pump, with

$$d(\Delta\tilde{\alpha}) = (-\Delta\tilde{\alpha} + \lambda\Delta\tilde{\alpha}_*)d\tau + \sqrt{\lambda}dW_1 \,, \qquad (4.4a)$$

$$d(\Delta\tilde{\alpha}_*) = (-\Delta\tilde{\alpha}_* + \lambda\Delta\tilde{\alpha})d\tau + \sqrt{\lambda}dW_2 \,. \qquad (4.4b)$$

This complex Ornstein-Uhlenbeck process corresponds to the Fokker-Planck equation for the degenerate parametric oscillator without pump depletion solved by us in Ref. [20]; here λ plays the role of the parameter ε/κ in our earlier work. The steady-state quadrature variances below threshold are ($\lambda < 1$)

$$\langle \Delta X_1^2 \rangle = \frac{1}{4}(1 - \lambda)^{-1} \,, \qquad (4.5a)$$

$$\langle \Delta X_2^2 \rangle = \frac{1}{4}(1 + \lambda)^{-1} \,. \qquad (4.5b)$$

These are plotted in Fig. 1. Squeezing increases with λ to the maximum $\langle \Delta X_2^2 \rangle = 1/8$ at threshold, as found by Milburn and Walls [19]. We have also solved for the time-dependent quadrature variances above threshold in Ref. [20]. This corresponds to a linearized analysis about the unstable steady state. Equation (4.5b) still holds in the long-time limit while fluctuations in the amplified X_1 quadrature diverge. The prediction of Eq.(4.5b) is extended above threshold as the dashed curve in Fig. 1(b). We might expect this result to hold during an initial interval for a system beginning in the unstable steady state; then, eventually the pump will be depleted and the system will approach one of the stable steady states defined by Eqs.(3.3). This was our earlier suggestion [20]. We look at the decay from the unstable state in the next section. The level of squeezing expected in the final stable steady state can again be calculated in a linearized treatment.

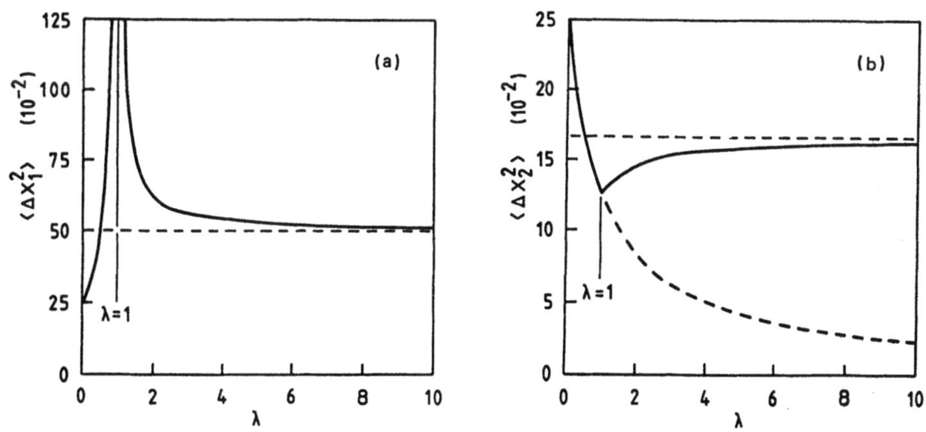

Fig. 1. Quadrature variances for the subharmonic mode from the linearized treatment of quantum fluctuations. Solid curves are plotted from: (a) Eqs.(4.5a) and (4.12a); (b) Eqs.(4.5b) and (4.12b). The dashed curve is for the unstable state and is plotted from Eq.(4.5b)

Above threshold the linearized version of Eqs.(2.15) couples the subharmonic and pump modes; we find the equations

$$d(\Delta\tilde{\alpha}) = [-\Delta\tilde{\alpha} + \Delta\tilde{\alpha}_* \pm \sqrt{2}(\lambda - 1)^{1/2}\Delta\tilde{\beta}]d\tau + dW_1 \ , \tag{4.6a}$$

$$d(\Delta\tilde{\alpha}_*) = [-\Delta\tilde{\alpha}_* + \Delta\tilde{\alpha} \pm \sqrt{2}(\lambda - 1)^{1/2}\Delta\tilde{\beta}_*]d\tau + dW_2 \ , \tag{4.6b}$$

$$d(\Delta\tilde{\beta}) = [-\Delta\tilde{\beta} \pm \sqrt{2}(\lambda - 1)^{1/2}\Delta\tilde{\alpha}]d\tau \ , \tag{4.6c}$$

$$d(\Delta\tilde{\beta}_*) = [-\Delta\tilde{\beta}_* \pm \sqrt{2}(\lambda - 1)^{1/2}\Delta\tilde{\alpha}_*]d\tau \ , \tag{4.6d}$$

where the upper and lower signs correspond to the upper and lower signs in Eq.(3.3a). The steady-state quadrature variances are calculated from elements of the covariance matrix

$$G = (\Delta\tilde{\alpha},\Delta\tilde{\alpha}_*,\Delta\tilde{\beta},\Delta\tilde{\beta}_*)^T(\Delta\tilde{\alpha},\Delta\tilde{\alpha}_*,\Delta\tilde{\beta},\Delta\tilde{\beta}_*) \ , \tag{4.7}$$

where T denotes the transpose. This satisfies the equation [6,23]

$$AG + GA^T = -D \ , \tag{4.8}$$

where A and D are drift and diffusion matrices for the stochastic process defined by Eqs.(4.6). (These are 4x4 matrices, although the stochastic differential equations are defined in eight real dimensions. See Ref.[6] for details.) Equation (4.8) is readily solved if we first transform to variables $1/2(\Delta\tilde{\alpha} + \Delta\tilde{\alpha}_*)$ and $-i\ 1/2(\Delta\tilde{\alpha} - \Delta\tilde{\alpha}_*)$ corresponding to the quadrature operators defined by Eqs.(4.2), and $1/2(\Delta\tilde{\alpha} + \Delta\tilde{\alpha}_*)$ and $-i\ 1/2(\Delta\tilde{\beta} - \Delta\tilde{\beta}_*)$ corresponding to the quadrature operators

$$Y_1 = \tfrac{1}{2}(\tilde{b} + \tilde{b}^\dagger) \ , \tag{4.9a}$$

$$Y_2 = -i\tfrac{1}{2}(\tilde{b} - \tilde{b}^\dagger) \ . \tag{4.9b}$$

Here $\tilde{b}$ and $\tilde{b}^\dagger$ are related to b and $b^\dagger$ as $\tilde{\beta}$ and $\tilde{\beta}_*$ are related to β and β_*. We find the following steady-state correlations:

$$\langle:\Delta X_1^2:\rangle = (1/8)(\lambda - 1)^{-1}(2\lambda - 1) \ , \tag{4.10a}$$

$$\langle:\Delta X_2^2:\rangle = -(1/8)(3\lambda)^{-1}(2\lambda + 1) \ , \tag{4.10b}$$

$$\langle:\Delta Y_1^2:\rangle = 1/2 \ , \tag{4.10c}$$

$$\langle:\Delta Y_2^2:\rangle = -(1/2)(3\lambda)^{-1}(\lambda - 1) \ , \tag{4.10d}$$

and

$$\langle:\Delta X_1\Delta Y_1:\rangle = \mp(1/4)(\lambda - 1)^{-1/2} \ , \tag{4.11a}$$

$$\langle:\Delta X_2\Delta Y_2:\rangle = \pm(1/4)(3\lambda)^{-1}(\lambda - 1)^{1/2} \ , \tag{4.11b}$$

$$\langle:\Delta X_1\Delta X_2:\rangle = \langle:\Delta Y_1\Delta Y_2:\rangle = 0 \ , \tag{4.11c}$$

$$\langle:\Delta X_1\Delta Y_2:\rangle = \langle:\Delta X_2\Delta Y_1:\rangle = 0 \ . \tag{4.11d}$$

For the subharmonic we then have ($\lambda > 1$)

$$\langle\Delta X_1^2\rangle = (1/8)(\lambda - 1)^{-1}(4\lambda - 3) \ , \tag{4.12a}$$

$$\langle\Delta X_2^2\rangle = (1/8)(3\lambda)^{-1}(4\lambda - 1) \ . \tag{4.12b}$$

These give the solid curves plotted above threshold in Fig. 1. Note that squeezing in the stable steady state is much reduced from that in the unstable state in correspondence with Fig. 1 of Milburn and Walls [19]. Limited squeezing remains, however, even far above threshold. We also find that the pump mode is squeezed above threshold. We have ($\lambda > 1$)

$$\langle \Delta Y_1^2 \rangle = 3/4 \,, \tag{4.13a}$$

$$\langle \Delta Y_2^2 \rangle = (1/4)(3\lambda)^{-1}(\lambda + 2) \,. \tag{4.13b}$$

The pump is actually squeezed more than the subharmonic for $\lambda > 2.5$.

It remains to see whether the squeezing obtained along the dashed curve in Fig. 1(b) is really attainable before the unstable state decays. For this we turn to a numerical solution of the nonlinear stochastic differential equations.

5. Numerical Simulation of the Nonlinear Equations

We have solved Eqs.(2.15) numerically using a simple Euler algorithm. Averages have been computed from ensembles of 25,000 trajectories. Our results are presented in Figs. 2-4. Each figure follows the evolution of one of the subharmonic quadrature variances as a function of the dimensionless time τ from an initial state with zero subharmonic amplitude. In all figures, with the exception of Fig. 4(b), we have taken a threshold pump photon number $n_{thr} = 10^{10}$ (Fig. 4(b) has $n_{thr} = 10^6$). This photon number has been calculated from Eqs.(2.3) and (2.11) for KDP with $L \sim 1$ cm in a cavity with $T = (1 - R) \sim 10^{-2}$ and $V_Q \sim 10^{-8}\ m^3$.

Figure 2 provides a single result for comparison with the linearized treatment below threshold. Here the pump mode begins with $\overline{\beta}(0) = \overline{\beta}_*(0) = 0$ and so its amplitude grows to $\overline{\beta} = \overline{\beta}_* = \lambda = 0.5$ over the first two or three cavity lifetimes. The subharmonic amplitude simply shows small fluctuations about the $\overline{\alpha}^s = \overline{\alpha}_*^s = 0$ steady state. The steady-state quadrature variances are in good agreement with Eqs.(4.5).

We are more interested in results above threshold. Figure 3 shows that the level of squeezing predicted by the dashed curve in Fig. 1 is achieved

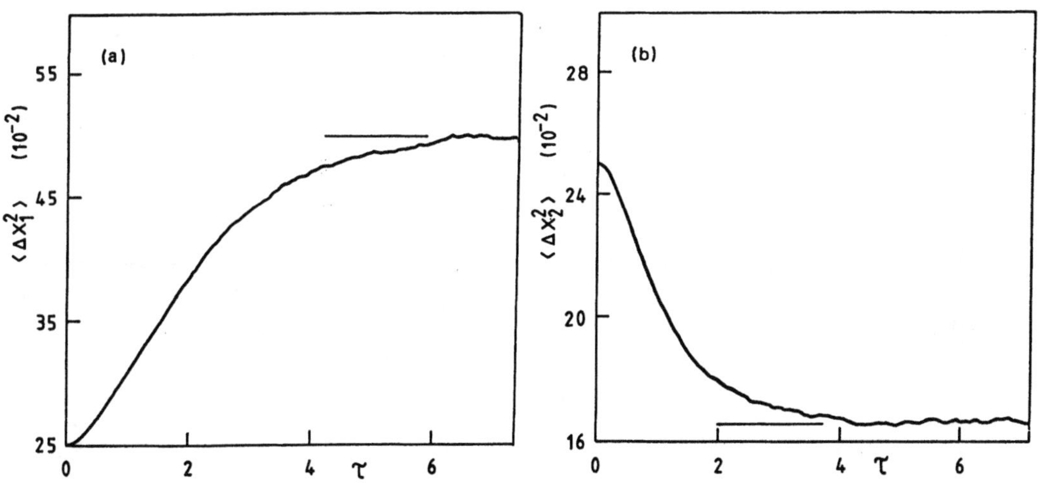

Fig. 2. Quadrature variances as a function of τ from the numerical simulation of Eqs.(2.15): $\lambda = 0.5$, $n_{thr} = 10^{10}$, $\overline{\alpha}(0) = \overline{\alpha}_*(0) = 0$, $\overline{\beta}(0) = \overline{\beta}_*(0) = 0$.

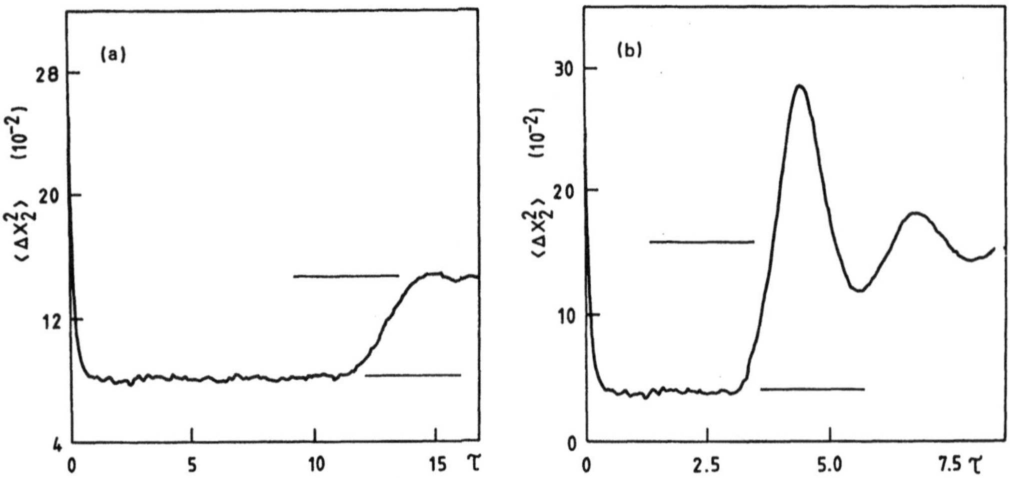

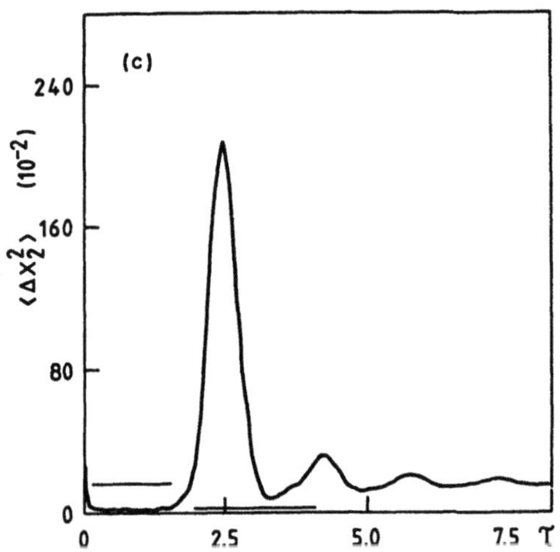

Fig. 3. Transient squeezing above threshold from numerical simulation of Eqs.(2.15): $n_{thr} = 10^{10}$, $\bar{\alpha}(0) = \bar{\alpha}_{*}(0) = 0$, $\bar{\beta}(0) = \bar{\beta}_{*}(0) = 0$; and (a) $\lambda = 2$, (b) $\lambda = 5$, (c) $\lambda = 10$. The straight lines indicate the predictions from linearized theory for squeezing in the unstable and stable steady states

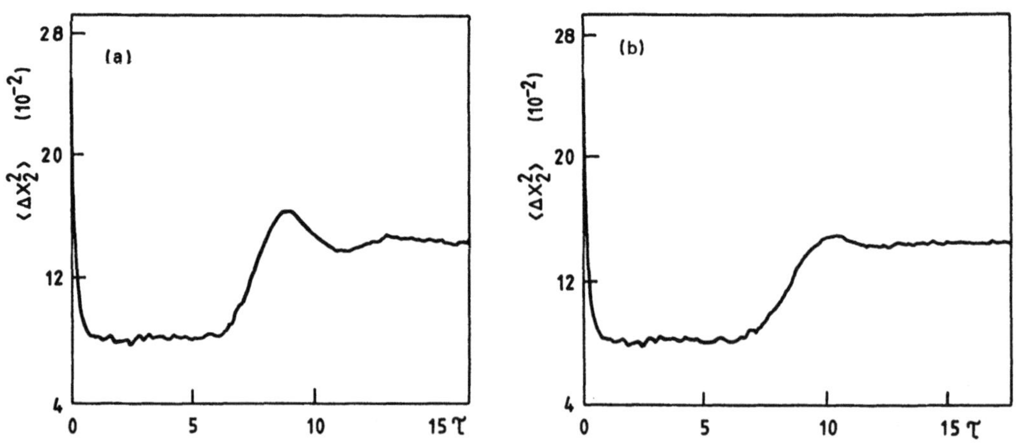

Fig. 4. Transient squeezing above threshold from numerical simulation of Eqs.(2.15): $\lambda = 2$, $\bar{\beta}(0) = \bar{\beta}_{*}(0) = 2$, and (a) $n_{thr} = 10^{10}$, $\bar{\alpha}(0) = \bar{\alpha}_{*}(0) = \pm 10^{-3}$; (b) $n_{thr} = 10^{6}$, $\bar{\alpha}(0) = \alpha_{*}(0) = 0$

initially, as we suggested [20], before the decay of the unstable state. The duration of this transient squeezing decreases as we move further above threshold to obtain larger squeezing. The squeezing attained before and after decay of the unstable steady state shows good agreement with Eqs. (4.5b) and (4.12b), respectively. We should mention that these results have all been obtained with the pump mode initially excited, $\overline{\beta}(0) = \overline{\beta}_*(0) = \lambda$. If the cavity fills with pump photons from $\tau = 0$ we can expect the squeezing to be degraded over the first few lifetimes. This will have a significant effect for the parameters of Figs. 3(b) and 3(c).

The duration of the transient squeezing is also affected by the presence of a small initial subharmonic amplitude, Fig. 4(a), and by the threshold pump photon number, Fig. 4(b). When $\overline{\alpha}(0) = \overline{\alpha}_*(0) = 0$ the mean subharmonic amplitude in the steady state is zero, since each stochastic trajectory has equal probability for terminating in either of the steady states given by Eqs.(3.3). When $\overline{\alpha}(0)$ and $\overline{\alpha}_*(0)$ are not zero one or other of these steady states is selected, depending on the phase of the initial subharmonic amplitude. If the initial perturbation from the unstable steady state is significantly larger than the size of the quantum noise,we see a more rapid decay to the stable steady state (Fig. 4(a) compared with Fig. 3(a)). Larger quantum noise, corresponding to a smaller n_{thr}, also gives a faster decay, as illustrated by Fig. 4(b).

Our results suggest that the degenerate parametric oscillator operated above threshold might still be capable of generating a significant amount of squeezing. There is more work to be done to provide results for a specific experimental measurement scheme. Here we have calculated quadrature variances for the intracavity field, not for the field at a detector outside the cavity. In the future we hope this work can be extended to investigate specific experimental possibilities.

6. Summary and Conclusions

We have modeled the degenerate parametric oscillator by a set of stochastic differential equations derived using the positive P-representation. We treat quantum fluctuations by a linearized analysis below and above threshold and also solve the full nonlinear problem by simulating the stochastic differential equations on a computer. A linearized treatment about the unstable steady state above threshold is equivalent to our earlier calculation for the degenerate parametric oscillator without pump depletion [20]; it predicts the continued increase of squeezing in the subharmonic mode above threshold with increasing pump intensity, while squeezing in the stable steady state decreases above threshold. Our numerical results show that the level of squeezing predicted in the vicinity of the unstable steady state can be maintained over a few to many cavity lifetimes, depending on: (1) the distance reached above threshold, (2) the size of any initial perturbation from the unstable steady state, (3) the threshold photon number in the pump mode.

The positive P-representation is required for our formulation in terms of a classical stochastic process,because the quantum dynamics responsible for squeezing give non-positive-definite diffusion if the traditional Glauber-Sudershan representation is used. We have highlighted the fact that the positive P-representation adds new steady-state solutions to those given by the usual classical equations. It is important that the role of these additional steady states be carefully studied in the future. They have recently been associated with anomalous behavior in the numerical simulation of quantum-stochastic models for absorptive optical bistability [17] and the single-mode laser [22].

Acknowledgment

This paper is based upon work supported by the National Science Foundation under Grant No. 8418070.

References

1. D.F. Walls, Nature 306, 141 (1983), and references therein.
2. R.J. Glauber, Phys. Rev. Lett. 10, 84 (1963); Phys. Rev. 130, 2529 (1963); ibid. 131, 2766 (1963).
3. E.C.G. Sudarshan, Phys. Rev. Lett. 10, 277 (1963).
4. W.H. Louisell, Quantum Statistical Properties of Radiation (Wiley, New York, 1973), pp. 331 - 417.
5. G.S. Agarwal, in Progress in Optics, Vol. XI, ed. E. Wolf (North Holland, 1973), pp. 1 - 75.
6. H.J. Carmichael, Quantum-Statistical Methods in Quantum Optics (Springer, Berlin, in Press).
7. F. Casagrande and L.A. Lugiato, Nuovo Cim. B55, 173 (1980).
8. P.D. Drummond and D.F. Walls, Phys. Rev. A 23, 2563 (1981).
9. H.J. Carmichael, D.F. Walls, P.D. Drummond, and S.S. Hassan, Phys. Rev. A 27, 3112 (1983).
10. H. Takahashi, Adv. Comm. System. 1, 227 (1965).
11. M.T. Raiford, Phys. Rev. A 2, 1541 (1970); ibid. 9, 2060 (1974).
12. E.Y.C. Lu, Lett. Nuovo Cim. 3, 585 (1972).
13. D. Stoler, Phys. Rev. Lett. 33, 1397 (1974).
14. L. Mista, V. Perinová, J. Peřina, and Z. Braunerová, Acta Phys. Pol. A 51, 739 (1977).
15. H.J. Carmichael, G.J. Milburn, and D.F. Walls, J. Phys. A 17, 469 (1984).
16. H. Risken, The Fokker-Planck Equation (Springer, Berlin, 1984).
17. H.J. Carmichael, J.S. Satchell, and S. Sarkar, "Nonlinear Analysis of Quantum Fluctuations in Absorptive Optical Bistability," submitted to Phys. Rev. A.
18. P.D. Drummond and C.W. Gardiner, J. Phys. A 13, 2353 (1980).
19. G. Milburn and D.F. Walls, Optics Commun. 39, 401 (1981).
20. M. Wolinsky and H.J. Carmichael, Optics Commun. 55, 138 (1985).
21. P.D. Drummond, "Nonequilibrium Transitions in Quantum Optical Systems," D. Phil. thesis, University of Waikato (1979).
22. S. Sarkar, J.S. Satchell, and H.J. Carmichael, "The Positive P-Representation and the Laser Equations," submitted to J. Phys. A.
23. C.W. Gardiner, Handbook of Stochastic Methods (Springer, Berlin, 1983) p. 110.

Quantum Brownian Motion: Recent Developments

A.O. Caldeira

Instituto de Física "Gleb Wataghin", Universidade Estadual de Campinas, Cidade Universitária "Zeferino Vaz", 13100 Campinas, SP, Brazil

This is a short report reviewing the progress achieved in Quantum Brownian Motion over the last five years. The emphasis will be, mainly, on basic ideas and results. I shall not try to invest any time in lengthy deductions or mathematical details of the approach employed here.

1) Introduction

Last century, the botanist R. Brown observed that small particles immersed in a fluid presented a very irregular motion. When no forces were acting on the particle, he noticed that its average velocity $\langle\vec{v}\rangle$ was zero and it had a finite value for the variance $\langle v^2\rangle$. This phenomenon was called "Brownian Motion" from then onwards.

The most successful theoretical approach to describe this sort of motion is the well-known "Langevin equation" which was first proposed by P. Langevin in the beginning of the twentieth century. This reads [1]

$$M\ddot{q} + \eta\dot{q} + V'(q) = f(t) \qquad , \qquad (1.1a)$$

where M is the mass of the particle, η is a phenomenological damping constant, V(q) is the external potential to which the particle is subject and f(t) is a fluctuating force obeying the relations

$$\begin{aligned}\langle f(t)\rangle &= 0\\ \langle f(t)f(t')\rangle &= 2\eta kT\ \delta(t-t')\end{aligned} \qquad (1.1b)$$

Like any other phenomenological equation, (1.1) has a restricted range of validity. It is reasonable to use Langevin equations when;

- the mass, M, of the Brownian particle is such that $M \gg m$, where m is the mass of the molecules of the viscous fluid and
- we are interested in the long-time behaviour of the system. By this we mean times long compared to the typical collision time of the particles components of the fluid.

An entirely equivalent way to study Brownian Motion is to solve the differential equation which describes the time evolution of W(q,p,t), the density of points in the phase space of the Brownian particle. This is the Fokker-Planck equation [1]

$$\frac{\partial W}{\partial t} = -\frac{\partial}{\partial q}(pW) + \frac{\partial}{\partial p}(V'(q)W) + 2\gamma\frac{\partial}{\partial p}(pW) + D\frac{\partial^2 W}{\partial p^2} \qquad (1.2)$$

where $\gamma \equiv \frac{\eta}{2M}$ and $D = \eta kT$. Once one knows the solution of this equation, averages of any physical quantity of interest can be computed.

Although I have used the classical example to introduce the idea of Brownian Motion there are many other systems that could serve as equally good paradigms of this phenomenon. Only to name a very special example (actually, this will be the sort of system specially interesting in quantum theory) we could think off the flux ϕ inside a LCR circuit, which obeys the equation

$$C\ddot{\phi} + \frac{RC}{L}\dot{\phi} + \frac{\phi}{L} = I_f \tag{1.3}$$

where I_f is a fluctuating current satisfying $\langle I_f \rangle = 0$ and $\langle I_f(t) I_f(t') \rangle = \frac{2RCkT}{L}\,\delta(t-t')$.

At this point one should start to worry about the necessity of the quantum limit in any of the systems I have mentioned above. In the first example, although the particle is very small it is still much heavier than the molecules components of the fluid where it is immersed. Therefore, observation of quantum effects involving the Brownian particle is hopeless. The same happens in the second example where the typical frequency of the circuit, $\sim 1/\sqrt{LC}$, is such that $\hbar/\sqrt{LC}\,kT \ll 1$ even for extremely low temperatures. However, there is a very special kind of circuit in which quantum effects are important. This is a superconducting ring closed by a weak link [2] which is found in the literature as SQUIDS (superconducting quantum interference devices).

The flux ϕ trapped inside a SQUID ring obeys the following Langevin equation [3]

$$C\ddot{\phi} + \frac{1}{R_n}\dot{\phi} + \frac{dU}{d\phi} = I_f \tag{1.4}$$

where $\langle I_f \rangle = 0$, $\langle I_f(t) I_f(t') \rangle = \frac{2kT}{R_n}\,\delta(t-t')$,

$$U(\phi) = \frac{(\phi-\phi_x)^2}{2L} - I_c\phi_o \cos\frac{2\pi\phi}{\phi_o} \tag{1.5}$$

and the quantities introduced above are:

C - capacitance of the weak link
R_n - normal resistance of the link
I_c - critical current of the link
ϕ_o = h/2e (flux quantum)
ϕ_x - flux due to the application of an external magnetic field perpendicular to the ring.

The equation of motion (1.4) is obviously analogous to the one of a particle of coordinate ϕ subject to a potential $U(\phi)$. Changing the external field to which the SQUID ring is subject, one can distort the shape of $U(\phi)$ (see fig. 1) and, consequently, vary the stability conditions of the system. The typical frequency, ω, of oscillation at one of the possible metastable equilibrium positions of $U(\phi)$ is of the order of $2\,(I_c/C\phi_o)^{1/2}$ which can be quite high depending on the capacitance and critical current of the link. For example, if $I_c \sim 10^{-5}$ A and $C \sim 10^{-12}$ F one has $\hbar\omega/k \sim 1^oK$ and quantum effects would be observable at easily attainable temperatures. It should be emphasized here that the values I have just mentioned for I_c and C are quite reasonable with the new technology of building weak links and junctions in general [4].

If one now neglects the influence of damping and quantizes this system in the canonical way, one is faced with the possibility of studying a rich variety of quantum effects just by changing the external field. One can study the quantum behaviour of the SQUID in the classically

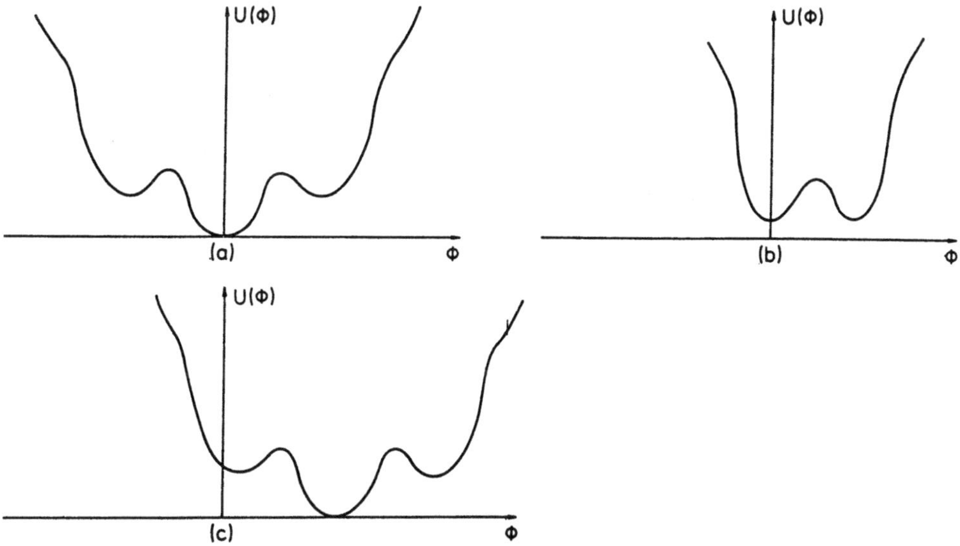

fig. 1: The potential $U(\phi)$ for; a) $\phi_x=0$, b) $\phi_x=\frac{\phi_0}{2}$, c) $\phi_x=\phi_0$

accessible region (fig. 1a) or coherent tunnelling (fig. 1b) or still decay by quantum tunnelling (fig. 1c). There are, however, two important questions about these considerations:

1) What is the meaning of this secondary quantization about which I have been talking?
2) How can one reconcile damped equations of motion with the process of quantization?

The answer for the first one is not that hard for the SQUID ring. As we know, $\phi = \oint \vec{A}.d\vec{\ell}$ and, therefore, I have been talking about quantization of the electromagnetic field generated by this superconducting circuit in some indirect way. The second question is somewhat harder and is partially the goal of next section.

2) The Basic Approach

Many people have tried to answer the second question above by very different methods. However, all these attempts fall into two main categories; [5]

- new schemes of quantization for non-isolated systems, or
- dynamics of subsystems (system of interest-plus-reservoir approach).

In this work I shall concentrate on the second possibility. The reason for this choice is simply because I know of no dissipative system in nature which is not coupled to an external reservoir.

Once one has made this choice there are still two possibilities;

- start with a first principle Hamiltonian of the system-plus-reservoir and solve a possibly complicated problem, or
- start with a simple Hamiltonian of the system-plus-reservoir and impose some conditions on the model in such a way that the appropriate classical limit is recovered.

Since in the great majority of cases one does not know enough about the reservoir to which the Brownian variable (the system of interest) is coupled, the second possibility seems again to be the most attractive one and the specific model I shall employ is defined in what follows.

Let us suppose that the Hamiltonian of the complete system is given by

$$H = H_S + H_{int} + H_R + C.T. \tag{2.1}$$

where

$$H_S = \frac{p^2}{2M} + V(q) \quad ; \quad H_{int} = \sum_k C_k q_k q \quad ;$$

$$H_R = \sum_k \frac{p_k^2}{2m_k} + \frac{1}{2} m_k \omega_k^2 q_k^2$$

and C.T. is a counter term [5,6]; $C.T. = \sum_k \frac{C_k^2}{2m_k\omega_k^2} q^2$. Let us also introduce a spectral function $J(\omega)$ defined as

$$J(\omega) \equiv \sum_k \frac{\pi C_k^2}{2m_k\omega_k} \delta(\omega-\omega_k) = \begin{cases} \eta\omega & \text{if} \quad \omega < \Omega \\ 0 & \text{if} \quad \omega > \Omega \end{cases} \tag{2.2}$$

where Ω is a very high frequency.

In general one is interested in studying average values of operators of the system only, or,

$$\langle \hat{O}(p,q) \rangle = tr_{RS}[\rho\hat{O}] \tag{2.3}$$

where ρ is the density operator of the system-plus-reservoir. Therefore, since $\hat{O}$ does not depend on reservoir operators, one has

$$\langle \hat{O} \rangle = tr_S[(tr_R\rho)\hat{O}] = tr_S[\tilde{\rho}\hat{O}] \tag{2.4}$$

where $\tilde{\rho} \equiv tr_R\rho$ is the reduced density operator of the system. This is the quantity in which I shall be interested from now on. In particular, I shall work with the Feynman path integral [7,8] representation of $\tilde{\rho}$ because in many cases of interest these integrals are either exactly soluble or very well approximated by the saddle point method. On top of this, I shall be avoiding the use of N+1 dimensional wave functions.

In equilibrium it can be shown [6,8] that

$$\tilde{\rho}_{eq}(x,y,\beta) = \tilde{\rho}_0 \int_y^x Dq\, e^{-\frac{1}{\hbar}\{S_E[q]+\Lambda_E[q]\}} \tag{2.5a}$$

$$\text{where } S_E[q] = \int_0^{\hbar\beta} \{\tfrac{1}{2} M\dot{q}^2 + V(q)\} d\tau \tag{2.5b}$$

$$\text{and } \Lambda_E[q] = \frac{\eta}{4\pi}\int_{-\infty}^{\infty} d\tau \int_0^{\hbar\beta} d\tau' \frac{\{q(\tau)-q(\tau')\}^2}{(\tau-\tau')^2} \tag{2.5c}$$

while the full dynamical expression of $\tilde{\rho}(t)$ has a more complicated form [5,7]

$$\tilde{\rho}(x,y,t) = \int dx'dy'J(x,y,t;x',y',0)\tilde{\rho}(x',y',0) \tag{2.6a}$$

where

$$J(x,y,t;x',y',0) = \int_{x'}^{x}\int_{y'}^{y} Dx(t)Dy(t)e^{i/\hbar S_0[x(t)]}e^{-i/\hbar S_0[y(t)]}F[x(t),y(t)]; \tag{2.6b}$$

$$S_0[x(t)] = \int_0^t \{\tfrac{1}{2}M\dot{x}^2 - V(x)\}dt' \quad ; \quad F = e^{i/\hbar\,\Phi[x(t),y(t)]} \tag{2.6c}$$

and $F[x(t),y(t)]$ is the influence functional [5,7] with Φ quadratic in $x(t)$ and $y(t)$ when the total Hamiltonian is the one in (2.1) with $J(\omega)$ as in (2.2).

It is an easy (although tedious) task to show that in the classical limit the Wigner transform, $w(q,p,t)$, of $\tilde{\rho}(t)$ obeys the Fokker-Planck equation (1.2) for the model proposed in (2.1) and (2.2) [5]. The full quantum mechanical equation of $w(q,p,t)$ is, in this case, a non trivial generalization of (1.2) (see ref. [9]).

Finally, I wish to mention something about the problem of the initial condition in (2.6a). In general, one chooses $\rho(0) = \rho_R(0)\rho_s(0)$ [5,7] where $\rho_R(0)$ refers only to the reservoir and $\rho_s(0)$ only to the system (obviously, $\tilde{\rho}(0) = \rho_s(0)$). Although this choice gives us almost all the expected results, some transients appear with a bizarre dependence on the cutoff frequency Ω. In order to overcome this problem one can choose a different initial condition as $\rho(0) = \rho_{RS}(0)\rho_s(0)$ where $\rho_{RS}(0)$ is the density operator of the reservoir in the presence of the Brownian particle and $\rho_s(0) = \tilde{\rho}(0)$ is the result of a measurement on that particle only [10]. When this happens there is a slight modification in the influence functional, F, but it still retains its quadratic form. The great advantage of this new initial condition is the fact that all transients are now independent of Ω. (cf. below)

3) Applications

In this section I shall mention the basic results of the application of the formalism introduced in the last section in many interesting problems.

3a) Classically accessible region: harmonic oscillator and free particle

Here I want to evaluate $\tilde{\rho}(x,x,t)$ using (2.6a), given that $\tilde{\rho}(0)$ is Gaussian. Since I shall be interested either in the harmonic oscillator or in the free particle and Φ is a quadratic functional, the final result will obviously be of the form [5]

$$\tilde{\rho}(x,x,t) = \frac{1}{\sqrt{2\pi\sigma^2(t)}} \exp -\frac{(x-x_0(t))^2}{2\sigma^2(t)} \tag{3.1}$$

where $x_0(t)$ and $\sigma(t)$ are analyzed below.

- harmonic oscillator: in this case, $x_0(t)$ is the solution of $M\ddot{q}+\eta\dot{q}+M\omega_0^2 q = 0$ with initial conditions compatible with $\tilde{\rho}(0)$ and $\sigma(t)$ is such that

$$\sigma^2(\infty) = \frac{\hbar}{\pi}\int_0^\infty d\omega \coth\frac{\hbar\omega}{2kT}\,\chi''(\omega) \tag{3.2a}$$

where $\chi''(\omega) = \frac{1}{M} \frac{2\gamma\omega}{(\omega^2-\omega_0^2)^2+4\gamma^2\omega^2}$ (3.2b)

which is the well-known "fluctuation dissipation theorem". The complete $\sigma(t)$ depends on the initial condition one chooses. Actually, they differ from one another only by long-lived transients.

Notice that (3.2a,b) could also have been obtained by the so-called "quantum Langevin equation" [11,12], which is an operator equation given by

$$M\ddot{q} + \eta\dot{q} + M\omega_0^2 q = \hat{f}(t) \tag{3.3a}$$

with $\langle\hat{f}(t)\rangle = 0$ and $\langle\{\hat{f}(t),\hat{f}(t)\}\rangle = \frac{2\hbar}{\pi}\int_0^\infty d\omega\eta\omega \coth\frac{\hbar\omega}{2kT} \cos\omega(t-t')$

(3.3b)

where { } is an anti-commutator. However, the validity of the quantum Langevin equation for potentials other than the harmonic one is still a moot question.

- free particle: now, $x_0(t)$ obeys $M\ddot{q}+\eta\dot{q} = 0$ with appropriate boundary conditions and the asymptotic form of $\sigma(t)$ depends on the choice of initial condition $\rho(0)$. It reads

$$\sigma(t) \sim \begin{cases} \ln \sqrt{\frac{\gamma\Omega}{2}}\, t & \text{if} \quad \rho(0) = \rho_R(0)\rho_S(0) \\ \ln \gamma t & \text{if} \quad \rho(0) = \rho_{RS}(0)\rho_S(0) \end{cases} \tag{3.4}$$

3b) Destruction of quantum interference: [12]

Now I wish to study the same problem of the harmonic potential with the only difference that $\tilde{\rho}(x',y',0)$ is given by

$$\tilde{\rho}(x',y',0) = \psi(x')\psi^*(y') \tag{3.5a}$$

where

$$\psi(x) - \psi_1(x)+\psi_2(x) = \bar{N}\,[\exp - \frac{x^2}{4\sigma^2} + \exp - \frac{(x-x_0)^2}{4\sigma^2}] \tag{3.5b}$$

$\bar{N}$ being a normalization constant. Therefore, we initially have a superposition of two localized packets in a harmonic well.

When there is no damping ($\eta = 0$) everyone knows that

$$\rho(x,t) = \rho_1(xt) + \rho_2(xt) + 2\sqrt{\rho_1(xt)\rho_2(xt)} \cos\varphi(x,t) \tag{3.6}$$

where I have defined $\rho(x,t) \equiv \rho(x,x,t)$. This means that the total probability density is the sum of probability densities corresponding to the first and second packets and an interference term. If one now makes η finite (3.6) changes to

$$\rho(x,t) = \rho_1(x,t)+\rho_2(x,t) + 2\sqrt{\rho_1(x,t)\rho_2(x,t)} \cos\varphi(x,t) \exp - \Gamma t \tag{3.7}$$

where Γ can be analyzed as below for very different conditions.

- High temperatures ($kT \gg \hbar\omega_0$)

$$\Gamma = \begin{cases} \frac{2NkT}{\hbar\omega_o}\gamma & \text{if} \quad \omega_o \gg \gamma \quad \text{and} \quad \hbar\omega_o^2/2kT\gamma \gg 1 & (3.8a) \\ \left(\frac{4}{3}\frac{NkT\gamma}{\hbar\omega_o^2}\right)^{1/3}\omega_o & \text{if} \quad \omega_o \gg \gamma \quad \text{and} \quad \hbar\omega_o^2/2kT\gamma \ll 1 & (3.8b) \\ \frac{2NkT}{\hbar\omega_o}\frac{\omega_o^2}{2\gamma} & \text{if} \quad \omega_o \ll \gamma & (3.8c) \end{cases}$$

- Low temperatures ($kT \ll \hbar\omega_o$)

$$\Gamma = \begin{cases} N\gamma & \text{if} \quad \gamma \ll \omega_o & (3.9a) \\ \frac{N\omega_o^2}{2\gamma} & \text{if} \quad \gamma \gg \omega_o & (3.9b) \end{cases}$$

The quantity N is a measure of the mean number of energy quanta present in the system at $t = 0$; $N \equiv x_o^2/2\sigma^2$. Therefore, one can say that if the two initial wave packets are well separated ($N \gg 1$) $\Gamma^{-1} \ll \gamma^{-1}$. In other words, the interference fringes are destroyed much faster than the relaxation time of the system (γ^{-1}). This is due to the fact that once the displaced part of the initial state loses one quantum of energy (which takes place within a time interval of the order of $\gamma^{-1}/N \ll \gamma^{-1}$) it is correlated to a state of the reservoir which is orthogonal to the state correlated to the remaining part of $\psi(x)$. This only happens because of the particular form of the interaction Hamiltonian we have in (2.1). Consequently, tracing over environmental variables will destroy off-diagonal blocks of the reduced density operator transforming it in a mixture of well-localized damped packets.

3c) Quantum Tunnelling

Suppose one is studying the motion of a Brownian particle in the metastable region of the potential

$$V(q) = \frac{1}{2} M\omega_o^2 q^2 - \lambda q^3 \equiv \frac{27}{4} V_o\left\{\left(\frac{q}{q_o}\right)^2 - \left(\frac{q}{q_o}\right)^3\right\} \qquad (3.10)$$

where $q_o \equiv \frac{1}{2}\frac{M\omega_o^2}{\lambda}$ is the coordinate of the exit point (fig. 2). A very natural question is then: what is the influence of damping on quantum mechanical tunnelling?

First, let us try to answer this question at zero temperature [6]. Using the reduced density operator of the system in equilibrium (2.5) it is possible to show [6] that the probability for the particle to stay in the metastable well is of the form

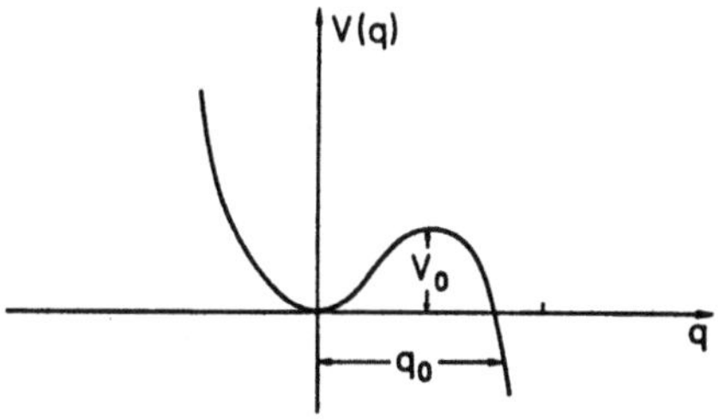

Fig. 2: The form of V(q) (3.10)

$$P(t) = \exp -\Gamma t \qquad (3.11)$$

where $\Gamma = \frac{2\,\mathrm{Im}E_o}{\hbar}$; E_o being the ground state energy of the system-plus-reservoir. When $\eta \cong 0$ one has:

$$\Gamma_o = A_o\, e^{-B_o/\hbar} \qquad (3.12)$$

where $B_o = \frac{36}{5}\frac{V_o}{\omega_o}$ and $A_o = \omega_o\sqrt{\frac{30B_o}{\pi\hbar}}$

while for finite η (3.12) is modified to

$$\Gamma = A\, e^{-B/\hbar} \qquad (3.13)$$

where

$$B \cong \begin{cases} B_o\left(1 + \frac{45}{\pi^3}\zeta(3)\frac{\gamma}{\omega_o}\right) & \text{if} \quad \frac{\gamma}{\omega_o} \ll 1 \\ \text{and} & \\ \frac{5\pi}{6}\frac{\gamma}{\omega_o}B_o & \text{if} \quad \frac{\gamma}{\omega_o} \gg 1 \end{cases} \qquad (3.14)$$

and

$$A \cong \begin{cases} \sqrt{\frac{30B}{\pi\hbar}} & \text{if} \quad \frac{\gamma}{\omega_o} \ll 1 \\ 0.29\left(\frac{\gamma}{\omega_o}\right)^{7/2} A_o & \text{if} \quad \frac{\gamma}{\omega_o} \gg 1 \end{cases} \qquad (3.15)$$

which allows us to conclude that damping always tends to supress quantum tunnelling. This is easily understood if we remember that applying the fluctuation-dissipation theorem (3.2) to the particle about $q = 0$ one has $(T = 0)$

$$\langle x^2\rangle = \frac{\hbar}{\pi}\int_0^\infty d\omega\, \chi''(\omega) < \langle x^2\rangle_{\gamma=0} \qquad (3.16)$$

Therefore, the probability density penetrates less under the barrier and this causes inhibition of the tunnelling process.

At finite temperatures the matter is more complicated. The natural extension of the method employed in [6] for $T \neq 0$ is criticized by many people. I shall mention here two of the usual objections; in the first place the expression for Γ in this case is given by [13,14,15]

$$\Gamma = \frac{2\,\mathrm{Im}F}{\hbar} \qquad (3.17)$$

where F is the free energy of the complete system. However, to attribute a physical meaning to the imaginary part of the free energy does not seem to be very trivial. Secondly, even if one accepts expression (3.17) the hypothesis of equilibrium distribution in the metastable well must be used. Nevertheless, one knows that when $\eta \to 0$ this is not true. [16]

On the other hand there are many alternative approaches to this problem (see e.g. [17,18]) but these also have a restricted range of validity and agree with (3.17) when the damping constant is not too small.

Therefore, taking into account only those systems to which (3.17) could be applied, we have

$$\Gamma(T) = A(T) \exp - \frac{B(T)}{\hbar} \tag{3.18}$$

where

$$B(T) = \begin{cases} B(0) - KT^2 & \text{if} \quad T \ll T_0 \\ -\frac{V_0 \hbar}{kT} & \text{if} \quad T \gg T_0 \end{cases} \tag{3.19}$$

and

$$T_0 \equiv \frac{\hbar}{2k\pi} \left[\sqrt{\omega_0^2 + \frac{\gamma^2}{4}} - \frac{\gamma}{2}\right]$$

showing that the increase in temperature would also increase the tunnelling rate. This is an expected result since the average height of the barrier to be crossed starts to diminish as T increases.

3d) Quantum Coherence

As a final example of possible applications of quantum Brownian motion, let us study the motion of a damped particle in a potential [19] (see fig. 3)

$$V(q) = -\frac{1}{2} M\omega_0^2 q^2 + \frac{1}{4} \lambda q^4 \tag{3.20}$$

when $kT \ll \hbar\tilde{\omega}$ (see fig. 3). Then, we are basically interested in a two level system coupled to the bath of oscillators and H_s and H_{int} in (2.1) can be replaced by [20]

$$H_s = -\frac{\hbar}{2} \Delta_0 \sigma_x \qquad \text{and} \qquad H_{int} = \frac{q_0}{2} \sigma_z \sum_k C_k q_k \tag{3.21}$$

where σ_i are Pauli matrices. What we want to compute now is $P(t) \equiv \langle\sigma_z(t)\rangle$. Defining a dimensionless damping constant as

$$\alpha \equiv \frac{\eta q_0^2}{2\pi\hbar} \tag{3.22}$$

and using (2.6) it was shown that [20]:

$$\text{If } \alpha > 1, \text{ for any } T; \; P(t) = \exp - \frac{t}{\tau} \tag{3.23a}$$

where

$$\tau^{-1} = \Delta_0^2 \frac{\sqrt{\pi}}{2} \frac{\Gamma(\alpha)}{\Gamma(\alpha+\frac{1}{2})} \left(\frac{\pi kT}{\hbar\Omega}\right)^{2\alpha-1} \tag{3.23b}$$

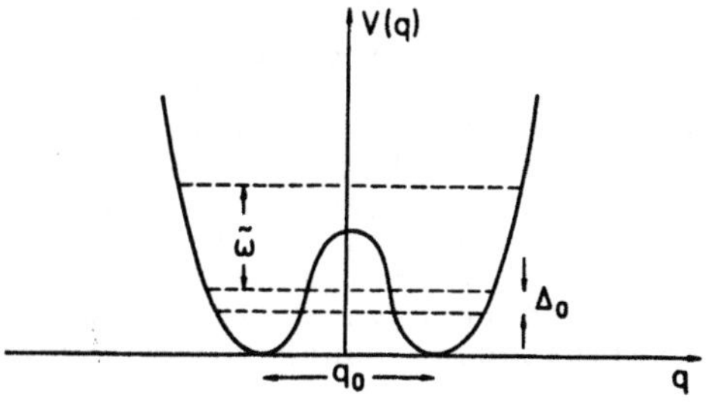

Fig. 3: The form of V(q) (3.20)

If $\frac{1}{2} < \alpha < 1$ and $T = 0$

$$P(t) = \exp - \frac{t}{\tau} \tag{3.24a}$$

where $\tau^{-1} \propto \Delta_r$

and

$$\Delta_r = \begin{cases} \Delta_o \left(\frac{\Delta_o}{\Omega}\right)^{\alpha/1-\alpha} & \text{if} \quad \alpha < 1 \\ 0 & \text{if} \quad \alpha > 1 \end{cases} \tag{3.24b}$$

however it should be emphasized that there might be in addition to (3.24a) a power law background algebraically identical to that in (3.25a) below [20].

If $0 < \alpha < \frac{1}{2}$ and $T = 0$

$$P(t) \cong \frac{A(\alpha)}{1-\alpha} \cos\{[\Delta_{eff} \cos(\frac{\pi\alpha}{2(1-\alpha)})]t\} \exp-[\Delta_{eff} \sin \frac{\pi\alpha}{2(1-\alpha)}]t$$

$$- \frac{1}{\pi} \sin 2\pi\alpha \; \Gamma(2-2\alpha)(\Delta_{eff}t)^{-2(1-\alpha)} \tag{3.25a}$$

where

$$\Delta_{eff} = \{[1+q(\alpha)] \cos \pi\alpha \; \Gamma(1-2\alpha)\}^{1/2(1-\alpha)} \; \Delta_r \tag{3.25b}$$

Here, one easily sees (3.24b) that there is a critical damping constant ($\alpha = 1$) above which the energy splitting at $T = 0$ renormalizes to zero. Below $\alpha = 1$, but still at $T = 0$, one can either have coherent or incoherent relaxation to $\langle\sigma_z\rangle = 0$. A very careful and complete treatment of this problem is about to appear in the literature [21].

4) Discussions and Conclusions

After we have seen all these applications based on the specific model (2.1), (2.2), a very basic question still remains, namely; what is the physical meaning of those oscillators?

They are nothing but low-energy excitations of the reservoir [6] (assuming that it is only weakly disturbed by the Brownian particle) and $J(\omega)$ is their spectral density. Consequently, in order to have a term of the form $\eta\dot{x}$ in the classical equation of motion we need the spectral density of excitations of the bath vanishing linearly with ω.

The success of this model in applications to SQUIDS and current-driven Josephson junctions (supported by many experimental results, see, e.g. [22,23] led people to think it could also be applied to many other problems, such as diffusion of heavy particles in solids. However, a detailed treatment of this sort of problem [24,25,26] shows that in the case the Brownian variable is really a geometric coordinate the concept of bath of oscillators is valid only locally. In other words, the motion must be restricted to a region of linear dimensions smaller than a typical length provided by the reservoir. The quantum mechanical expressions are now more complicated, although the classical limit is the same as before.

At this point one concludes that

- Quantum effects in local Brownian motion can be investigated by convenient couplings to a single bath of oscillators.
- Quantum effects tend to be destroyed by ohmic damping $(\alpha \eta \dot{q})$
- Quantum effects in non-local Brownian motion have to be studied by a detailed analysis of the reservoir or sophistications of the bath of oscillators
- The above-mentioned considerations tend to invalidate the idea of quantizing dissipative systems in a direct way.

Acknowledgements

I am extremely thankful to the organizing committee of the "Fourth International Symposium on Quantum Optics" for their invitation and hospitality, and to IBM - Brasil for having made my presence at the symposium possible.

References

1) Selected Papers on Noise and Stochastic Processes, Nelson Wax, ed. (Dover, NY, 1954)
2) K.K. Likharev, Rev. Mod. Phys., vol. 51, No. 1 (1979) 101
3) A.O. Caldeira, D. Phil. Thesis, Univ. of Sussex (1980) unpublished
4) R.F. Voss and R.A. Webb, Phys. Rev. Lett. 47 (1981) 697
5) A.O. Caldeira and A.J. Leggett, Physica 121A (1983) 587 and references therein*
6) A.O. Caldeira and A.J. Leggett, Annals of Physics 149 (1983) 374**
7) R.P. Feynman and A.R. Hibbs, "Quantum Mechanics and Path Integrals" (McGraw Hill, New York, 1965)
8) R.P. Feynman, "Statistical Mechanics" (Benjamin, New York, 1972)
9) L.D. Chang and D. Waxman, preprint of the SUNY at Stonybrook (1985)
10) V. Hakim and V. Ambegaokar, Phys. Rev. A32 (1985) 423
11) R. Benguria and M. Kac, Phys. Rev. Lett. 46 (1981) 1
12) A.O. Caldeira and A.J. Leggett, Phys. Rev. A31 (1985) 1059
13) H. Grabert, U. Weiss and P. Hänggi, Phys. Rev. Lett. 52 (1984) 2193
14) H. Grabert and U. Weiss, Z. Phys. B 56 (1984) 171
15) A. Larkin and Yu. N. Ovchinikov, J.E.T.P. Lett. 37 (1983) 382
16) M. Büttiker and R. Landauer, Phys. Rev. B 30 (1984) 1551
17) D. Waxman, D. Phil. Thesis, Univ. of Sussex (1984) unpublished
18) W. Zwerger, preprint of the Univ. of Illinois at Urbana (1984)
19) A.J. Bray and M.A. Moore, Phys. Rev. Lett. 49 (1982) 1545
20) S. Chakravarty and A.J. Leggett, Phys. Rev. Lett. 52 (1984) 5
21) S. Chakravarty, A.T. Dorsey, M.P.A. Fisher, A. Garg, A.J. Leggett and W. Zwerger, to appear in Rev. Mod. Phys.
22) S. Washburn, R.A. Webb, R.F. Voss and S.M. Faris, preprint of the IBM - Thomas J. Watson Research Center (1985)
23) D.B. Schwartz, B. Sen, C.N. Archie and J.E. Lukens, preprint of the SUNY at Stonybrook (1985)
24) F. Guinea, Phys. Rev. Lett. 53 (1984) 154
25) P. Hedegard and A.O. Caldeira - Research Report, IBM Thomas J. Watson Research Center (1985) and submitted to Phys. Rev. B.
26) P. Hedegard and A.O. Caldeira - Research Report, IBM Thomas J. Watson Research Center (1985) and submitted to Phys. Rev. B.

*) see also erratum, Physica 130A (1985) 374
**) see also erratum, Annals of Physics 153 (1984) 445

Quantum and Many-Body Effects in a Compton Regime Free-Electron Laser

M. Orszag

Facultad de Fisica, Pontificia Universidad Católica de Chile, Casilla 6177, Santiago 22, Chile

1 Introduction

When an electron travels through a periodic transversal magnetic field, emission of radiation occurs. The first experiments were done at Stanford [1] and infrared radiation was obtained. More recently, there have been also experiments in the visible [2].

2 The model

The model for 1 electron presented here is based on the work of Bambini and Renieri [3]. In this model, one performs a Lorentz transformation where the laser frequency and the wiggler (magnetic) quasi-photon have the same frequency, and therefore, the lasing process can be described as a wiggler quasi-photon that is turned around into a laser photon by the electron. The electron recoils to conserve momentum. Since the two frequencies are the same, this is an elastic scattering event. (Figure 1).

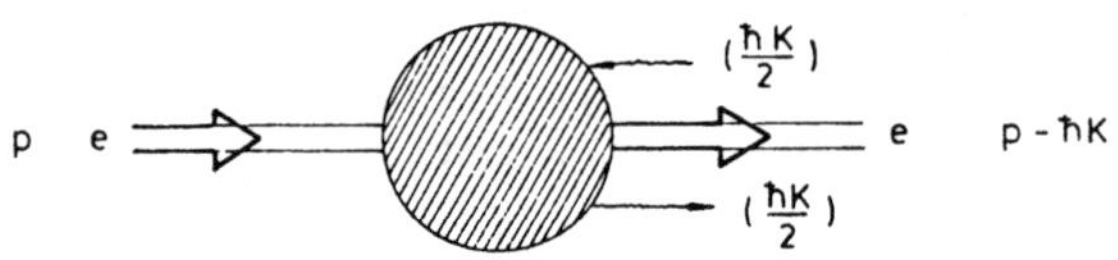

Fig. 1. A wiggler quasiphoton is converted into a laser photon in an elastic scattering process, where the momentum is provided by the electron's recoil.

In the Bambini-frame, the problem becomes non-relativistic and the full quantum Hamiltonian for a one-electron model can be written as:

$$H = \hbar\omega(a_W^+ a_W + a_L^+ a_L) + \hbar\Omega S_z^2 + i\hbar g(a_L^+ a_W S^- - a_L a_W^+ S^+), \quad (1)$$

where $\hbar K$ is the momentum recoil of the electron, $S_z = (\frac{p}{\hbar K})$ is the electron's scaled momentum, $S^\pm = e^{\pm ikz}$ and $g = \frac{4\pi c}{KV} r_o$. ($r_o$ = classical electron radius, V : interaction volume). a_W and a_L are the annihilation operators for the wiggler and laser fields, ω their common frequency (in the Bambini frame) and $\hbar\Omega = \frac{(\hbar k)^2}{2m}$ is the recoil energy of the electron in every emission event.

In this problem, there are two constants of motion:

$$a_L^+ a_L + a_W^+ a_W = \text{const.} ,$$
$$S_z + \frac{1}{2}(a_L^+ a_L - a_W^+ a_W) = \text{const}, \quad (2)$$

and combining these, we get

$$S_z + a_L^+ a_L = c = \text{constant}. \quad (3)$$

It is simple to show that the S_z, $S^\pm$ matrices obey the following commutation rules:

$$[S^+, S^-] = 0 ,$$
$$[S_z, S^\pm] = \pm S_\pm . \quad (4)$$

Now, we can assume that the wiggler-field is much stronger than the laser field

$$|A_w|^2 >> |A_L|^2 ,$$

and therefore consider the wiggler classical. We can re-write the Hamiltonian as:

$$H = \hbar\Omega S_z^2 + i\hbar g\alpha_w(a^+S^- - aS^+), \quad (5)$$

where α_w is the classical wiggler amplitude. In eq. (5) we have eliminated the constant number of photons, and a represents the laser field.

We can use a $|m>|n>$ product basis, with

$$a^+a|n> = n|n> ,$$
$$S_z|m> = m|m> , \quad (6)$$

We can always expand the eigenstate of $H, |\phi>$ in this basis as:

$$|\phi> = \sum_{n=0}^{\infty} Cn|n>|m = C-n>. \quad (7)$$

By direct application of Schrödinger's equation [4], we can write:

$$i\hbar \sum_{n=0}^{\infty} \dot{C}_n(t)|n>|m = c-n> = \sum_{n=0}^{\infty} \{\hbar\Omega(c-n)^2 C_n|n>|m = c-n>$$
$$+ i\hbar g\alpha_w C_n(\sqrt{n}|n-1>|m = C-n+1>- \sqrt{n+1}|n+1>) \quad (8)$$

Because of the constant of motion, the problem has been reduced to one degree of freedom and we can write a fictitious Hamiltonian

$$H_{fict} = \hbar\Omega(c-a^+a)^2 + i\hbar g\alpha_w(a-a^+) \quad (9)$$

that reproduces exactly the difference differential equation (8)

3 Normal ordering

One can always write the Hamiltonian in normal ordered form:

$$H = \sum_{\ell,m} h_{\ell,m}(t)(a^+)^\ell a^m , \quad (10)$$

in such a way that the differential equation for the evolution operator:

$$i\hbar\frac{\partial U}{\partial t} = HU , \tag{11}$$

can be written as a c-number differential equation [4]:

$$i\hbar \frac{\partial U^{(n)}}{\partial t} = \sum_{\ell,m} h_{\ell,m}(t)(\alpha^*)^{\ell}\left(\alpha + \frac{\partial}{\partial\alpha^*}\right)^m U^n(\alpha,\alpha^*,t) \tag{12}$$

In our particular case:

$$i\hbar\frac{\partial U^{(n)}(\alpha,\alpha^*,t)}{\partial t} = [\ \hbar\Omega(1-2c)\alpha^*\left(\alpha + \frac{\partial}{\partial\alpha^*}\right)$$

$$+ \hbar\Omega(\alpha^*)^2\left(\alpha + \frac{\partial}{\partial\alpha^*}\right)^2 + i\hbar g\alpha_W\left(\alpha + \frac{\partial}{\partial\alpha^*} - \alpha^*\right)]U^n(\alpha,\alpha^*,t) \tag{13}$$

Now, if we define

$$U^{(n)}(\alpha,\alpha^*,t) = \exp G(\alpha,\alpha^*,t) , \quad \text{and} \tag{14}$$

$$G(\alpha,\alpha^*,t) = \sum_{\ell,m=0}^{\infty} f_{\ell m}(t)\alpha^{\ell}(\alpha^*)^m, \tag{15}$$

then inserting the equations (14) and (15) into the eq. (13), after some algebra, one gets [5]:

$$\begin{aligned}\frac{df_{\ell m}}{d\tau'} = & - \frac{i(1-2c)}{c}\delta_{\ell,1}\delta_{m,1} - \frac{i}{c}\delta_{\ell,2}\delta_{m,2} \\ & - \frac{im(m-2c)}{c} f_{\ell,m} - 2i\frac{(m-1)}{c} f_{\ell-1,m-1} \\ & - \frac{i}{c}\sum_{\ell',m'} f_{\ell' m'} f_{\ell-\ell',m-m'} \cdot m'(m-m') \\ & + \frac{G}{c}(\delta_{\ell,1}\delta_{m,0} - \delta_{\ell,0}\delta_{m,1}) + \frac{G}{c}(m+1)f_{\ell,m+1},\end{aligned} \tag{16}$$

with $\tau' = \Omega ct$ and $G = \alpha_W g/\Omega$.

This is an infinite set of non-linear difference differential equations.

A particular simple case is that of no-initial photons. Setting $\ell = 0$ for this case, we get a simpler version of eq. (16):

$$\begin{aligned}\frac{df_m}{d\tau} = & -W\delta_{m,1} + W(m+1)f_{m+1} - ih_m f_m \\ & - \frac{i}{c^{3/2}}\sum_{m'=0}^{m} f_{m'}f_{m-m'}(m')(m-m')\end{aligned} \tag{17}$$

with

$$\tau = \Omega c^{3/2} t, \quad W = \frac{G}{c^{3/2}} = \frac{\alpha_w g}{\Omega c^{3/2}}, \quad h_m = \frac{m(m-2c)}{c^{3/2}} \tag{18}$$

4 Numerical results showing quantum effects

We show here some of the numerical results by solving over 100 equations, by truncating the set of eqs. (17).

In figure 2 we show the average number of photons as a function of W (wiggler parameter) with $\tau = 0.5$ and $c = 10$. This type of oscillatory behavior was shown before by Fedorov [6].

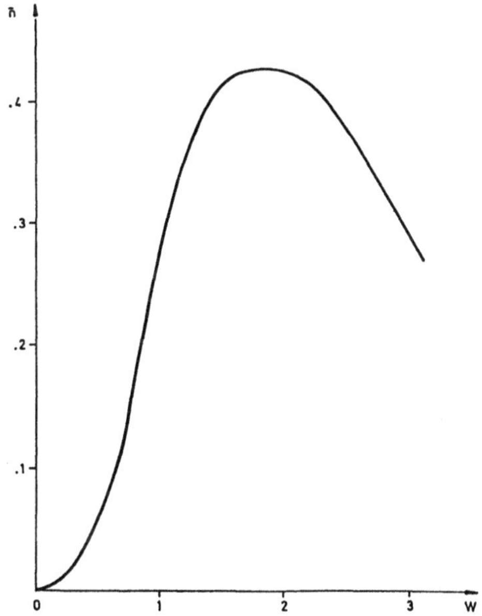

Fig. 2 Average number of photons versus the wiggler parameter for $\tau = 0.5$ and $C = 10$.

We are also interested in quantum effects such as squeezing and antibunching. For a single-mode laser radiation, one defines two hermitian operators

$$A_1 = \frac{a+a^+}{2}, \qquad A_2 = \frac{a-a^+}{2i}, \tag{19}$$

which are the 2 quadratures of the electromagnetic field, which obey the commutation and uncertainty relations:

$$[A_1, A_2] = \frac{i}{2}, \qquad (\Delta A_1)(\Delta A_2) \geq 1/4 . \tag{20}$$

Squeezing, of course, implies that $(\Delta A_1)^2$ or $(\Delta A_2)^2$ is less than 0.25, one quadrature at the expense of the other one.

For the light statistics, we define the second-order coherence function $g^{(2)}(\tau)$:

$$g^2(\tau) = \frac{\langle n^2\rangle - \langle n\rangle}{\langle n\rangle^2} . \qquad (21)$$

$g^{(2)} = 1$, $g^{(2)} < 1$ and $g^{(2)} > 1$ correspond to coherence, antibunching and bunching respectively. In Fig. (3) we show $g^{(2)}$ as a function of (τ) for various values of the wiggler parameter W. Antibunching is observed for certain ranges of τ

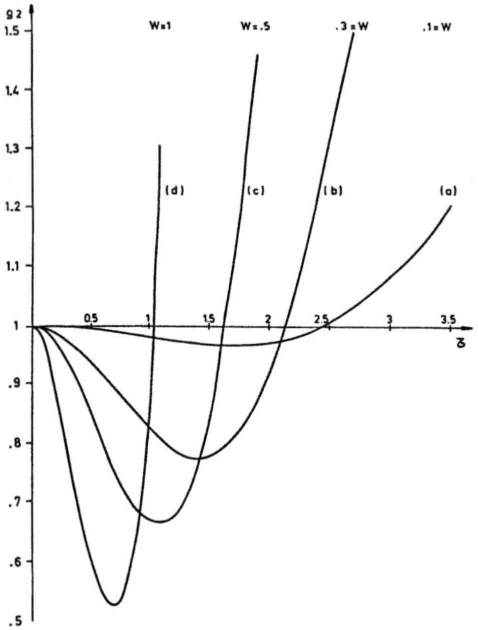

Figure 3. Second-order coherence function $g^{(2)}(\tau)$ for W = 0.1 (curve a), W = 0.3 (curve b), W = 0.5 (curve c) and W = 1 (curve d). In all the curves C = 10.

It is apparent from these curves that we get more antibunching for larger wiggler parameters W. In fig. (4) we show $(\Delta A_2)^2 - 1/4$ (curve a) and $(\Delta A_1)^2 - 1/4$ (curve b) versus τ for W = 0.2 and C = 10

It is worthwhile mentioning that Becker et al [7] obtained symmetrical results for the two quadratures of the field in a first order theory on the electrons recoil. We see that this is true for small values of τ only. Although these results show a definite squeezing, we have to remark that this squeezing is not on the electromagnetic states of the FEL radiation, but rather on the combined electron-photon states.

5 Many-body effects

By simple extension of the 1-electron model, we can write the Ne-electron Hamiltonian (per unit $\hbar$) as:

$$H = \omega \sum_{i=1}^{Ne} (S_z^i)^2 + ig \sum_{i=1}^{Ne} (S_i^- a^+ - a S_i^+) \qquad (22)$$

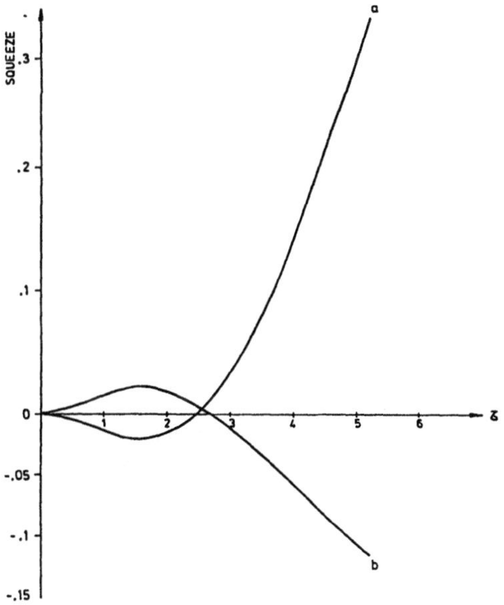

Figure 4. $(\Delta A_2)^2 - 1/4$ (curve a) and $(\Delta A_1)^2 - 1/4$ (curve b) as a function of τ for W = 0.2 and C = 10.

(we have included α_W in g and ω is the recoil frequency) S_z^i, $S_i^{\pm}$ are the operators for the ith electron, obeying the same commutation rules as before

Define the Green's Function G (Ω)

$$G(\Omega) = \frac{1}{\Omega - H} = \sum_\nu \frac{|\nu\rangle\langle\nu|}{\Omega - \Omega_\nu} , \qquad (23)$$

where $|\nu\rangle$ and Ω_ν the eigenstates and eigenvalues of H.

It is simple to show [8] that the probability amplitude for a transition from an initial state $|i\rangle$ to a final state $|f\rangle$ is given by:

$$a_{fi} = \sum_\nu \text{Res}[e^{-i\Omega t} G_{fi}(\Omega)]_{\Omega_\nu} , \qquad (24)$$

with $G_{fi} = \langle f|G|i\rangle$, and Res means the residue evaluated at the Green's function pole Ω_ν.

Our aim in this section is to calculate the probability of emission of 1, 2 or more photons (starting from vacuum) in order to be able to say something about the many-body effects in the light statistics.

We use a basis $|m_1, m_2 \ldots m_{Ne}, n\rangle$, such that

$$\begin{aligned} S_z^i|m_1,m_2\ldots m_{Ne},n\rangle &= m_i|m_1\ldots n\rangle, \\ a^+a|m_1,m_2\ldots m_{Ne}n\rangle &= n|m_1\ldots n\rangle, \end{aligned} \qquad (25)$$

and

$$S_{\pm}^{i}|m_1,m_2\ldots,\ m_i,\ldots\ m_{Ne}\rangle \ = |m_1,\ldots m_i\pm 1,\ldots n\rangle \tag{26}$$

Since

$$[S_z^i, S_i^{\pm}] \ = \ \pm S_i^{\pm}\ . \tag{27}$$

Now, let's re write the definition of the Green's function as:

$$(\Omega - H)G(\Omega) \ = \ 1, \tag{28}$$

and take matrix elements, multiplying by $\langle m_1\ldots m_i\ldots n|$ from the left and by $|m_1'\ldots m_{Ne}'n'\rangle$ from the right. Also, introducing a $1 = \sum_{\{m_i''\}} |m''\ldots\ m_i''\ldots n''\rangle\langle m_1''\ldots m_i''\ldots n''||$ and taking the matrix elements of H:

$$(H)_{m_1\ldots m_{Ne},n;m_1''\ldots m_{Ne}''n''}$$

$$= \ \omega\ \sum_i (m_i)^2 \delta_{m_i''m_i}\delta_{nn''} \ + \ ig[\sum_i \delta_{m_i''-1,m_i}\prod_{j\neq i}\delta_{m_j'',m_j}\delta_{n''+1,n}\sqrt{n''+1}$$

$$- \sum_i \delta_{m_i''+1,m_i}\prod_{j\neq i}\delta_{m_j'',m_j}\delta_{n''-1,n}\sqrt{n''}]\ , \tag{29}$$

we can readily obtain the following equation:

$$Q\{m_i\} - \sum_i X_{m_i}^{n} \ - g^2(n+1)\sum_i \frac{1}{X_{m_i-1}^{n+1}} \ = \ \frac{\delta_{n,n'}\prod_j \delta m_j,m_j'}{G_{m_i,n;m_i',n'}}\ , \text{ where} \tag{30}$$

$$Q\{m_i\} \ \equiv \ \sum\ (\frac{\Omega}{Ne} - \omega m_i^2), \tag{31}$$

$$X_{m_i}^{n} \ \equiv \ \frac{ig\sqrt{n}\ G_{m_{i+1},n-1;m_i',n'}}{G_{m_i n,m_i' n'}}\ ,$$

and the short hand notation $G_{m_i+1,n-1;m_i'n'}$ means $G_{m_i,m_2\ldots,m_i+1\ldots m_{Ne},n-1,m_i',m_2'\ldots m_{Ne}',n'}$. We will assume now that the primed quantum numbers will represent our initial state and furthermore, we will give the same initial momentum to all the electrons. Therefore, the initial state of our system will be

$|m,m,\ldots n_0\rangle$ or $m_i' = m,\ \ n' = n_0$.

Consider different situations for the final quantum numbers $m_i n$.

a) $m_i = m \; ; \; n = n_0$

In this case, we can write the eq. (30) as:

$$G_{m_i n, m_i n} = \frac{1}{\sum_{i=1}^{Ne} \left(\frac{\Omega}{Ne} - \omega m_i^2 - X_{m_i}^{N_0} - \frac{g^2(n_0+1)}{X_{m_i-1}^{n_0+1}}\right)} . \tag{32}$$

Since all the m_i's are equal, the sum in the eq. (32) reduces to a factor Ne. The poles of $G_{m_i n, m_i n}$ corresponds to:

$$\frac{\Omega}{Ne} - \omega m^2 - X_m^{n_0} - g^2 \frac{(n_0+1)}{X_{m-1}^{n_0+1}} = 0. \tag{33}$$

The eq. (33) give us a recursion relation for the X_m^n, whose solution is given in terms of continued fractions [8]:

$$X_m^{n_0} = \cfrac{g^2 n_0}{q_{m+1} - g^2 \cfrac{(n_0-1)}{q_{m+2} \ldots}} \; ; \; \frac{g^2(n_0+1)}{X_{m-1}^{n_0+1}} = \cfrac{g^2(n_0+1)}{q_{m-1} - \cfrac{g^2(n_0+2)}{q_{m-2} \ldots}} \tag{34}$$

with: $q_m = \frac{\Omega}{Ne} - \omega m^2$.

Notice that we have given two different functional forms for $X_m^{n_0}$, one ascending and the other one descending in the index of q_m. Furthermore, when we will have a power series expansion in g, we will treat the two expressions as independent. Of course these two expressions coincide at the poles when taking all orders of g(exact).

b) $m_i = m \; (i \neq j) \; ; \; m_j = m-1, \; n = n_0+1$

One can easily check that

$a^+a + \sum_i (S_z^i) =$ constant of motion, therefore if the ith electron will recoil in a unit of momentum, it will also emit an extra photon.

In this case, the eq. (30) can be written as:

$$\sum_{i \neq j} \left\{ \left(\frac{\Omega}{Ne} - \omega m^2\right) - X_{m_i}^{n_0+1}(m_j = m = 1) - \frac{g^2(n_0+2)}{X_{m_i-1}^{n_0+2}(m_j = m-1)} \right\}$$

$$+ \left[\frac{\Omega}{Ne} - \omega(m-1)^2 - X_{m-1}^{n_0+1} - \frac{g^2(n_0+2)}{X_{m-2}^{n_0+2}}\right] = 0. \tag{35}$$

The term in square brackets vanishes since it's just the eq. (33) with $m \to m-1$ and $n_0 \to n_0+1$.

Now, since all the m_i's are equal, the eq. (35) reduces to

$$q_m - X_m^{n_0+1}(m-1) - \frac{g^2(n_0+2)}{X_{m-1}^{n_0+2}(m-1)} = 0. \qquad (36)$$

However, the eq. (36) is identical to eq. (33) if $X_m^{n_0} \to X_m^{n_0+1}(m-1)$ and $n_0 \to n_0+1$. Therefore:

$$X_m^{n_0+1}(m-1) = X_m^{n_0+1}. \qquad (37)$$

In a similar way, it is simple to show that:

$$X_m^{n_0+1}(m+1) = X_m^{n_0+1},$$

$$X_m^{n_0+2}(m-1,m-1) = X_m^{n_0+2}(m-2) = X_m^{n_0+2}, \qquad (38)$$

$$X_m^{n_0+3}(m-1,m-1,m-1) = X_m^{n_0+3}.$$

and so forth.

The notation used in the eq. (38) is the following

$$X_m^{n_0+2}(m-1,m-1) \equiv X_{m_i}^{n_0+2}(m_j = m-1,\ m_k = m-1)$$

$$\equiv \frac{ig\sqrt{n_0+2}\,G_{m_i+1 m_j=m-1, m_k=m-1, n_0+2, mn_0}}{G_{m_i, m_j=m-1, m_k=m-1, n_0+2; mn_0}}$$

As we can see, for instance, a three-body function $X_{m_i}^{n_0+2}(m-1.m-1)$ can be expressed as a one-body function $X_{m_i}^{n_0+2}$, just setting the right number of photons. Therefore, we will consider Ne=1 in all the expressions for X_m^n and G. The Ne-body effects will appear, considering two simple quantum rules:

a) For a fixed initial state, if we have several paths to arrive to a common final state, we have to add the probability amplitudes of the various processes.

b) For a fixed initial state, if we have different final states, we add the various probabilities.

There are two fundamental processes that we will call i^+ and i^-

i^+ : The electron i recoils and emits a photon

i^- : The electron i absorbs a photon and gains a unit of momentum.

From the definition of X^n_m, we can write:

$$G_{m_i-1,n_0+1;\, m_i,n_0} = \frac{ig\sqrt{n_0+1}}{X^{n_0+1}_{m_i-1}} G_{m_i,n_0;m_i n_0} : (i^+) , \quad (39)$$

and

$$G_{m_i+1,n_0-1;m_i,n_0} = \frac{X^{n_0}_{m_i}}{ig\sqrt{n_0}} G_{m_i,n_0;m_i,n_0} : (i^-) \quad (40)$$

We can re-write both factors appearing in the eqs. (39) and (40):

$$\frac{ig\sqrt{n_0+1}}{X^{n_0+1}_{m_i-1}} = \frac{ig\sqrt{n_0+1}}{q_{m-1}-\frac{g^2(n_0+2)}{q_{m-2} \ddots}}, \quad \frac{X^{n_0}_{m_i}}{ig\sqrt{n_0}} = \frac{-ig\sqrt{n_0}}{q_{m_i+1}-\frac{g^2(n_0-1)}{q_{m_i+2} \ddots}}, \quad (41)$$

and notice that they are of the order of g (or higher). Everytime an emission or absorption occurs, there is a factor of the order of g(or higher). Let us take a higher order process leading to the emission of one photon, like $j^+i^+j^-$. This corresponds to the emission of a photon by the electron j^- followed by the emission by electron i and the absorption by the electron j. We will consider this scattering process, step by step.

Let us start by j^+. The corresponding factor is $\frac{ig\sqrt{n_0+1}}{X^{n_0+1}_{m_j-1}}$. The next one is i^+, and the electron (i) is in the presence of an extra photon and the momentum of the jth electron is m_j-1 and the factor is $\frac{ig\sqrt{n_0+2}}{X^{n_0+2}_{m_i-1}(m_j-1)}$ and finally j^- is an absorption and the electron j is in the presence of 2 extra photons and m_j-1, m_i-1, the factor is $\frac{X^{n_0+2}_{m_j-1}(m_i-1)}{ig\sqrt{n_0+2}}$. With this procedure, we can write, for example, the various terms corresponding to a given final state as well as the various final states contributing to the probability of emission of one photon P_1 (Assume we are studying the spontaneous emission, $n_0=0$)

$\underline{P_1}$: The simplest final state is:

$$|f\rangle_1 = |m_i-1,1\rangle$$

$$i^+ : G^{(1)}_{f_i} = \frac{ig\sqrt{1}}{X^{(1)}_{m-1}} G_{m_0,m_0} ,$$

$$i^+j^+j^- \; : \; G_{f_i}^{(2)} = \frac{X_{m-1}^{(2)}}{ig\sqrt{2}} \; \frac{ig\sqrt{2}}{X_{m-1}^{(2)}} \; \frac{ig\sqrt{1}}{X_{m-1}^{(1)}} \; G_{m0,m0}$$

$$j^+i^+j^- \; : \; G_{f_i}^{(3)} = \frac{X_{m-1}^{(2)}}{ig\sqrt{2}} \; \frac{ig\sqrt{2}}{X_{m-1}^{(2)}} \; \frac{ig\sqrt{1}}{X_{m-1}^{(1)}} \; G_{m0,m0} \; ,$$

$$i^+j^-j^+ \; : \; G_{f_i}^{(4)} = \frac{ig\sqrt{1}}{X_{m}^{(1)}} \; \frac{X_{m}^{(1)}}{ig\sqrt{1}} \; \frac{ig\sqrt{1}}{X_{m-1}^{(1)}} \; G_{m0,m0} \; ,$$

$$j^+j^-i^+ \; : \; G_{f_i}^{(5)} = \frac{ig\sqrt{1}}{X_{m-1}^{(1)}} \; \frac{X_{m-1}^{(1)}}{ig\sqrt{1}} \; \frac{ig\sqrt{1}}{X_{m-1}^{(1)}} \; G_{m0,m0} \; .$$

We immediately notice that $G_{f_i}^{(2)}$ and $G_{f_i}^{(3)}$ are identical, since (i^+) is associated with the operator $S_i^- a^+$ and (j^+) with $S_j^- a^+$ and they commute, therefore we consider it one and put a (2!) factor. In higher order terms like $i^+j^+k^+j^-k^-$ we have a (3!)(2!) factor, and so forth. Other possible final state is:

$$|f>_2 \; = \; |m_i-2, m_j+1, \; 1>$$

$$i^+i^+j^- \; : \; G_{f_i}^{(6)} = \frac{X_{m}^{(2)}}{ig\sqrt{2}} \; \frac{ig\sqrt{2}}{X_{m-2}^{(2)}} \; \frac{ig\sqrt{1}}{X_{m-1}^{(1)}} \; G_{m0,m0}$$

$$i^+j^-i^+ \; : \; G_{f_i}^{(7)} = \frac{ig\sqrt{1}}{X_{m-2}^{(1)}} \; \frac{X_{m}^{(1)}}{ig\sqrt{1}} \; \frac{ig\sqrt{1}}{X_{m-1}^{(1)}} \; G_{m0,m0} \; .$$

$$|f>_3 \; = \; |m_i-1, \; m_j-1, \; m_k+1, \; 1>$$

$i^+j^+K^-$ (factor 2)

$$: \; G_{f_i}^{(8)} = \frac{X_{m}^{(2)}}{ig\sqrt{2}} \; \frac{ig\sqrt{2}}{X_{m-1}^{(2)}} \; \frac{ig\sqrt{1}}{X_{m-1}^{(1)}} \; G_{m0,m0} \; ,$$

$i^+K^-j^+$ (factor 2)

$$: \; G_{f_i}^{(9)} = \frac{ig\sqrt{1}}{X_{m-1}^{(1)}} \; \frac{X_{m}^{(1)}}{ig\sqrt{1}} \; \frac{ig\sqrt{1}}{X_{m-1}^{(1)}} \; G_{m_0,m_0}$$

There are of course higher order terms (in $g^{(8)}$ or higher) contributing to P_1 which have not been included here. If we associate the amplitude $a_\mu \to G_{f_i}^{(\mu)}$, we can write

$$P_1 = \{|a_1 + (Ne-1)(a_2+a_3+a_4+a_5) + (Ne-1)(Ne-2)(\ldots) + \ldots|^2 \; Ne + \frac{Ne(Ne-1)}{2!} \; |(a_6 + a_7)+\ldots|^2 + \frac{Ne(Ne-1)(Ne-2)}{3!} \; |2a_8+2a_9+\ldots|^2 + \ldots \quad (42)$$

In a similar way, we can write for P_2:

$$|f\rangle_4 = |m_i-2,2\rangle$$

$$i^+i^+ = G_{f_i}^{(10)} = \frac{ig\sqrt{2}}{X_{m-2}^{(2)}} \frac{ig\sqrt{1}}{X_{m-1}^{(1)}} G_{m_0,m_0}$$

$$|f\rangle_5 = |m_i-1,m_j-1,2\rangle$$

$$i^+j^+ \text{ (factor 2)} : G_{f_i}^{(11)} = \frac{ig\sqrt{2}}{X_{m-1}^{(2)}} \frac{ig\sqrt{1}}{X_{m-1}^{(1)}} G_{m_0,m_0}$$

and

$$P_2 = \{Ne|a_{10}+\dots|^2 + \frac{Ne(Ne-1)}{2} |2a_{11}+\dots|^2\dots\} \tag{43}$$

P_3:

$$|f\rangle_6 = |m_i-3,3\rangle$$

$$i^+i^+i^+ : G_{f_i}^{(12)} = \frac{ig\sqrt{3}}{X_{m-3}^{(3)}} \frac{ig\sqrt{2}}{X_{m-2}^{(1)}} \frac{ig\sqrt{1}}{X_{m-1}^{(1)}} G_{m_0,m_0}$$

$$|f\rangle_7 = |m_i-2,m_j-1,3\rangle$$

$$i^+i^+j^+ : G_{f_i}^{(13)} = \frac{ig\sqrt{3}}{X_{m-1}^{(3)}} \frac{ig\sqrt{2}}{X_{m-2}^{(2)}} \frac{ig\sqrt{1}}{X_{m-1}^{(1)}} G_{m_0,m_0},$$

$$i^+j^+i^+ \text{ (factor 2)} : G_{f_i}^{(14)} = \frac{ig\sqrt{3}}{X_{m-2}^{(3)}} \frac{ig\sqrt{2}}{X_{m-1}^{(2)}} \frac{ig\sqrt{1}}{X_{m-1}^{(1)}} G_{m_0,m}$$

$$|f\rangle_8 = |m_i-1,\ m_j-1,\ m_{k-1},3\rangle$$

$$i^+j^+K^+ \text{ (factor 3!)} \quad G_{f_i}^{(15)} = \frac{ig\sqrt{3}}{X_{m-1}^{(3)}} \frac{ig\sqrt{2}}{X_{m-1}^{(2)}} \frac{ig\sqrt{1}}{X_{m-1}^{(1)}} G_{m_0,m_0}$$

and

$$P_3 = \{|a_{12}+\dots|^2 Ne + \frac{Ne(Ne-1)}{2} |a_{13} + 2a_{14}+\dots|^2 + \frac{Ne(Ne-1)(Ne-2)}{3} |a_{15}(3!) + \dots|^2 + \dots\} \tag{44}$$

This calculation allows us, in principle, to calculate $g^{(2)}(\tau)$, $\langle n \rangle$ $\langle n^2 \rangle$ to any order in (g). A preliminary calculation shows that other previous perturbative calculations[7] can be reproduced. A detailed numerical analysis and results obtained by exploiting this method will be the subject of a near future publication.

References

[1] D. A. G. Deacon et al, Phys. Rev. Lett. 38, 892 (1977)

[2] J. Billardon et al, Phys. Rev. Lett. 51, 1652 (1983)

[3] A. Bambini and A. Renieri, Lett. Nuovo Cimento 31, 399 (1978)

[4] M. Orszag, Optica Acta, 31, 267 (1984)

[5] M. Orszag, R. Ramírez, Report IC/85/25 Trieste 1985, also M. Orszag, R. Ramírez (to be published JOSA)

[6] M. V. Fedorov, Progress in Quant. Electr. 7, 73 (1981)

[7] W. Becker and J. K. McIver, Phys. Rev. A27, 1030 (1983)

[8] J. Roessler, M. Orszag, R. Ramírez, J. Phys. A, 15, 1353 (1982)

Time-Symmetric Electrodynamics and Quantum Measurement

D.T. Pegg

School of Science, Griffith University, Nathan, Brisbane 4111, Australia

1. Introduction

A phenomenon such as the instantaneous collapse of the wavefunction associated with the measurement process seems necessary in order to predict results from quantum mechanics. That such a postulate gives correct results for spatially extended quantum optical EPR type experiments, as has been experimentally verified, makes it somewhat of an enigma, because an instantaneous collapse does not have a Lorentz covariant description. Philosophical arguments based on logical positivism, which involve abandoning the concept of physical reality, have been used to circumvent this difficulty, but to many these are not satisfying and attempts are still being made to understand the experiments in terms of real physical processes which permit a specification of the quantum state during all evolution stages before the measurement takes place. The simplest type of experiment concerns the detection of a photon from a two-level atom a by a detector at a distance r away. A question which might be asked is - does the detection or measurement have a real physical influence which causes a to jump to the ground state and, if so, when does this occur, at the same time as detection, or a time r/c before or at r/c after detection?

It is the purpose of this paper to show how the Wheeler-Feynman theory of time-symmetric electrodynamics gives definite answers to such questions and to more complicated questions concerning the intermediate states in a quantum optical EPR type experiment. The essential feature of this theory is the non-rejection of the advanced potential component of the solution of the electrodynamic equations. To achieve this purpose, the paper has been written with a minimum of mathematical detail which might otherwise obscure the essential arguments. The explicit mathematical derivations can be found in the references given, and in the references they contain.

2. Time-Symmetric Electrodynamics

Maxwell's electrodynamic equations, both in classical and operator form, are time symmetric and permit both retarded and advanced potential solutions and normalized linear combinations of the two. Conventionally the advanced potential solution, which represents radiation propagating backward in time, is immediately rejected on the grounds that its retention might lead to an observable violation of retarded causality, for example, it might be possible for an accelerated charge to affect a distant test charge before the acceleration takes place. In time-symmetric electrodynamics, on the other hand, the advanced component of the solution is not immediately rejected and instead the potential from an isolated charge is taken as half-retarded and half-advanced. This symmetry implies that, in an otherwise charge-free universe, an isolated accelerated charge would neither lose nor gain energy or, quantum mechanically, an isolated excited atom would not spontaneously decay. At first sight this seems to indicate that rejection of the advanced component is required to obtain agreement with observation, but WHEELER and FEYNMAN [1] were able to show, in their absorber theory, that the action of the other charges in the universe is just sufficient to cause an accelerated charge to lose energy at the correct rate and the ad hoc rejection of the advanced potential of these grounds is not justified. They showed that it is the absorber which is the source of the radiative reaction.

The formal description of the action of the classical absorber is given in Ref. [1], here we illustrate the mechanism with reference to Fig.1. The time-symmetric field from the accelerated charge a is $\frac{1}{2}F(ret,a) + \frac{1}{2}F(adv,a)$ where the brackets refer to the retarded and advanced components of the field from a. In Fig. 1 we consider only one particular directional component of the field and call the retarded and advanced parts of this R and A respectively. R leaves a at time t_0, travels into the future and interacts with the absorber, causing the charges in the absorber to move. For complete absorption these charges must produce a field -R on the outside (here the right side) of the absorber. But in doing so, they must also produce a corresponding advanced field which, if the retarded field is -R, must be equal to -A. This advanced field -A travels back into the past and interacts with a at the same instant t_0 as a produces R and A, and cancels A on the left side of a. The final non-zero fields are R from a acting on the absorber, giving energy to it, and the advanced response field -A from the absorber acting on a and removing energy from a. The sum of R over all directions is $\frac{1}{2}F(ret,a)$. The corresponding sum of -A is a field continuous near and at a, i.e. it exists on both the past and future light cones of a, which is equal to $\frac{1}{2}F(adv,a)$ on the past light cone. The only such continuous field is $\frac{1}{2}F(ret,a) - \frac{1}{2}F(adv,a)$, which is a solution of the homogeneous Maxwell equations. Thus the sum of -A is just Dirac's radiative reaction field. Also the total field between a and the absorber reduces to just $\frac{1}{2}F(ret,a) + \frac{1}{2}F(ret,a)$, that is, to the fully retarded field of experience.

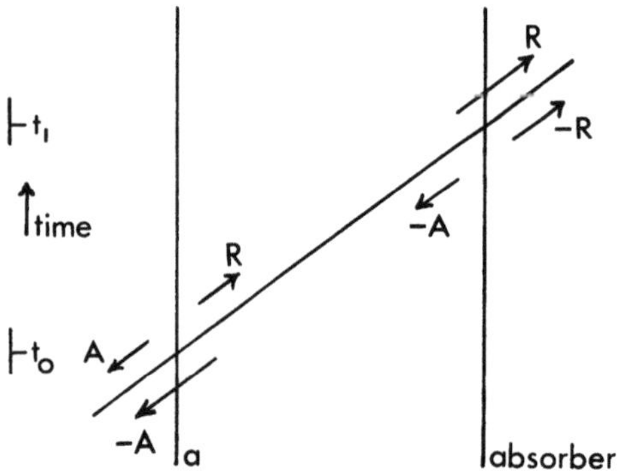

Figure 1. Mechanism for energy transfer in absorber theory.

To sum up, the advanced response field from the absorber provides the radiative reaction $\frac{1}{2}F(ret,a) - \frac{1}{2}F(adv,a)$ which (a) acts on a at t_0, removing energy at the correct rate, (b) cancels the half-advanced field from a, reducing the total field on the past light cone of a to zero and (c) adds to the half-retarded component $\frac{1}{2}F(ret,a)$ from a on the future light cone, increasing its value to the full F(ret,a). Thus an introduced test charge will only be affected after the time t_0

and retarded causality is not violated. For example, the advanced field from a can not be used to propagate information backwards in time. Again there is no justification for the rejection of the advanced component on these grounds, as is done in the conventional theory.

3. Quantum Absorber Theory

In a universe which can completely absorb all retarded radiation, classical absorber theory gives results indistinguishable from those of conventional classical electrodynamics, with the advanced effects being completely hidden. Are there differences in the quantum domain, that is, is the difference between quantum and classical electrodynamics sufficiently great for advanced effects to be apparent in quantum absorber theory?

Different approaches to a quantum form of absorber theory have been adopted by different authors (see, for example Refs. [2,3,4] and the references therein). All these approaches predict consistent final results of experiments which, disappointingly, agree with those obtained from conventional electrodynamics. Some of the effects calculated are the spontaneous emission rate, Lamb shift, dynamic Stark spectrum, photoelectron antibunching and violation of Bell's inequalities [5,6]. Although the final results are the same as for the conventional theory, absorber theory is a physically different theory in that the radiative reaction field is an advanced field whose source is the absorber, so the intermediate mechanisms should be different from those of conventional theory. To understand why we might expect some advanced effects to be manifest in the intermediate processes of quantum absorber theory, while being hidden in classical absorber theory, we recall that in the classical case different parts of the outgoing retarded field from the emitter are absorbed at different, well separated, parts of the absorber and so the total radiative reaction is independent of the properties of any particular part of the total absorber, and is completely isotropic, just as it is the conventional theory. In the quantum domain, however, when the emitter is a single excited atom, the absorption in the absorber is very localized, that is, the energy lost can appear as an excitation in a single absorber atom, with the other absorber atoms being unaffected. This means that the radiative reaction is not only advanced but is also highly directional and can indeed by polarized, with a polarization corresponding to a property of the absorbing atom. This is very different from conventional quantum electrodynamics where the radiative reaction has no particular preferred direction, whether it is due to vacuum fluctuations or self-action, and can have no correlation with any state of the absorbing atoms. Thus the physical description of detection events involving the absorption of a single photon will, in terms of absorber theory, be different from that in terms of conventional theory.

The time symmetry of absorber theory permits the use of a direct-action formalism [3,5] for which the interaction of two atoms a and b is described by an interaction Hamiltonian [3] which, to avoid unnecessary geometric factors, can be written as

$$\mathcal{H}_{ab} = -\mu_o e^2 \underset{\sim}{v}_a \cdot \underset{\sim}{v}_b / (4\pi r) \tag{1}$$

where $\underset{\sim}{v}_b$ is perpendicular to the displacement $\underset{\sim}{r}$ of b from a. With this interaction Hamiltonian the evolution of the states of a and b is described in terms of the retarded time variable, which is equal to the normal time minus r/c, where r is the distance of the particular atom from a[3]. We are interested in the probability amplitude that the system initially (in terms of retarded time) in state $|a_e\rangle|b_g\rangle$, where these refer to the excited and ground states of a and b, makes a transition. For r sufficiently large, first order perturbation theory can be used. We find [5] that the only non-vanishing transition amplitude is to the state $|a_g\rangle|b_e\rangle$, which is only appreciable when the energy gap of b is close to that of a. It follows that if b is found in $|b_e\rangle$ then a must be found in $|a_g\rangle$ at the same retarded time. It is the retarded action of a on b which excites b and the corresponding advanced action of b on a, the radiative reaction, which pushes a from $|a_e\rangle$ to $|a_g\rangle$. Thus, if b is an atom in a photodetector, it is the detection event, that is the upward transition in b, which is the physical cause, via the advanced action, of the "jump" of a from its excited to its ground state. We note that a is forced into its ground state immediately "after" the detection event in retarded time, which in fact is r/c before the detection event in real time. This is the simplest manifestation of an advanced effect in absorber theory, and there can be no similar effect in conventional theory, in which the advanced potential is rejected.

We can also include the interaction of a with other detector atoms c, d ... by introducing the appropriate extra terms into the Hamiltonian. We then find [5] that if any one of these atoms is excited, a is put into its ground state at the same retarded time, which immediately makes it impossible for a to excite any other atom, unless a is then itself re-excited. This, of course, is just the photoelectron antibunching effect. In addition, the advanced radiative reaction from the detector atom has another effect, besides forcing a to $|a_g\rangle$, which arises because of its directionality. Because it is highly directional, the radiative reaction exerts a radiation pressure on a causing a to recoil in the direction opposite to that in which the detector atom lies. Again this recoil takes place at time r/c before the detection event, but is a physical consequence of the absorption which will take place in a particular atom at a future time. As such, even though absorption by any one of a number of atoms isotropically distributed around a may be equally likely, the direction of the atom which will eventually be excited can be determined in advance by measuring the direction of the recoil which the detection event causes.

4. Kocher-Commins Experiment

While the atomic recoil is a consequence of the directionality of the radiative reaction, a more complicated and interesting effect is that due to the polarization of the radiative reaction. This involves the fresh interpretation which absorber theory offers for the Kocher-Commins experiment [7], which has recently been repeated in a more rigorous form [8]. In this experiment an atom a can undergo cascade decay from a J=0 state $|e\rangle$ to a J=1 state $|m\rangle$, emitting photon B say, then to a J=0 state $|g\rangle$, emitting photon C. Photons B and C are detected by polarized detectors c and b in opposite directions, along the z-axis, with polarizers at angles ϕ_c and ϕ_b to x-axis. Given a sufficient detector-emitter distance, the angles can be set after the transitions in atom a have taken place. The joint probability of detection is found to be proportional to $\cos^2(\phi_b-\phi_c)$. A possible space-time configuration for such an experiment is shown in Fig.2, in which photon C is detected first. The conventional interpretation is that, after emission and before detection, the field, in a linear x,y polarization basis, is in the state

$$|2\ \text{photons}\rangle = 2^{-\frac{1}{2}}(|x_B\rangle|x_C\rangle + |y_B\rangle|y_C\rangle) \tag{2}$$

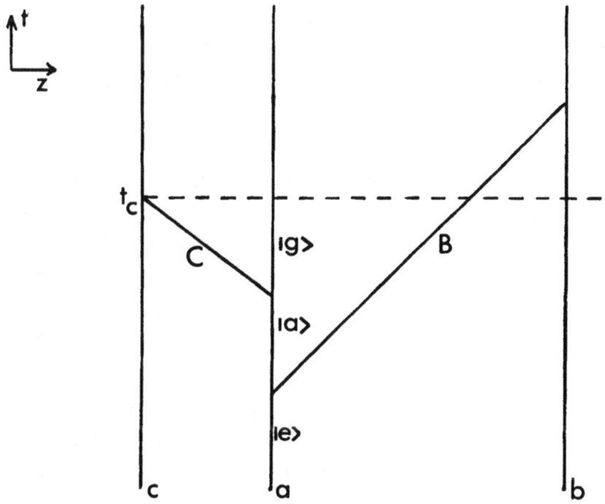

Figure 2. Space-time diagram for Kocher-Commins experiment.

The detection of C at time t_C projects state (2) onto $\cos\phi_c|x_C\rangle + \sin\phi_c|y_C\rangle$, immediately leaving the field in the one-photon state $\cos\phi_c|x_B\rangle + \sin\phi_c|x_B\rangle$. Thus at time t_c the photon B is given the polarization angle ϕ_c. This result does seem strange when we remember that in principle photon B could be, for example, light days away from a before photon C is even emitted by a and before the experimenter has even chosen the value at which to set his adjustable angle ϕ_c. Nevertheless this conventional recipe does predict the correct result. An obvious difficulty with the recipe is that it is not Lorentz covariant, because it invokes an effect on B which is simultaneous with the detection of C. Different inertial observers will disagree when this takes place.

Absorber theory offers an alternative mechanism. All the interactions,advanced and retarded, take a place along the light cone so covariance is preserved. We can let polarized detectors b and c be single atoms with $\underset{\sim}{y}_b$ and $\underset{\sim}{y}_c$ at angles ϕ_b or ϕ_c to the x-axis. If photon B is absorbed by b, then b is the source of the first transition advanced radiative reaction, which will thus have polarization ϕ_b. It is not difficult to see that, under the action of radiation with this definite polarization, a will make a transition to a corresponding definite superposition of the m=1, m=0, and m=-1 states of the J=1 intermediate state. We can calculate this superposition [6] indeed to be simply

$$|a\rangle = -\exp(-i\phi_b)|+1\rangle + \exp(i\phi_b)|-1\rangle \tag{3}$$

The light emitted by the subsequent transition $|a\rangle$ to $|g\rangle$ will have its electric vector in the direction $\langle g|\underset{\sim}{y}_a|a\rangle$ which is just [6] $\cos\phi_b\underset{\sim}{i} + \sin\phi_b\underset{\sim}{j}$, i.e. the light from the second transition is polarized at the time of emission in the direction ϕ_b. Consequently the probability of absorption by c, if there is absorption at b, is just $\cos^2(\phi_b-\phi_c)$.

In absorber theory, therefore, we can trace the individual dynamical steps in the overall process and assign a definite state to the system at all times. The advanced nature of the action of b on a is crucial for this. Our final result is the same as that calculated with quantum electrodynamics and it is worthwhile noting that we have obtained it without having to invoke any overall conservation laws. These follow automatically from the detailed dynamics, as they should in good theories. The atom exists in a real, definite intermediate state after the first transition, which is determined by the polarization of the absorbing atom at the later time of detection, i.e. the atom possesses a memory of a future event. It is quite clear that the results are independent of precisely when the polarizers are set before detection.

Thus, according to absorber theory, the results of the Kocher-Commins experiment are a manifestation of the influence of the future on the past by means of an advanced potential as an intermediate part of the dynamics. The question naturally arises, can we make use of this violation of retarded causality at the intermediate level to send useful, that is controllable, information at a speed greater than that of light, or perhaps even into the past? We might envisage an ensemble of such experimental arrangements, or the same arrangement used at successive times, in which we alter the polarizer in front of detector b in order to control the intermediate state of a and hence affect the polarization of the light reaching detector c, thereby sending a controllable message. Unfortunately, this would not work for the following reasons. Suppose the polarizer at b is set at ϕ_b. If the photon heading towards b passes through the polarizer and is thus absorbed by b, this polarization will be transmitted to a and hence to the photon heading towards c. If, however, the photon B is absorbed by the polarizer, the polarizer becomes the source of radiative reaction which will then have polarization direction $\phi_b + \pi/2$. The important point is that each photon has at best a probability of one-half of being absorbed by the detector and one-half of being absorbed by the polarizer, and we have no control over which one will do the absorbing. Thus the photons heading towards c will have a 50% probability of being polarized at ϕ_b and a 50% probability of being polarized at $\phi_b + \pi/2$, and will thus appear as unpolarized light, carrying no useful information. We see therefore that while the quantization of absorber theory gives rise to localized absorption and directional polarized advanced radiative reaction, it is also the quantum probabilistic nature of the theory which prevents us from making use of these properties to violate retarded causality in the transmission of controllable information.

5. Conclusion

Although not designed by Wheeler and Feynman specifically for the purpose, time-symmetric electrodynamics allows an additional factor to be introduced into attempts to understand measurement processes involving the absorption of photons. This factor is the advanced nature of the physical reaction of the detector back on the atomic system being measured, a reaction which can not occur in conventional quantum electrodynamics. This gives an entirely fresh understanding of experiments of the Kocher-Commins type. Indeed the explanation in terms of absorber theory is so natural that it is tempting to regard such experiments as support for absorber theory.

References

1. J.A. Wheeler and R.P. Feynman: Rev.Mod.Phys. 17, 157 (1945); Rev.Mod.Phys. 21, 425 (1949).

2. F. Hoyle and J. Narlikar: Action at a Distance in Physics and Cosmology (Freeman, San Francisco, 1974).

3. D.T. Pegg: Ann.Phys.(N.Y) 118, 1 (1979); Rep.Prog.Phys. 38, 1339 (1975); Physica Scripta, in press (1986).

4. A.V. Durrant: Proc.R.Soc.A. 370, 41 (1980); Eur.J.Phys. 1, 106 (1980).

5. D.T. Pegg: J.Phys.A: Math.Gen. 13, 1389 (1980).

6. D.T. Pegg: Eur.J.Phys. 3, 44 (1982).

7. C.A. Kocher and E.D. Commins: Phys.Rev.Lett. 18, 575 (1967).

8. A. Aspect, J. Dalibard and G. Roger: Phys.Rev.Lett. 49, 1804 (1982).

Part IV

Optical Tests of General Relativity and Other Contributions

Quantum-Noise Quenching in Correlated Spontaneous Emission Lasers, and Their Application to High-Precision Measurements

J. Gea-Banacloche and M.O. Scully

Max-Planck-Institute für Quantenoptik,
D-8046 Garching bei München, Fed. Rep. of Germany and
Center for Advanced Studies and Department of Physics and Astronomy,
University of New Mexico, Albuquerque, NM 87131, USA

1. Introduction: Quenching Spontaneous Emission Noise

The concept of "correlated spontaneous emission" lasers has been introduced recently [1]. In these devices, the spontaneous emission events occurring in two different laser modes are correlated in such a way that, under appropriate conditions, the relative phase between the two modes may be freed from random noise. Ordinarily, of course, the phase of a laser undergoes a diffusion process [2] such that, after a time t, its value has accrued an uncertainty

$$\Delta\phi = \sqrt{Dt} , \tag{1}$$

where D is a diffusion constant intimately related to the laser linewidth. In terms of the cavity damping rate γ and the number of photons in the cavity $\bar{n}$ one has

$$D = \frac{\gamma}{2\bar{n}} . \tag{2}$$

Given then any two lasers, their phases will drift independently, so that the relative phase will diffuse away at twice the diffusion rate (2). This is a restriction on the accuracy achievable in measurements that involve comparing the relative phase (or the frequencies; see Ref. [3]) of two lasers; such as, e.g., in gravity-wave detection [4] or ring laser gyros [5]. The latter hold promise as a probe to test the predictions of different metric theories of gravitation [5]. Thus it is of interest to see if their sensitivity can be improved beyond the current limits, one of which is precisely this "quantum noise" just described.

In a correlated spontaneous emission laser (CEL) [1], one common atomic medium is used to radiate two laser modes whose relative phase will be free of noise. The need for a common atomic medium is clear, since only in this way could one correlate the random spontaneous emission events. In addition, some applications (e.g., gravity-wave detection) require the two modes to oscillate in different cavities. One must then be able to separate them, which may be achieved if they have sufficiently different frequencies, or different polarizations. This leads one naturally to consider three-level schemes such as in Fig. 1.

On the other hand, some other applications do not require two physically separated cavities; for example, the ring laser gyroscope. As shown in Ref. [1b], this system does not require a three-level medium, since the same atomic transition can support two oppositely-directed traveling waves. In this case the phase correlation between the two waves is ensured by modulating the gain medium, so that the two traveling waves will tend to add to form an interference pattern with minimum intensity at the points of smaller gain (the pattern doesn't necessarily have to be a standing wave; it could be a moving grating, if the two counterpropagating waves have slightly different frequencies).

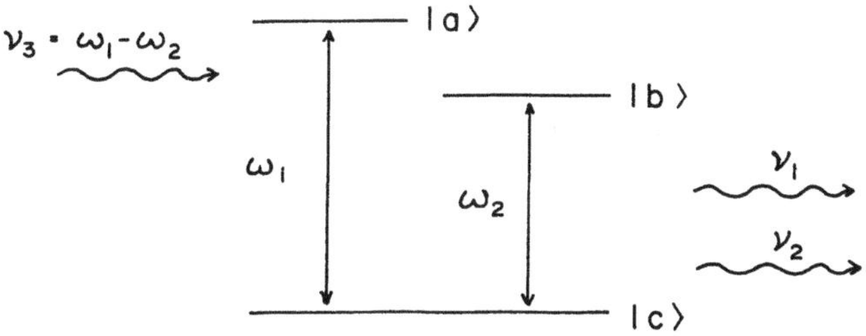

Fig. 1. Three-level medium used in the quantum-beat CEL. The laser operates simultaneously on the two transitions shown (ν_1, ν_2 = laser frequencies) while the upper levels $|a\rangle$ and $|b\rangle$ are mixed coherently by a strong external field at frequency ν_3. For the Hanle-effect CEL, the levels $|a\rangle$ and $|b\rangle$ are degenerate, and the two transitions differ by their polarization. In this case no external mixing field is used; rather, the atom is excited into a coherent superposition of $|a\rangle$ and $|b\rangle$ by the pumping mechanism.

The simplest example of the above, a ring laser operating in a "standing wave" configuration, shows in a special case how this perfect phase correlation is possible; in effect, the two oppositely-directed traveling waves carry the same phase information and no diffusion in the relative phase results when the relative phase angle ψ vanishes. A similar situation obtains in the other schemes mentioned (quantum-beat laser, Hanle-effect laser [1]): when correlation is perfect (i.e., when $\psi = 0$), the laser may be seen as operating in a single, generalized mode. However, one is in general interested in the situation wherein ψ is not equal to zero, since the result of the measurement one wants to perform is precisely given by the phase difference. Hence, the relative phase must be allowed to change in order to register the information of interest, and the correlation cannot be perfect.

2. Theory of the CEL Detector

In general, the phase angle between the two modes is found to obey an equation of the form [6]

$$\dot{\psi} = a - b \sin \psi . \tag{3}$$

Here "a" is a term proportional to the physical quantity that we want to measure (e.g., gravity-wave amplitude, rotation rate...). In the absence of the second term, it would lead to a phase difference ψ increasing linearly with time (i.e., a frequency difference between the two lasers). The second term has the typical form of a mode-locking term (see, for example, [2], Chapter 9, or [5]) and arises from the correlation imposed on the atomic medium. Under the circumstances of interest (those leading to maximum correlation, or vanishing relative-phase diffusion) one has $b = 2\alpha$, where α is the (saturated) gain per unit time, in each of the two modes.

As for the relative-phase diffusion itself, it is given by the diffusion constant [1,6,7]

$$D(\psi) = \frac{\gamma}{\bar{n}} (1 - \cos\psi) . \tag{4}$$

The meaning of $D(\psi)$ is clearer when one considers the Fokker-Planck equation for the probability density $P(\psi,t)$ (the probability that the relative phase has the value ψ at time t). One finds [7] that $P(\psi,t)$ obeys

$$\frac{\partial P}{\partial t} = -\frac{\partial}{\partial \psi}\left[(a - b \sin\psi)P\right] + \frac{1}{2}\frac{\partial^2}{\partial \psi^2}(D(\psi)P) . \tag{5}$$

In the absence of an external perturbation ($a = 0$) (3) shows that ψ locks at the value $\psi = 0$, and (4) then gives a vanishing diffusion constant, i.e., the relative phase is then completely free from noise. This corresponds to perfectly correlated operation. On the other hand, when $a = 0$, but $a < b$ (as will often be the case) one is still in a locked regime, but the lock-in angle ψ now is given by

$$\psi = \sin^{-1}\left(\frac{a}{b}\right) \simeq \frac{a}{b} , \tag{6}$$

(the approximation being good for $a \ll b$); in this case $D(\psi)$ (see (4)) is not exactly zero, and the correlation is not perfect, although the diffusion will be very small if $a \ll b$ (this will be shown below; for now, note only that $D(\psi)$ is of second order in a/b).

Equation (6) shows that information concerning the perturbation (the quantity a) is still available in the locked regime. However, the size of the signal is not very large. One can see that the free-running lasers would produce (in the absence of the second term in (3)) a phase difference $\psi(t) = at$ in a time t, while the size of (6) is only of the order of $a\tau$, where τ is the photon lifetime in the cavity (since $\tau = 1/\gamma$, and $\gamma = \alpha$, the saturated gain, and, as we noted above, $b = 2\alpha$). In the case of a gravity-wave detector, [4] this completely offsets the potential advantage of this "active" device over the usual passive-cavity interferometric techniques.

Indeed, if a measurement time t_m is available, an ideal situation would be obtained if the locking coefficient b could be reduced to be of the order of, say, $1/t_m$. Then the signal (6) could be as large as $a\,t_m$ (provided that this is still < 1), just as in the free-running case with the given measurement time. Two questions may be asked at this point: First, is such a locking reduction in fact possible? Second, will such a procedure spoil the correlation between the modes to the point of negating the phase-diffusion quenching?

The first question has been addressed in Ref. [4]. It is shown there that if one extracts some light from one of the cavities and injects it into the other, a mode coupling results that leads to the addition of a term $c \sin\psi$ to the right-hand side of (4):

$$\dot{\psi} = a - b \sin\psi + c \sin\psi . \tag{7}$$

This is of course the same as reducing b. It is shown in [4] that c may be made as large as $2\gamma_c$, where γ_c represents the loss rate of each cavity due to precisely this extraction-reinjection coupling (a different port needs to be used for injection than for extraction, to avoid turning the problem into one of three coupled cavities). The size of the signal is now given by

$$\psi = \sin^{-1}\left(\frac{a}{b - c}\right) \simeq \frac{a}{b - c} , \tag{8}$$

(where as before the last equality will hold only provided that a is still $\ll b-c$, which will generally be the case). Clearly the signal of (8) is larger than that given by (6). Since ψ will not lock unless the measurement time t_m is larger than $1/(b-c)$ ($b-c$ is the new locking coefficient), (8) is in general not valid if $b-c < 1/t_m$, and in general one can see from (7) that one cannot get $\psi > a\,t_m$ in a measurement lasting a time t_m. But when $b-c \gtrsim 1/t_m$, (8) shows that one can in principle obtain as large a phase shift as with free-running lasers.

Let us also note in passing that for the ring laser one could think of alternative mechanisms to yield the desired coupling, $c \sin\psi$, in (7): e.g., backscattering from one traveling wave into the other. This will be discussed in more length elsewhere [8].

3. Noise when $\psi \neq 0$.

Having seen that a respectable signal can be achieved we next address the second question, namely, will this increase in signal be accompanied by a corresponding increase in noise? The extraction-reinjection technique by itself is noiseless; however, one may be concerned that the perfect correlation that obtains when $\psi = 0$ is spoiled, and therefore $D(\psi)$ (see (4)) no longer vanishes, and the information contained in the relative phase diffuses away. As discussed below, this is not the case.

The diffusion of ψ will in fact be given by (5), but with b-c in place of b. We may use this equation to follow the evolution of both $\langle\psi(t)\rangle$, the average value of the signal, and $\langle(\psi - \langle\psi\rangle)^2\rangle \equiv \langle\Delta\psi^2\rangle$, the average random diffusion error.

Assume that the signal we want to detect is very small, and that we start from the locked ($\psi = 0$) configuration, so that $\psi \ll 1$ always. Then we may expand the trigonometric functions in (4) and (5) to yield

$$\frac{\partial P}{\partial t} = -\frac{\partial}{\partial\psi}\left[(a - (b-c)\psi)P\right] + \frac{\gamma}{4\bar{n}}\frac{\partial^2}{\partial\psi^2}(\psi^2 P). \tag{9}$$

This yields the moment equations

$$\frac{d}{dt}\langle\psi\rangle = a - (b-c)\langle\psi\rangle, \tag{10a}$$

$$\frac{d}{dt}\langle\Delta\psi^2\rangle = -\left(2(b-c) - \frac{\gamma}{2\bar{n}}\right)\langle\Delta\psi^2\rangle + \frac{\gamma}{2\bar{n}}\langle\psi\rangle^2 . \tag{10b}$$

According to (1) and (2), over a measurement time t_m the error in the relative phase would be, for the free-running (not CEL) lasers,

$$\Delta\psi = \sqrt{2Dt_m} = \sqrt{\frac{\gamma t_m}{\bar{n}}} . \tag{11}$$

On the other hand, equation (10b) yields, for the CEL,

$$\Delta\psi_{CEL}(t_m) = \left[\frac{\gamma}{2\bar{n}}\int_0^{t_m} e^{-(2(b-c)-\gamma/2\bar{n})(t-t')}\langle\psi(t')\rangle^2 dt'\right]^{1/2} , \tag{12}$$

which is, at most, of the order of

$$\Delta\psi_{CEL}(t_m) < \sqrt{\frac{\gamma t_m}{2\bar{n}}}\langle\psi(t_m)\rangle , \tag{13}$$

which is smaller than (11) when the signal $\langle\psi(t_m)\rangle \simeq a/(b-c)$ is small. We find therefore that the noise-quenched configuration of the CEL offers an advantage in terms of the signal-to-noise ratio, since the signal phase difference may be made comparable to that obtained from two independent lasers, while the CEL noise is much smaller.

Finally, we point out that some complications arise in the gravity-wave detector case having to do with the extraction of "signal" beams from the cavity, which lead to a somewhat reduced effective signal [4], although still greater than the passive interferometers. Thus we find an increased signal-to-noise ratio for our envisioned CEL gravity wave detector, as well as the CEL laser gyro [8].

4. Acknowledgments

This work was supported by the ONR and the AFOSR.

5. References

1. (a) M. O. Scully: Phys. Rev, Lett., 55, 2802 (1985); (b) M. O. Scully, (Phys. Rev.), to be published.
2. M. Sargent III, M. O. Scully, W. E. Lamb Jr.: Laser Physics (Addison-Wesley, Reading, MA, 1974).
3. J. Gea-Banacloche, M. O. Scully, and D. Z. Anderson: Opt. Commun., to be published.
4. M. O. Scully and J. Gea-Banacloche: to be published.
5. W. W. Chow, J. Gea-Banacloche, L. M. Pedrotti, V. E. Sanders, W. Schleich and M. O. Scully: Rev. Mod. Phys. 57., 61 (1985).
6. L. M. Pedrotti and M. O. Scully: R. J. Glauber Festschrift, to be published.
7. M. O. Scully, et al.: to be published.
8. M. O. Scully: Phys. Rev. A, to be published.

Ring Interferometric Tests of Nonclassical Gravitational Effects

G.E. Stedman

Department of Physics, University of Canterbury, Christchurch 1, New Zealand

The use of various detectors, and particularly ring interferometers of various types, in testing newtonian gravitational effects in the quantum limit, or post-newtonian effects at the classical or quantum level, is reviewed.

1. INTRODUCTION

First we describe the essential physical characteristics of a variety of ring interferometers: the optical ring interferometer, including the ring laser gyroscope; the SQUID (superconducting quantum interference device, based on Josephson interferometry); and the neutron interferometer, in connection with the Sagnac effect.

We then discuss their relative merits in testing theories of gravity, and the extent to which quantum properties are of fundamental importance in these tests. This follows my recent review [1] with some updating and extension, including in particular a fuller discussion of the range and type of experiments of interest in such a "quantum gravity" context. As a general rule, references will be given only for such updated topics. There were two reasons for my earlier review. Each of the above types of interferometry has its own enthusiastic and highly expert advocates, some of whom are represented at this Symposium. However, a comparison at an elementary level had not been previously published, as far as I am aware. Second, Professor Hans R Bilger of the School of Electrical and Computer Engineering, University of Oklahoma, Stillwater, Oklahoma has suggested that we in Christchurch, New Zealand could be interested in collaboration on a global network of optical ring interferometers for measurement of gravitational angular momentum - or gravitomagnetic effects - however caused, at the level at which general relativistic effects become significant. It was necessary to gain some perspective on the present state of the art to evaluate this.

2. RING INTERFEROMETERS

2.1 INTRODUCTION: SAGNAC EFFECT

As a sensor of noninertial effects, ring interferometers are best known as gyrometers, or rotation detectors, in connection with the Sagnac effect. Counterpropagating waves in a ring (or polygonal) cavity of area N rotating at angular speed $\underline{\Omega}$ suffer a relative phase shift on recombination of

$$\phi = 8\pi \underline{N}.\underline{\Omega}/\lambda v \qquad (1)$$

which with the de Broglie relation $\lambda = h/mv$ implies that

$$\phi = 4m\underline{N}.\underline{\Omega}/\hbar. \qquad (2)$$

This may be proved by any of the following methods:

(a) A classical (ether-theoretic!) time-of-flight argument;

(b) via the special relativistic Doppler effect at the mirrors defining the cavity;

(c) by analogy with the Aharonov-Bohm effect, in which the Coriolis force is the analogue of the Lorentz force;

(d) by the general theory of relativity, in which at least for light waves the device measures the contour integral around the cavity of the mixed space-time component h of the metric tensor ($\underline{h} = \Omega \times \underline{r}/c$) in the rotating frame. This generalises to all sources of $\underline{h}$, which is called the gravitomagnetic field (note the analogy mentioned in (c) in which $\underline{h}$ parallels the vector potential $\underline{A}$, and also the essentially dipole-dipole character of the coupling between a ring interferometer and the source of the noninertial effect).

The experimental interest in our topic lies in the fact that gravitomagnetic effects special to general relativity and to other post-newtonian theories have not yet been detected directly.

On the face of it, the most sensitive gyrometer will utilize particle-waves for the interference which have minimal speed-wavelength product, or equally maximal mass. This would appear to favour the neutron and electron over the optical photon. However, the reverse is the case.

2.2 OPTICAL RINGS

Early experiments include the remarkable work of MICHELSON, GALE and PEARSON [2] who measured the rotation rate of the earth via the fringe shift in a passive ring interferometer with a perimeter of 1900 m!

A dramatic improvement became possible when ring lasers were devised. If the counterpropagating modes are made to lase, they acquire a frequency difference compensating the phase difference given by $\delta\omega = \phi/t$, where t is the transit time. This gives an increase in sensitivity by a factor of the velocity (of light!) over the perimeter P, so that

$$\delta\omega = 8\pi \underline{N}.\underline{\Omega}/\lambda P \tag{3}$$

This has caused great interest, and triggered intensive effort, in the commercial development of ring laser gyroscopes. Some of this progress has been recorded in the engineering literature. The technology is now deployed in aircraft (Boeing 757, 767, Airbus) for inertial guidance, using triangular ring laser cavities of a few cm size for sensitivities of 10^{-3} ^{o}/h. Other applications are commercial or even national secrets (witness the spy scandal of April 1985, when five Soviet officials were expelled from Britain, reportedly [3] because of concern over the security of ring laser technology as developed for guidance of nuclear submarines). Still others have not been documented since they have not been thought significant in the commercial context, though in the context of fundamental science they have some importance, as we discuss.

Initially the performance of ring lasers was limited by mode locking (setting a lower limit on the beat frequency). This has been overcome in a variety of ways [4]; anisotropic Faraday rotators will split the frequencies of counter-propagating modes, but introduce stability problems, which in turn are overcome by using dither, for example; different polarizations are frequency split by using a non-planar loop; bias errors are then eliminated by detecting the difference of two beat frequencies. Passive rings with frequency-locked external lasers have equal potential. Fibre optic rings show great promise and are under extensive development for commercial application, but are likely to be more noisy for the purposes discussed here.

Accuracies have reached the level dubbed the "quantum limit" at which photon shot noise is a constraint. This limit is itself reducible by signal averaging, by increasing the area N, and by increasing the Q factor of the cavity. Signal averaging will not lessen the 1/f noise; however, increasing N and Q will, and recently SAYEH and BILGER [5] have demonstrated the $1/Q^4$ dependence of the 1/f noise. The area is ultimately limited by questions of mechanical and optical stability, such as the need for single mode operation. At the time of writing planned systems have areas $N \sim 50$ m^2, $Q \sim 10^{12}$, and signal averaging times of the order of days. These should permit easily the detection of effects at the level of

10^{-8}-10^{-9} of the rotation rate of the earth. A pilot ring of this size is being built at the Air Force Base, Seiler, Colorado, with similar parameters, using a pyrex ring (ultimately this would be zerodur) and a threshold frequency resolution of 100 nHz, compared to the 677 Hz beat frequency induced by the earth's rotation itself.

2.3 SQUIDS

A Josephson junction transmits a current I governed by the phase difference ϕ between the wavefunctions of the Cooper pairs on each of the superconductors, according to $I = I \sin \phi$. If the superconductors are joined into a circuit, the phase difference within the resulting SQUID is governed by the line integral of the canonical pair momentum $m\underline{v} + q\underline{A}$. The Josephson current thus has an interferometric sensitivity to noninertial effects (through the term $m\underline{v}$) as well as to magnetic fields, its more familiar measurement function. It may therefore function as a gyrometer, the phase shift being generated by $v = \Omega \times \underline{r}$, and as such being the analogue of the Sagnac effect. This was demonstrated by ZIMMERMAN and MERCEREAU [6] using SQUIDs with an area of the order of 1 cm^2, and rotation rates of the order of a few hertz.

While the sensitivity of SQUIDs has also reached their quantum limit (a more ambiguous reference than the term may suggest - shot noise is very prominent in the neutron interferometer), their optimisation has been mainly in connection with magnetic field measurement, for which the self-inductance and self-capacitance need to be minimised. This is achieved by going to thin film fabrication techniques, producing devices with areas of the order of 10^{-5} m^2, and which are decidedly not optimal for gyrometers. One basic problem is that the particles participating in the quantum interference are charged, and so their motion induces magnetic fields which oppose the Sagnac effect on the phase. Other such cancellations from this source, some complete, bedevil other measurements of noninertial effects. Nevertheless, suggestions for the use of SQUIDs for the detection of gravitational effects have been made (Section 3.2).

2.4 NEUTRON INTERFEROMETER

In a neutron interferometer three ears are sculptured from a single crystal of silicon, and act as Bragg reflectors for a beam of thermal neutrons with wavelengths of the order of 10^{-10} m and speeds typically 10^{-5} c. The beams are split at the first ear, and reflected at the second, forming a parallelogram geometry. On recombination at the third ear, any source of relative phase shift (due to the different gravitational or electromagnetic histories of the beams) influences the resulting interference. The Bragg reflection serves to align the beams to an accuracy of typically 10^{-6} rad, and to impart a coherence length of typically 10^9 wavelengths, i.e. of the order of cm (the typical size of zone refined single crystal silicon). Many striking experiments have been performed to illustrate fundamental principles of relativistic and of quantum physics.

A dramatic improvement on the optical experiment of MICHELSON et al [2] was possible with this device; the mass factor advantage to the neutron (Section 2.1) permitted WERNER et al [7] to measure the earth's rotation rate in the laboratory in a very beautiful experiment. However, two claims of WERNER et al [7] should be amended. This was not the first experiment to test equation (2) for matter waves; the experiment of ZIMMERMAN and MERCEREAU [6] was much earlier. Nor was it the first laboratory measurement of the earth's rotation rate. Ring laser gyroscopes of similar size had been developed for many years, and were being tested for commercial reliability; they have a sensitivity of 10^{-3} of the earth's rotational frequency. The latter is routinely subtracted as a bias in aircraft inertial navigation. Curiously this has never been explicitly reported in the scientific literature. A measurement of the earth's rotation rate to an accuracy of 10% using ring lasers is implicit in BILGER and STOWELL [8], for example.

The comparison between optical and other ring interferometers is then not only a comparison between masses and speeds of the relevant particles as they affect a universal formula; the observable is different. This invalidates the comment by ANANDAN

[9]: "there seems to be no hope of detecting gravitational effects in optical interference," in a general setting.

3. TESTS OF GRAVITY IN A QUANTUM LIMIT

3.1 INTRODUCTION: THE COW EXPERIMENT

First we discuss one experiment which is based on ring interferometry and has special interest as a test of quantum gravity.

In principle, interferometry could also find application in accelerometers, i.e. for measuring the acceleration relative to the local inertial frame (LIF). For this one would need the dual-beam setup of the neutron interferometer in which the wave paths are physically different, not just oppositely directed. The gravitational redshift implied by all PPN theories, or equivalently the twin paradox effect for the two beams, will generate a phase difference ϕ on recombination of

$$\phi = 2\pi mg\, N_0/hv = 2\pi g N_0/\lambda v^2 \qquad (4)$$

where N_0 is the component of the area perpendicular to the vertical.

The prototype of such experiments is the elegant experiment of COLELLA, OVERHAUSER and WERNER [10], hereafter called the COW experiment. This was performed in a neutron interferometer, which because of the velocity factor in equation (4) has a proportionately greater advantage re the Sagnac effect. Hence in the COW experiment the effect of mechanical sag on rotation (to change N_0) was found by using X rays in place of neutrons; the difference in fringe shifts between the two experiments was taken as the COW signal. Hence also the laser is not as suitable for an accelerometer as for a gyrometer in inertial guidance.

COLELLA et al [10] regard their experiment to be pioneering, in that it was a test of gravitation theory (in particular of the equivalence principle) in a quantum limit; it was the first whose outcome depended on both Newton's and Planck's constant. This claim is less than compelling for several reasons.

First, for many years, experiments in gravitation have been conducted in which a quantum effect has been used for readout - an NMR frequency, a SQUID detection of a London moment and so a rotation axis, etc. This has been done with the totally pragmatic philosophy that the quantum nature of the observable does not introduce a fundamental change of interpretation, but is merely a matter of convenience and feasibility, the quantum properties of the measuring device having been sufficiently explored in independent calibration experiments. This widespread philosophy is in sharp contrast to that of, for example, ANANDAN [11], [12]. He argues that since one may discern some elegance in a quantum description of general relativity, general relativity is itself naturally a quantum theory, at least in the senses that all gravitational effects have a quantum origin, and that information about metrical geometry is obtained ultimately from the phase of the quantum mechanical wavefunction. He concludes that tests of any theory of gravity which involve quantum interference are of more fundamental significance than those which do not. We shall return to this shortly.

Second, it is difficult to see what value the COW experiment has in the Popperian sense of falsifying possible alternatives (compare the comment of HALL [13] in another context). We lack "straw man" theories as alternatives to be discredited by such experiments as the COW experiment, and indeed a theory of theories (analogous to the PPN or TH$\varepsilon\mu$ formalisms of post-newtonian gravity) within which the COW experiment could put bounds on parameters. GREENBERGER [14] has outlined ways in which theories contradicting the COW result might be obtained, and the failure to develop these fully might reasonably be attributed to good sense in the avoidance of much perverse ingenuity. The COW experiment certainly checks that a correct description of the neutron interferometer is obtained by adding a newtonian potential energy term to Schrodinger's equation.

A third problem is that the quantum nature of such experiments is ephemeral from some viewpoints. In the case of the COW experiment, Planck's constant appears in only one of the equations (4), and its presence or absence is to some extent a matter of historical perspective. For the optical interferometer, the form omitting Planck's constant is more natural. Why then should lowering the speed of a particle, and endowing it with rest mass, have the supposedly decisive effect of taking us from a situation well described by classical (general) relativity to one where the subtleties and unsolved problems of quantum gravity are necessarily involved, especially when the same equation suffices for the experimental analysis? A similar problem of interpretation holds for the Sagnac effect: Planck's constant appears in (2), but not (1); both classical and quantal proofs, and indeed ether-theoretic and general relativistic proofs are possible, as noted in Section 2.1. A correspondence principle is clearly forcing these different approaches to coincide. To defend the novelty of the COW experiment, one has to argue strongly for a fundamental distinction between the interpretation of experiments with particles of zero and of finite rest mass. (Such distinction involves not only the question as to whether quantum mechanics is overtly relevant, but also whether the gravitational effect is newtonian or not [9].) However, this creates its own problems of interpretation. This resolution seems artificial in view of the pragmatic experimental philosophy mentioned above, and particularly so in view of the equivalence principle, that rest and kinetic mass should gravitate equally. I conclude that the equivalence principle is not readily reconciled with the philosophy of COLELLA et al [10], and is not as natural an outgrowth from a quantum basis for general relativity as the philosophy of ANANDAN [11], [12] would imply.

Fourth, if one ignores the role of correspondence principles in dictating the dependence of the observable on g and h, one may argue a case for a quantum gravity interpretation of virtually any experiment. Newtonian gravity satisfies an equivalence principle in which a gravitational field may be removed by an acceleration. Hence, the newtonian potential entering the description of the COW experiment might be regarded as derived from the noninertial effect of its acceleration relative to the freely falling frame. The Sagnac effect is equally a manifestation of noninertial effects in quantum interference; the ZIMMERMAN and MERCEREAU experiment [6] could then be regarded as antedating the COW experiment. Any shift of spectral lines with temperature could be attributed to noninertial effects via the Unruh effect and so to a gravitational field via the equivalence principle, but measurements of phonon-induced spectral shifts - a commonplace of solid-state physics - are not usually regarded as test of gravitation theory in the quantum limit. What this points to is that it has to be shown for any experiment ostensibly testing gravitation in the quantum limit that the physical content of a quantum gravitational description is greater than the minimum set by correspondence principles. This is unconvincing for the COW experiment in view of the alternative forms of equation (4).

While the above reservations are equally open to counter-argument, it is difficult to see any limit to the ingenuity with which they might be refined. As a result, proposed experiments in this context lack a convincing rationale.

3.2 REPRESENTATIVE EXPERIMENTS

If we accept that the COW experiment is striking for the obvious way in which gravity and quantum mechanics are involved, we may categorise other experiments by their corresponding interest.

Null experiments may not have the same prestige in terms of obtaining an observable dependent on g and h, but may be argued to have a quantum-gravity significance. One could go back to the early work of DREVER and of HUGHES (see the more recent experiment of PRESTAGE et al [15] and GALLOP and PETLEY [16] and the analysis of HAUGHEN [17] and of FISCHBACH et al [18] for example, as well as the comments of ROBINSON [19]), in which measurement of an NMR (quantal) frequency monitored possible violations of the weak equivalence principle, or equivalently of local Lorentz invariance, or of a preferred-frame effect. When the results are interpreted as a measurement of the directional dependence of mass, including the binding energies as a function of excitation state, the quantum foundation is not conspicuous; however, an analysis of the wavefunction and energy for any quantum state will certainly involve Planck's constant in the observable, if a frequency. Since in practice a frequency ratio is measured, the observable does not depend on Planck's constant. Similarly the Mossbauer measurements, as indeed the more recent and accurate hydrogen maser measurements, of the gravitational redshift predicted by the equivalence principles are also non-quantal, despite the obviously quantum nature of the Mossbauer effect itself. Incidentally a curious, if null, test of one possible consequence of Weyl geometry by HARRIS et al [20] using Mossbauer technology must also be regarded as classical in this context.

Experiments in which nonvanishing newtonian gravitational effects are verified by a quantum process include the suggestion by NIETO and GOLDMAN [21] for a quantum measurement (by the Josephson effect) of the gravity-induced voltage within a battery whose terminals have a vertical separation, and in their argument anticipate objections to a quantum gravity interpretation not dissimilar to those voiced above about the COW experiment. Similarly we may include in the COW category those experiments testing the principle of equivalence, as much a characteristic of newtonian as of einsteinian gravity theory, in quantum interference. This includes such proposals of ANANDAN as that of a gravity-dependent frequency shift in microwave emission from two SQUIDs at different heights in a gravitational field [22]. Direct interference would not show this shift, since the microwave photons themselves suffer a compensating gravitational redshift, but the phase of their interference would reveal the effects of gravity [23].

Experiments to test non-newtonian effects by interferometry have obvious interest, though (as mentioned above) what is or is not newtonian may itself depend on our viewpoint.

Let us commence with effects which are not obviously of quantum significance. Laser interferometry in a 'skyhook' or in satellite or lunar tracking may detect the tidal components of the gravitational field [24], [25]. Such effects may be regarded as newtonian were a massive fundamental particle used for the detection [9], but then the interferometric detection might be regarded as essentially quantum mechanical and the experiment might be categorised with the COW and the NIETO-GOLDMAN experiments.

A more convincing demonstration of a non-newtonian effect is that of detecting gravitational waves. Optical ring interferometry is one possibility [26], but it might be argued to be essentially classical. SQUID interferometry seems to be feasible [27], and to have the advantage that the observed electromagnetic field changes are not limited by the speed of sound [28]. It is also of quantum character.

Another approach is to look for relativistic corrections to quantum processes measuring a gravitational effect. ANANDAN [9] discusses various relativistic corrections to the COW experiment, including tidal effects. CABRERA et al [29] discuss relativistic corrections in rotating superconductors (it may be noted that there are errors in their initial Hamiltonian [30], although these do not affect their conclusions). ANANDAN and collaborators [31], [32] discuss electromagnetic field-dependent effects on electron interference in the presence of a gravitational field. For magnetic effects at least this is arguably relativistic; this is not so obvious for electric field effects. FISCHBACH and collaborators [33] suggest a search for gravitational mixing of the hydrogen states split by the Lamb shift. Hydrogenic states are relativistic; hydrogen is abundant on stars with high gravitational fields; interference with the electroweak coupling will generate a characteristic geometric dependence for the circularly polarised transition radiation.

In the Schiff gyroscope experiment, the precession of spherical spinning balls in orbit is also to be monitored by SQUID technology (detecting changes in the orientation of the London field generated by a superconducting coating of the balls). This would mean not merely the first detection of the gravitomagnetic field of the earth, but the detection of a quantum gravity effect, if we endorse the philosophy of the COW experiment.

The role of ring laser experiments is that of detecting the gravitomagnetic field, or the angular momentum of the gravitational field relative to the local inertial frame. Apart from tidal effects, this is all that a ring interferometer of this type measures [34], and makes it essentially a gyrometer, as for the Sagnac effect. It is just the gravitomagnetic predictions of general relativity which have not yet been tested. This gives ring interferometry a special interest. Within a PPN formalism, three contributions to the gravitomagnetic field are of interest.

The geodetic effect is the consequence of parallel transport in a curved space-time by orbital or motional effects. It is comparable to the following effects for a laboratory-based device, but a factor 10^3 greater for a satellite experiment, because of the much greater speed (compared to the earth's rotational speed), and so need careful discrimination.

The Lense-Thirring effect is the local dragging of inertial frames caused by the rotation of the earth nearby. Its existence is suggested by Mach's Principle [35]; a pendulum at the North Pole will by that principle strike a compromise by swinging in a plane which is at rest with respect to neither the fixed stars nor the rotating earth, but somewhere in between. However, the Lense-Thirring effect is strictly at variance with Mach's Principle, although derived from general relativity, since the mass of the fixed stars plays no role in defining the frame dragging. The Schiff gyro experiment is better adapted to measuring the Lense-Thirring effect, since an earthbound interferometer is not readily able to distinguish the Sagnac effect, itself about 10^9 greater in magnitude.

Third, a search for a preferred-frame effect is the main goal of the proposed ring-laser experiments, since the diurnal variation implied by such an effect would be distinguishable by Fourier analysis. This will complement the earlier tests of a preferred-frame effect; the tests mentioned above do not monitor the gravitomagnetic field contributions.

Surprisingly, a gravity gradiometer could also detect a gravitomagnetic effect. Traditionally this device consists of a Michelson interferometer, in which a quadrupole distortion of the local Minkowski metric is measured by the associated change in the interference of beams along orthogonal directions. SQUID technology would allow a similar observable to be measured, by a quantum observable [36]. The components of the relevant tidal matrix K_{ij} are traditionally related to the "electric" components R_{0i0j} of the Riemann curvature tensor (by contraction with the relative position vector). However, under geodetic motion the situation is more complicated, and all components of the tidal matrix receive a contribution which is linear in the angular momentum of the rotating object, and indeed of relatively large size (see for example THEISS [37]). Hence, an orbiting SQUID gravity gradiometer, such as that planned for operation in the near future, would permit yet another "quantum" test of non-newtonian gravity.

Since the analysis of THEISS [37] is limited to general relativity, it would be of interest to cast his formalism in the wider context of PPN theory, say, to find to what extent this technique would complement the proposed experiments with ring lasers. Its ostensibly quantum character would certainly be one complementary feature.

REFERENCES

1 G.E. Stedman: Contemp. Phys. 26, 311 (1985)
2 A.A. Michelson, H.G. Gale and R.F. Pearson: Astrophys. J. 61, 140 (1925)
3 Daily Telegraph (London: U.K.) April 24 & 25 (1985)
4 W.W. Chow et al: Rev. Mod. Phys. 57, 61 (1985)
5 M.R. Sayeh and H.R. Bilger: Phys. Rev. Lett. 55, 700 (1985)
6 J.E. Zimmerman and J.E. Mervereau: Phys. Rev. Lett. 14, 887 (1965)
7 S.A. Werner, J.-L. Staudenmann, and R. Colella: Phys. Rev. Lett. 42, 1103 (1979)
8 H.R. Bilger and W.K. Stowell: Phys. Rev. A. 16, 313 (1977)
9 J. Anandan: Phys. Rev. D. 30, 1615 (1984)

10 R. Colella, A.W. Overhauser, and S.A. Werner: Phys. Rev. Lett. 34, 1472 (1975)
11 J. Anandan in Quantum theory and gravitation (Ed. A.R. Marlow)(Academic, N.Y., 1980)
12 J. Anandan in Quantum Optics: Experimental Gravitation and Measurement Theory (Ed. P. Meystre and M.O. Scully)(Plenum: N.Y. 1983)
13 J. Hall: in Quantum Optics, Experimental Gravitation and Measurement Theory (Ed. P. Meystre and M.O. Scully)(Plenum, N.Y., 1983) p. 347.
14 D.M. Greenberger: Rev. Mod. Phys. 55, 875 (1983)
15 J.D. Prestage et al: Phys. Rev. Lett. 54, 2387 (1985)
16 J.C. Gallop and B.W. Petley: Nature, 303, 53 (1983)
17 M.P. Haughen: Ann. Phys. (N.Y.) 118, 156 (1979)
18 E. Fischbach et al: Phys. Rev. D. 32, 154 (1985)
19 A.L. Robinson: Science 229, 245 (1985)
20 E.G. Harris et al: Phys. Rev. D 7, 2326 (1973)
21 M.M. Nieto and T. Goldman: Phys. Lett. 79A, 449 (1980)
22 J. Anandan: Class. Quant. Gravity, 1, L51 (1985)
23 J. Anandan: Phys. Lett. 105A, 280 (1984)
24 V.B. Braginsky and K.S. Thorne: Nature 316, 610 (1985)
25 N. Ashby and B. Bertotti: Phys. Rev. Lett. 52, 485 (1984)
26 A. Kulak: Acta Phys. Polonica B15, 3 (1984)
27 R. Schrader: Phys. Lett. 143B, 421 (1984)
28 J. Anandan: Phys. Lett. 110A, 446 (1985)
29 B. Cabrera, H. Gutfreund and W.A. Little: Phys. Rev. B 25, 6644 (1982)
30 S.H. Payne and G.E. Stedman: Phys. Lett., 50A, 415 (1975)
31 J. Anandan and L. Stodolsky: Phys. Rev. Lett. 50, 1730 (1983)
32 J. Anandan: Gen. Rel. Grav. 16, 33 (1984)
33 E. Fischbach and B.S. Freeman: Gen. Rel. Grav. 11, 377 (1979); E. Fischbach: in Cosmology and Gravitation (Ed. P.G. Bergmann and V. de Sabbata)(Plenum, N.Y. 1980) p. 359.
34 A. Kuriyagawa, M. Ihara and S. Mori: Phys. Rev. D 12, 2955 (1975)
35 W. Rindler: Essential Relativity (2nd Ed.)(Springer, N.Y., 1977)
36 J. Anandan: Phys. Lett. 105A, 280 (1984)
37 D.S. Theiss: Phys. Lett. 109A, 19 (1985)

Fluorescence Enhancement of Rhodamine 6G Molecules in TiO_2 Semiconductor Colloid

B. Zhang and C. Zunsheng*

University of Missouri-Columbia, Columbia, MO 65211, USA

1. Introduction

Since the discovery of Surface-Enhanced Raman Scattering[1] much effort has been devoted to understanding the nature of the interaction of molecules with rough surfaces. Several theoretical models have been suggested to explain it[2]. The authors turn special attention to the surface plasmon model[3]. According to this model, the enhancement is due to the amplification of the incident and radiated optical fields by plasma resonances of the metal.

It is evident from this model that other linear and nonlinear optical processes close to metal surface or metal particles should also be enhanced.

It has been demonstrated that the luminescence and two-photon fluorescence of dye molecules absorbed onto silver and gold particles are enhanced[4]. Enhanced second harmonic generation has also been demonstrated[5].

The purpose of our experiment:

(1) to confirm the surface plasmon model;

(2) to demonstrate a prediction that the enhancement of the optical processes close to small semiconductor particles or surface of semiconductor can occur in certain conditions.

2. Experiment

Sample: Rhodamine 6G dye was dissolved in the TiO_2 semiconductor colloid.

Advantages of the sample:

(1) High extinction coefficient.

(2) Long lifetime of carriers (>1ms).

(3) Suitable positioning of the valence band and conduction band.

(4) It's easy to make the sample.

Sample preparation:

TiO_2 colloid was prepared by hydrolysis of $TiCl_4$ in acidic (HCl) aqueous solution (PH<3). The concentration of TiO_2 is 2.86 g/L. The concentration of Rh.6G was $1x10^{-7}$ - $5x10^{-7}$mol.

The basic idea of the experiment: At the same time, shine UV light and 5145Å visible light on the sample. The colloid does not absorb visible light. A sharp

*Permanent Address: Physics Department, Zhengzhou University, Zhengzhou, P.R. China

absorption band rises toward the UV spectrum region at $\lambda \leqslant 3800\mathring{A}$, which corresponds to the band gap of TiO_2 semiconductor.

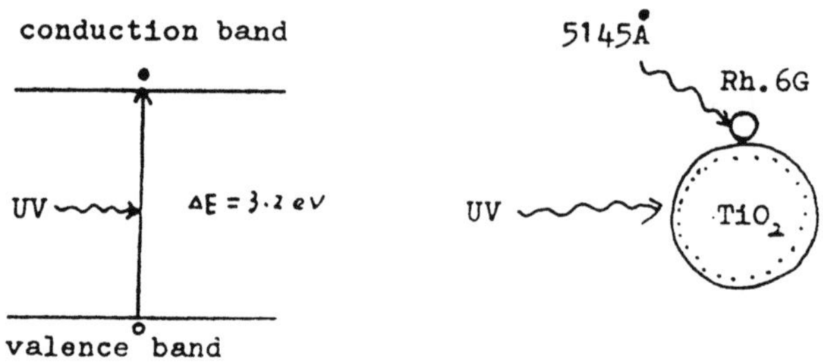

Fig.1. Schematic of TiO_2 semiconductor colloid excitation

When we shine UV light on the TiO_2 particle of the colloid, the electrons are excited from valence band to conduction band of TiO_2 semiconductor by the UV light to form the surface plasma.

The $5145\mathring{A}$ light is used as the exciting radiation to excite Rhodamine 6G molecules absorbed on the surface of TiO_2 particle. The dipole oscillation of the molecule irradiates the fluorescence.

Measure the spectrum in two different conditions (with UV light and without UV light) to see the difference between them.

Apparatus :

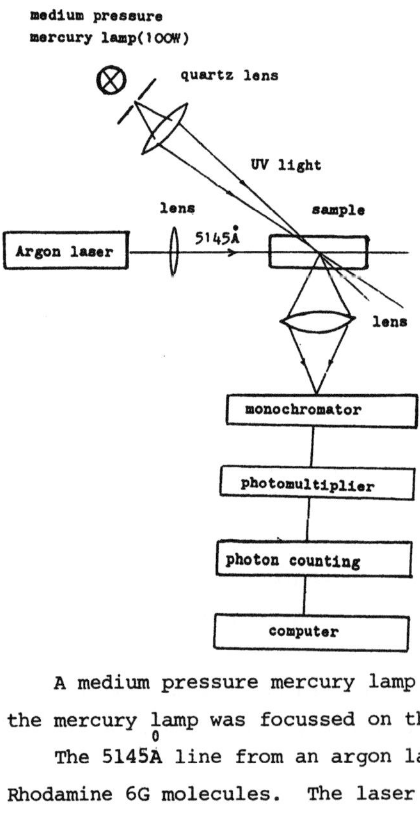

Fig.2. Apparatur schematic of the experiment

A medium pressure mercury lamp (100W) was used as UV source. UV light from the mercury lamp was focussed on the TiO_2 colloid by a quartz lens.

The $5145\mathring{A}$ line from an argon laser was used as the exciting source to excite Rhodamine 6G molecules. The laser power at the sample was 50 mW.

A 90° scattering geometry was employed for fluorescence measurements using a Spex double monochromator equipped with a cooled RCA photomultiplier tube, photon-counting electronics and a dedicated computer for data storage and analysis.

3. Results and Discussion

Discussion:

According to the surface plasmon model, the explanation of the enhancement is as follows:

(1) UV light excites the electrons from valence band to conduction band in TiO_2 particles, and produces surface plasmon.

(2) Surface plasmon can be excited by the incident beam (argon laser). When the incident laser frequency is close to the surface plasmon frequency, the optical electric local field near the surface can be enhanced by the resonant excitation of surface plasmon.

W.H. Weber and G.W. Ford calculated optical electric-field enhancement from surface plasmon excitation(3). They obtained the following formula.

$$\frac{|E_{SP}(0^+)|^2}{|E_0|^2} = \frac{2\cos\theta\epsilon_2'^2(1-R)}{\epsilon_1^{\frac{1}{2}}\epsilon_2''(-\epsilon_2'-\epsilon_1)^{\frac{1}{2}}} = f^2(\omega_i) = \frac{I_{Loc}(\omega_i)}{I(\omega_i)}$$

The local field factor $f(\omega_i)$ can become large on resonance where $-\epsilon_2' \sim \epsilon_1$

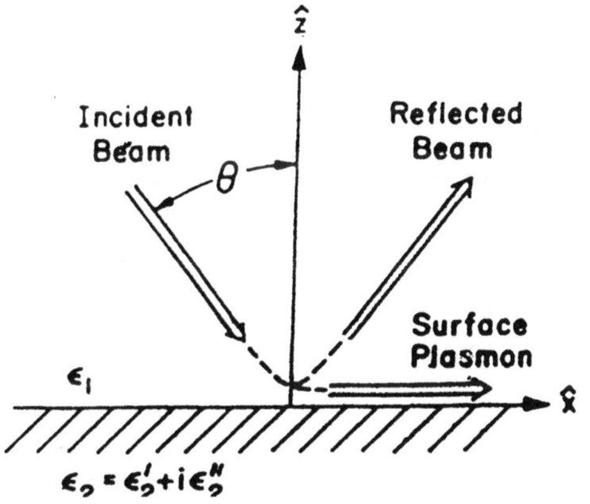

Fig.3. Schematic drawing of the excitation of a surface-plasmon wave with a plane-wave coupler, which is represented by the dashed lines. The coupler can be a prism, a grating or random roughness

(3) The rhodamine 6G molecules absorbed on the surface are excited by the optical local field. The excitation rate R can be written:

$$R = N\sigma I_{Loc}(\omega_i) = N\sigma f^2(\omega_i)I(\omega_i).$$

N is the density of absorbed molecules, σ is the absorption cross-section and $I_{Loc}(\omega_i)$, $I(\omega_i)$ are the local and incident intensities. The excitation rate is enhanced by the factor $f^2(\omega_i)$.

(4) The intensity of fluorescence radiated by Rh 6G molecules is

$$I(\omega_L) = qRf_L^2(\omega_L) = Nq\sigma f^2(\omega_i)f_L^2(\omega_L)I(\omega_i).$$

q is the radiative quantum efficiency of the molecules. $f_L(\omega_L)$ is the local field factor for fluorescence. It's enhanced twice: $f^2(\omega_i)f_L^2(\omega_L)$.

According to quantum theory, we suggest the surface plasmon is an intermediate state in surface-enhanced fluorescence.

The transition probability of the fluorescence can be written as:

$$R \sim \left| \sum_{sp} \frac{\langle n-1,1,1,g|\, \hat{H}_{sm}\, |sp\rangle\langle sp|\, \hat{H}_{ps}\ n,o,o,g\rangle}{\hbar\omega_{sp}(\vec{K}) - h\omega_i} \right|$$

where $|n,0,0,g\rangle$: initial state

$|n-1,1,1,g\rangle$: final state

$\hat{H}_{ps}$, $\hat{H}_{sm}$ are the Hamiltonians for the interaction of photons with plasmons and plasmons with molecules, respectively.

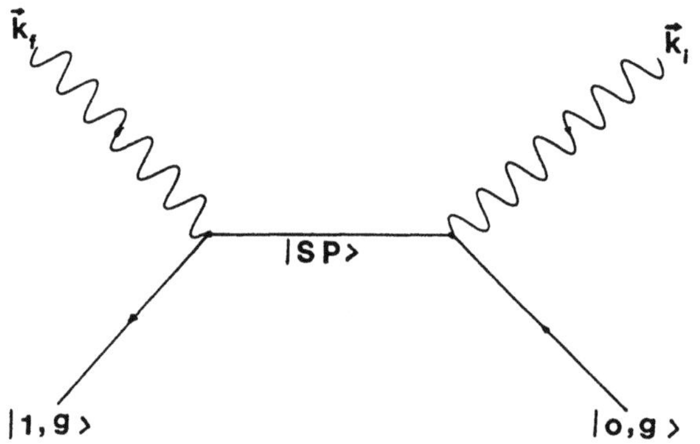

Fig.4. Diagrammatic representation of photon-surface plasmon model

If the fluorescence enhancement proceeds through resonant excitation of surface plasmons, we can drop the sum over all intermediate states and consider only the resonant state; the transition probability is greatly enlarged when incident light frequency is close to surface plasmon frequency.

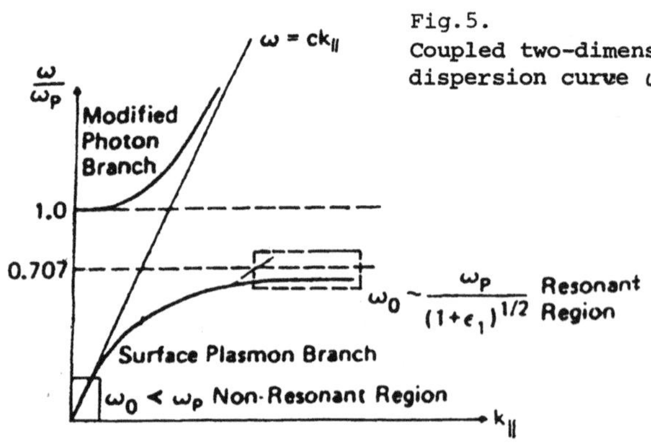

Fig.5.
Coupled two-dimensional photon-surface-plasmon dispersion curve $\omega_k^2 = \frac{1}{2}\omega_p^2 + k_\parallel^2 c^2 - (\frac{1}{4}\omega_p^4 + k_\parallel^4 c^4)^{\frac{1}{2}}$ (Ref.6)

In our situation, the calculated electron density on the surface of TiO_2 particles is about $5\times10^{18}/cm^3$. The plasmon frequency ω_{sp} is about 10^{13} - 10^{14}Hz. For 5145Å line of argon laser, ω_i is 5.8×10^{14}Hz. It satisfied near-resonant condition.

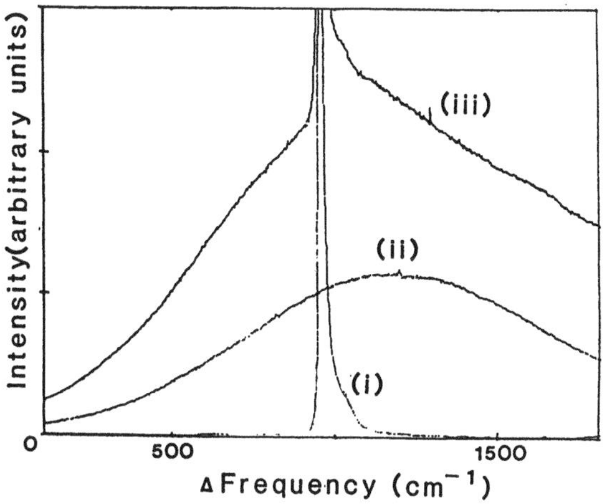

Fig.6. Fluorescence spectra of Rhodamine 6G in TiO_2 colloid
Colloid made by hydrolysis of titanium tetraisopropoxide; concentration of Rh 6G:1×10^{-7} mol
(i) mercury UV light only
(ii) 5145Å argon laser only
(iii) UV light and 5145Å argon laser

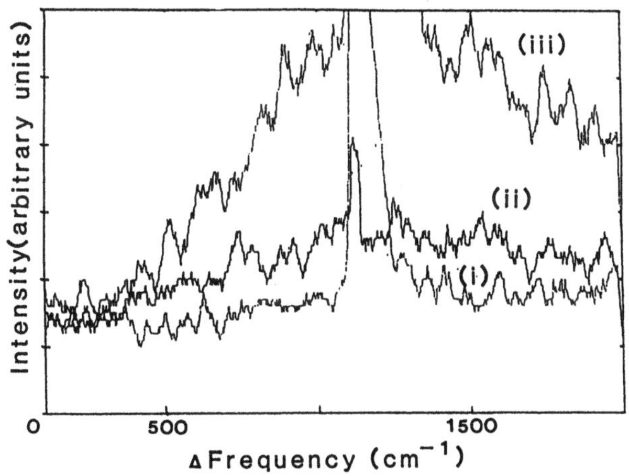

Fig.7. Fluorescence spectra of Rhodamine 6G in TiO_2 colloid
TiO_2 colloid made by hydrolysis of $TiCl_4$ concentration of Rh6G:5×10^{-7} mol
(i) mercury UV light only
(ii) 5145Å argon laser only
(iii) UV light and 5145Å argon laser

4. Conclusion

(1) The fluorescence intensity was three times enhanced for 5×10^{-7} mol concentration of Rh 6G in TiO_2 colloid using argon laser and mercury UV light as excitation sources at the same time.

(2) The fluorescence enhancement can be explained by the surface plasmon model.

(3) We predict that maybe surface-enhanced Raman scattering can be observed on the surface of semiconductor in certain condition.

Acknowledgment

We thank U. Venkateswaran, Dr. M. Chandrasekhar and H.R. Chandrasekhar for many exciting discussions and much support.

References

1 M. Fleischman, P.J. Hendra and A.J. McQuillan, Chem. Phys. Lett. 26, 163(1974)
D.L. Jeanmaire and R.P. Vanduyne, J. Electronanal Chem. 84, 1 (1977)
2 T.F. Furtak and J. Reyes, Surface Science 93, 361 (1980)
3 J.C. Tsang, J.R. Kirtley and J.A. Bradley, Phys. Rev. Lett. 43, 772 (1979)
W.H. Weber, G.W. Ford, Optics Lett. 6, 122 (1981)
4 A.M. Glass, P.F. Liao, J.G. Bergman, D.H. Olson, Optics Lett. 5, 368 (1980)
A.M. Glass et al, Phys. Rev. B, 24, 4906 (1981)
5 C.K. Chen, A.R.B. de Cestro, and Y.R. Shen, Phys. Rev. Lett. 46, 145 (1981)
6 H. Morawitz and M.R. Philpott, Phys. Rev. B, 10, 4863 (1974)
7 A. Wokaun, J.P. Gordon and P.F. Liao, Phys. Rev. Lett. 48, 957 (1982)

Stabilizing Lasers for Applications in Quantum Optics

*J.L. Hall**

Joint Institute for Laboratory Astrophysics, National Bureau of Standards and University of Colorado, Boulder, CO 80309, USA

In our field of quantum optics, it is now quite usual for us to discuss "squeezed" radiation states with sub-shot-noise levels of fluctuation. The pioneering experiments [1] of Slusher, Hollberg, Yurke, Mertz and Valley were the first to demonstrate an unambiguous level of squeezing, and it was natural for such a demonstration to prefer to squeeze the vacuum or zero-photon state. Indeed, one of the most exciting applications of squeezing — that of enhancing the sensitivity of gravity-wave interferometers — will be able to make powerful use of such a well-squeezed vacuum state input into the reference port, along the lines discussed by Caves [2], for example. Still, in some ways it is an even more exciting prospect to have a strongly-squeezed radiation field with substantial intensity. Such a field could be applied to a single cooled, trapped, two-level ion for some amusing and perhaps fundamental contact between subdisciplines of atomic physics, radiation physics, and quantum measurement theory. It may also be the field of choice for illuminating various types of sensitive interferometers, reducing the fluctuating radiation-pressure forces which lead to perturbation of the mirrors. Such a highly-squeezed strong field would also facilitate systematic investigation of the several apparently distinct physics models which discuss the origins of noise in heterodyne detection. In the contribution by H. J. Kimble and the present author in these proceedings, we give a progress report on our experimental approach to generation of such a squeezed, high-intensity field by doubly-resonant intracavity second harmonic generation. With this background of the quantum optics ideas then, let me explain the purpose of the present article: Here we try to give an indication of the performance level of absorption spectroscopy and laser stabilization, assuming that the laser fields employed are utilized "merely" with shot-noise limited performance. To make the discussion here compelling and self-contained, it is useful to begin with a few tutorial remarks about the nature and behavior of servo control systems, including for example those which stabilize a laser's intensity or frequency.

Let us consider first a laser servo system intended to produce a constant output intensity. As indicated in Fig. 1, the laser's output will contain some undesirable noise, schematically introduced in our discussion here by $N(\omega)$. This noise may contain both random and deterministic components. For the control system, the laser output is directed through an electronically-variable optical attenuator and thence to a beam-dividing mirror which sends the fraction 1-f on through it to form the useful, stabilized output. The output fraction f is directed to a low noise detector. Its electrical output in turn will be sent to a differencing amplifier where the electrical image of the fraction f of the laser system's output will be compared with a quiet dc level which represents the electrical equivalent of what we wished the system's output to be. If the output were indeed the "correct" value, this difference amplifier's output would be just zero. Thus we can design a lot of gain into the following amplifier which will drive the optical attenuator. The motivation for wanting high gain is that a higher gain will force the output to more closely correspond to our set-point condition, and be correspondingly less controlled by the intrinsic original noise characteristics, $N(\omega)$. We can illustrate the idea with the following argument: Suppose the

*Staff Member, Quantum Physics Division, National Bureau of Standards.

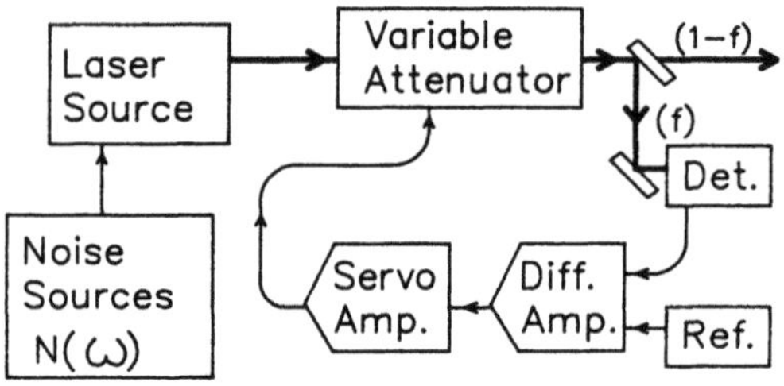

Fig. 1. Laser intensity stabilization. After passing through a fast electronic attenuator, a portion of the output beam is detected and compared with a reference or "target" set level for the output power. The difference or "error signal" is amplified strongly and fed to the attenuator, thus closing the loop. In general the laser contains intrinsic amplitude noise, here modeled by the explicit noise source N(ω).

optical attenuator/modulator requires 100 volts to be applied to cut the transmission coefficient to 10% from its usual 90% level obtained at 0 volts. Let the photodiode amplifier sensitivity be scaled to give us 5 V output from the 100% measurement level input. Suppose we are trying to stabilize the output at the 85% level, so as to maintain some margin for unexpected downward changes in the laser's output. We can guess that the necessary modulator voltage will therefore be about 15 V (the transmission varies as $\sin^2(V/V_p)$. If our servo amplifier has a gain of 1, even the full 5 V nominal output of the difference amplifier will hardly cause any effect on the transmitted light since the modulator really needs 100 V to have a powerful effect. Suppose instead that the servo amplifier has a gain of 100. Now only 3% of the full output of the difference amplifier, 0.15 V, when amplified 100-fold will give the 15 V needed to reduce the output to near the set level. If the servo amplifier gain were instead 1000x, the photodiode amplifier output of 10-fold smaller magnitude would suffice to drive the output to near the required 15 V level. Inclusion of an integrator in the servo amplifier is interesting since it will continue to accumulate even very small input error increments. Eventually we will reach the needed output level with zero error (=offset from the set level).

Stability analysis of such control systems has been a venerable subject since the topic was introduced by Bode [3] in the 1930's. (Parenthetically we may note that, following the suggestion by T. W. Hänsch [4], computer "spread-sheet" programs provide a qualitative revolution in the ease and power of such stability analyses. Rather than following my net financial worth in month seven as a consequence of investments, income and expenses in the preceding six months, we may instead choose to program the finite difference equations describing the voltage state of some integrating capacitor in the servo amplifier in terms of the charge input events associated with the six previous passes of an assumed input transient through our laser intensity control servo system. This new capability to instantly evaluate the consequences of a servo designer's parameter choices will undoubtedly lead to sharper understanding and more powerful designs. However, it is not necessary here for us to delve too deeply into such delicious details. Rather we should clearly state the global outlook of the conventional frequency-domain servo system analysis.)

It is effective to speak of servo system performance and gain in the Fourier plane. The laser source may have some effective noise distribution, say N(ω), representing the fractional noise per unit bandwidth center around the Fourier frequency ω. This noise is presumed to contain a random part R(ω) intrinsic to the laser (for example, spontaneous emission-caused intensity or frequency fluctuations), augmented by a deterministic noise distribution D(ω) (for example produced by inadequate filtering of the power supply or by laboratory vibrations, etc.). Similarly, the detector output will contain noise, even in addition to the expected electrical images of the laser's noise $N(\omega)$. The minimum possible additional detector noise, $S(\omega)$, may be ascribed to photoelectron shot noise in the detected light sample we plan to use for our stabilization procedure. We may think of these noise distributions as being spectral power densities representing laser intensity changes occurring in the bandwidth of one Hertz centered at the Fourier frequency ω. Finally, the closed servo loop will have some frequency distribution of servo gain, $G(\omega)$, which may be understood by (mentally) opening the closed loop at any point, sending a Fourier disturbance around, converting it to light changes, converting them back to electrical signals, and continuing through the electronics until reaching the location of the cut. The reponse at frequency ω for a unit input at ω is the gain $G(\omega)$.

As anticipated by our introductory discussion above, we will want to make the closed-loop servo gain $G(\omega)$ as large as possible. What indeed does fix the maximum usable gain? Physically, the limit is fixed by the inevitable time delays around the servo loop. If we have a sinusoidal input error, we will not be very effective in reducing it by feedback action unless the correction signal arrives "in time," say with a phase shift of 90 degrees or less. Thus if the closed servo loop is characterized by a transit time T, the frequency ω_1 at which the value of $G(\omega)$ reaches unity should be chosen so that $\omega_1 T < \pi/2$. If the frequency ω_1 were chosen to be twice this value, then the correction signal would be 180 degrees shifted from the original perturbation. The result would be to increase the error signal on each pass through — that is to say the system would be in self-sustained oscillation.

Even in the absence of time-delay phase shift, there is an intrinsic coupling between phase shift at frequency ω and the slope of the gain/attenuation function in the vicinity of frequency ω. These reciprocal coupling relationships are known in physics as dispersion relations or as the "Kramers-Kronig" relations. In the context of electronics, Bode may have been the first to emphasize the inescapable existence of a phase shift [5] in the presence of an amplitude response which varies as a function of frequency. The consequence for our present discussion is that the gain-vs-frequency curve, $G(\omega)$, cannot be allowed to rise too rapidly toward lower frequencies, particularly at the high frequencies near unity closed loop gain. The simulation technique referred to previously shows that a suitable $G(\omega)$ vs ω plot will have a slope of -1 to -1.5 in a log-log presentation. Eventually as we come to frequencies much lower than the unity gain frequency ω_1, the system can afford to have a steeper slope, say -2.0 for a decade or two in frequency. In this way we can approach — but only at low frequencies — the objective we understood as desirable from the initial discussion: more gain is better as far as servo system performance is concerned.

Now we must distinguish just which errors this enormous servo gain will suppress. Certainly the intrinsic problems of the laser will be better hidden by more gain. Under feedback we will have the effective noise density of the system $N'(\omega) = N(\omega)/G(\omega)$, i.e. the high gain can reduce the intrinsic noise of the laser to any small value we specify. We need only provide enough gain at the frequency of interest. Of course, the price for the high gain is that the frequency of unity gain must be pushed up accordingly, and the loop propagation time delay reduced appropriately. Usually, this procedure will have power dissipation as well as amplifier cost consequences. Also, increasingly subtle design tricks will be needed — as we undertake to increase and then increase again the bandwidth — to avoid loss of the intrinsic signal-to-noise ratio.

This last remark in turn brings us to face the signal/noise question directly. For example, our photodetector system sensing a fraction of the laser power for our intensity servo will certainly have a maximum signal/noise ratio: we will not be able to discern any genuine laser intensity changes smaller than the photoelectron shot noise of the measured light. With high servo gain, the small noise current of shot-noise origin will give rise to an actual light intensity variation -- negatively correlated — of just the scale needed so that the apparent fluctuation of the photodetector output vanishes. We obtain a laser system output noise power spectral density then which is

$$N''(\omega) = [N(\omega)/G(\omega)] + S(\omega)\times(1-f)/f \quad . \qquad (1)$$

Effectively, the native noise of the laser may be reduced by any factor we choose, being replaced by the noise of the measurement process we used to develop the error signal for the servo system.

In this way we have produced in the lab servo-controlled laser beams which are within a few percent of the shot-noise amplitude limit and/or which exhibit phase stability better than a radian for times approaching 10 seconds. These results take on certain significance when we recall that the broadband shot-noise level for a mW beam is below 1 ppm, while the input laser beam rarely has a power variation smaller than the percent level. Particularly the dye laser is a challenging beast -- its amplitude and frequency noise spectra do not approach the shot-noise level until we reach 2 or 3 MHz Fourier frequency. A jet-stream dye laser will ordinarily have a phase jitter of 1 radian on a microsecond time scale! It may be useful to discuss some of the actual performance limitations in detail.

Accurate stabilization of a dye laser was undertaken in 1975 in collaboration with J. J. Snyder and D. Frohlich. We built a servo system [6] that controlled out amplitude variations below about 4 MHz. Because of the stability-dictated finite gain-slope increase toward low frequencies — discussed above — our system had a small embarrassment near 150 kHz where the intrinsic laser noise had increased so much faster than our servo gain that a 6 dB noise hump was visible above the shot-noise measurement floor. Below this range a spectrum analyzer looking at the photodiode monitor point showed indeed a noise level below the shot-noise level. This is not an indication of squeezed light -- although recently there have been discussions of this interpretation [7]. Rather it merely indicates that all is well in servo-land, that a sensed input error is driven essentially to zero by a high gain, properly-designed servo control loop. (Parenthetically we note that servo design to include one or more stages of integration are interesting in this connection because the apparent servo error, measured at the null point, gets smaller and smaller as the gain rises toward low frequencies. Some of the recent quantum optics experimenters have restricted their work to proportional servo designs where less dramatic null point noise levels were obtainable.)

Of course, the real test of the achieved performance can only be accomplished with a second, independent detector system. Disappointment is the experimenters' first reward for this measurement since the laser beams' spatial purity is limited. The beam may be pure TEM_{00} on the 1% level, but we are talking 10^{-6} or 10^{-7} for a shot-noise level. Other spatial distributions have different diffraction characteristics and more-or-less rapidly diffract out of the beam differentially between the two detectors. (Surface roughness of the dye laser's jet stream is a natural way for frequency and intensity noise to develop spatial inhomogeneity.) Obviously small-scale scattering from component surfaces or high spatial frequency optical noise of any origin is particularly damaging. We found that with a decent-looking Gaussian beam, the test photodiode agreed that the laser's intrinsic noise had been suppressed by the servo as the discussion above would lead us to expect. Also, within the servo bandwidth the apparent base noise level observed in the second detector was indeed increased about two-fold as expected. Near the upper frequency limit of the servo system, the gain is smaller and the phase is not quite zero. This results in a few dB noise bump near the unity gain frequency, as viewed on the output of the independent test photodetector.

It was interesting to add spatial filtering either differentially or before the beam divider and two detectors. Of course, treating the detectors differentially, even to just having them with a few tens of cm different distance from the beam divider, rapidly decorrelated their outputs. With spatial filtering before beam division we measured noise suppression of >60 dB below 10 kHz with the proof detector. Correlation loss of 3 dB had a scale distance of ~1.5 m with spatial filtering before division.

With this extended introduction to the artform of laser servo-systems, the reader may be ready to learn of recent experiments mentioned in the introduction. However, we first need to discuss the frequency stabilization limits imposed by the quantum noise of the (frequency) measurement process, and relate them to the intensity stabilization concepts just discussed. To be concrete, we introduce in Fig. 2 a generic servo system configured to stabilize a laser frequency. The error signal is derived by measuring the deviation of the actual laser frequency from the central point of the frequency discriminator. This measurement has random photoelectron shot noise $S(\omega)$, just as in the intensity measurements of the earlier example. Further, we can be sure that the central frequency of the discriminator itself will have long-term drift and noise represented by $D(\omega)$. On the laser source side, we have idealized the noise sources into the categories of deterministic and stochastic noises, principally because environmental vibrations play such an ubiquitous role in causing deterministic laser frequency noise. Surface roughness of the dye laser jet stream is a rich source of fast stochastic frequency variations. It is important now to be more explicit about the laser frequency stabilization system shown in Fig. 2 and particularly about its frequency discriminator.

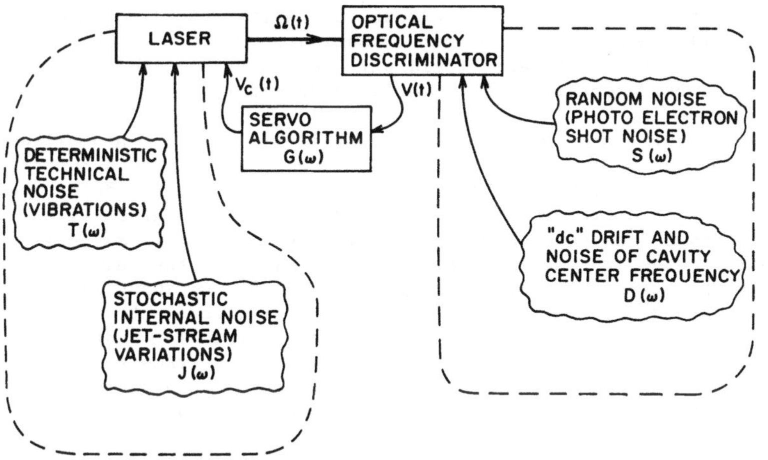

Fig. 2. Generic laser frequency stabilization system. The frequency of the laser emission is compared with the center frequency of the optical frequency discriminator, and the differences between "set level" and actual performance form the error signal input for the servo amplifier. Its output will change the laser's frequency. We idealize the actual laser frequency noises as those originating primarily from random processes $J(\omega)$ and deterministic technical and environmental processes, $T(\omega)$. The servo will essentially eliminate these from the output. However, the measurement process introduces random shot noise $S(\omega)$ and the frequency discriminator probably is subject to drift and vibrations, $D(\omega)$.

There are several techniques for obtaining a frequency error signal by illuminating a resonant cavity with laser light. Two of the most powerful schemes involve measurement of the optical phase shift near resonance by heterodyne of cavity-reflected light with imposed local oscillator sidebands (the "Pound/Drever" discriminator [8]) or by homodyne detection with a quadrature-phase local oscillator field (the "dc Pound stabilizer" [9]). However, it is initially more useful for us here to focus our attention on the simplest and most direct use of a resonant cavity to perform the frequency discriminator function. This scheme also has a certain aura of popularity since, in the dozen years since its introduction by Barger, Sorem, and the present author, this "fringe-side" frequency discriminator technique [10] has been employed in the >800 stabilized lasers produced commercially by two major laser firms in California.

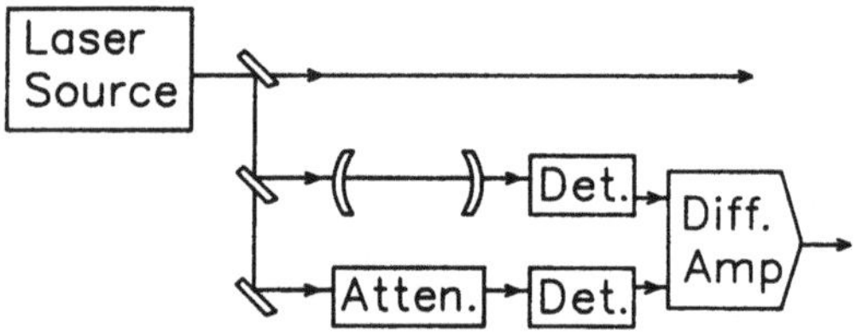

Fig. 3. "Fringe-side" optical frequency discriminator. A portion of the laser output is passed through a sharply frequency-selective filter, to be compared with another portion which is only attenuated. Frequency noise will be displayed at the output of the difference amplifier. When the servo is locked the average value of this output is zero. Thus amplitude noise of the laser is suppressed in first order.

The fringe-side stabilization technique may be understood with the help of Fig. 3, which shows the laser input being divided into two portions, passing through two separate channels, and being detected by two separate photodetectors. In one of these transmission channels we have placed a resonant cavity, tuned so that the input laser frequency falls near the 1/2 maximum transmission point. In this way a small frequency change of the input laser field will be converted into an amplitude change and thus will be detected by the associated detector.

Of course this detector will also have a fundamental sensitivity to an amplitude changes of the laser input as well. This explains the need for a second detector which monitors the laser intensity through an optical attenuator which reduces the light level presented to the reference detector. A physicist might naturally be inclined to take the ratio of the two photodetector signals, since this measure would be sensitive only to the frequency-dependent transmission of the resonant cavity, and would be completely independent of the magnitude of the input laser power. Unfortunately, as a practical matter, the wonderful little electronic (analog) modules which perform this ratio function are only marginally fast enough and accurate enough to respond in the microsecond or quicker time frame required by a dye laser. Thus in the realistic application of this scheme one finds it desirable to subtract rather than divide the two photodetector outputs.

In use one wants to adjust the optical attenuator in the "reference" photodetector branch so that its output current matches that of the "cavity fringe" detector when the laser is tuned for the 1/2 peak transmission condition. These two photocurrents can be organized to be of opposite sign and their difference directly amplified to form the "discriminator" output signal to be used for our laser frequency servo system. When the servo control loop is closed, the frequency of the laser will be dynamically adjusted so that the frequency-dependent photocurrent exactly matches and cancels the intensity reference signal. In this way one can see that small intensity variations of the laser will not affect the frequency lockpoint since they will give rise to matching -- and canceling -- variations in the difference current.

One lingering weakness of the system as just described is that the transfer function of the discriminator, in volts/MHz, still carries a linear dependence on the laser power level. Thus it is attractive to normalize the difference current so that this first-order power-dependence is suppressed. For this purpose one can use the analog divider modules mentioned above, as some of these units have rather good bandwidth for changes of the numerator input, even though the denominator response is often too sluggish. An even more attractive approach can be based on a fast attenuator in the output line of the discriminator, where the attenuator is driven by a signal approximating the inverse of the laser power. Numerous delicious electronic details enter into the design of such a system, but it is not necessary for us to address these issues here. The interested reader may find some further remarks in the paper by Helmcke et al. [11] and in Ref. [8].

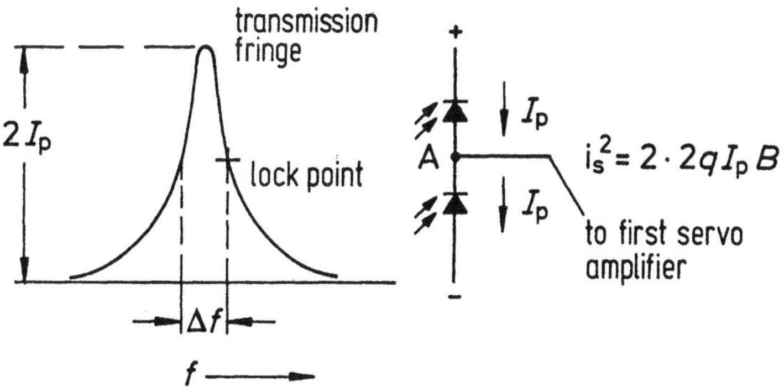

Fig. 4. Electronic aspects of the fringe-side discriminator. See Fig. 3 for operation. The dc average currents are balanced when the servo is locked. Note that the photodetection fluctuations of both detectors are passed to the error amplifier, along with the genuine signal due to frequency fluctuations.

So now we may turn to the question of the ultimate possible performance of a frequency servo. Basically this is our original question of the signal/noise ratio for frequency stabilization. Figure 2 represents the overall situation in a schematic way, while the diagram of Fig. 4 allows us to see the noise processes clearly. The two photocurrents are equal and so their dc values cancel, assuming that the frequency-control servo system is indeed locked and operating properly. Any small changes from equality of the two currents will be input to the frequency servo and will be interpreted as frequency errors. Certainly we know one minimum noise current to be so presented -- the rms sum of the photodetection shot-noise currents. As indicated in Fig. 3, we will obtain a noise current variance of

$$S^2 = 2\ i_s^{\ 2} = 2(I_o \cdot 2 \cdot e \cdot B) \quad . \tag{2}$$

where I_o is the dc photocurrent in each detector. With the high servo-gain assumption we have discussed before, we can be sure that the frequency servo electronics will interpret this voltage as a frequency error and dutifully attempt to correct it away. In this noise equation we may regard the bandwidth B as being approximately the bandwidth over which the servo has a gain of unity or greater. It is clear that the frequency noise produced by this photodetection shot-noise current will be smaller if the conversion slope of the discriminator is very high. If we write the cavity full line width at half-maximum response as $\Delta\Omega_c$, then the discriminator slope at the half-height will be

$$\frac{dI}{d\Omega} = \frac{2I_o}{\Delta\Omega_c} \quad , \tag{3}$$

where $2I_o$ represents the maximum fringe photocurrent when the laser is tuned to the cavity resonant frequency. So over the bandwidth B the servo system will accept the photodetection shot noise as its input and produce an output frequency variation essentially equal to its negative, when dimensionally converted by the discriminator. For the noise-induced frequency excursions we have

$$d\Omega = \frac{(S^2)^{1/2}}{(dI/d\Omega)} = \Delta\Omega_c(eB/I_o)^{1/2} \quad . \tag{4}$$

This formula expresses the rms frequency excursions within the bandwidth B due to the conversion of measurement noise into frequency by the servo action — it's only doing what we apparently requested! Clearly, if we were clumsy experimentally and let power mains currents leak into our sensitive circuits as well — one part of the D(ω) in Fig. 2 — we would be rewarded by seeing a nearly perfect image imposed onto the laser's frequency output.

A more insidious problem arises when the spurious signal is contained in the discriminator function itself. If it occurred for example as an added current in the photodetector buffer, its presence would be revealed as a deterministic output of this buffer when the light signal was blocked. However, if the cavity resonance frequency itself were being jiggled somehow, we would have no internal way of discerning it. It is remarkable to see papers being published even nowadays in which the author purports to estimate the performance of his system by looking at the discriminator error signal. This location is good to study as a function of servo gain to develop an appreciation of what kinds of noise the servo is "hiding" and of whether the present gain is indeed "enough" in the sense discussed previously. Good performance at this measurement point is a powerful diagnostic -- we may even say mandatory requirement — of the servo's good closed loop performance. But absolutely no diagnostic measurement of the closed loop error signal will teach one that his cavity's resonance frequency is being perturbed by vibrations or magnetic fields in its environment! It is quite necessary to have an independent system to prove the performance.

Suppose we have indeed avoided the pitfalls and built up a "shot-noise-limited" frequency-stabilized laser system. What typical performance can be expected? Evidently the performance is then controlled by our assumptions, particularly by the questions of reference cavity line width ($\Delta\Omega_c$), assumed light level (photocurrent I_o), and the necessary servo bandwidth B. Here it will be convenient to consider two relatively extreme cases — first the dye laser which changes very rapidly and needs therefore a quick servo. Later we will consider the case of the He-Ne laser that begins life with a drastically better intrinsic stability.

The dye laser designer may choose a cavity line width $\Delta\Omega_c/2\pi \approx 2$ MHz [11], and considering interferometric efficiency, he may finally have 300 μW detected at the fringe peaks. Looking at the 1/2 height where the slope is maximum then gives 150 μW transmitted light for detection. At 0.6 μA/μW we then obtain the current $I_o \approx 100$ μA. For a dye laser a servo bandwidth of ≈2 MHz is appropriate [11] and consistent with our earlier assumptions. Taking only shot noise as the noise source, from Eq. (2), we can calculate a noise current of $\sqrt{2}\, i_s = 2\sqrt{I_o eB} = 11.3$ nA. The measurement S/N is therefore 100 μA/11.3 nA ≈ 8850. With Eq. (4) we predict that a laser — operating with high loop gain to suppress its own noise — will be given an rms line width of 225 Hz by the finite S/N ratios due to photodetection shot noise.

So we have the advertised situation that the servo has suppressed the laser's natural noise almost completely and left us with a low level, broadband fm noise reaching toward 2 MHz and amounting to 225 Hz rms. This exchange can be seen to be attractive,since the fast random noise due to the laser jetstream [J(ω) in Fig. 2] is 1/2 MHz in a good case, and the low-frequency variations due to pressure surges can amount to several, even tens of megaHertz. Still it is interesting to press a little farther and to try to understand how a super-narrow absorber would respond to this modified laser field. To avoid unnecessary complexity we may look for the atom's linear response, and so we come to ask about the power spectrum of the laser under feedback. Neglecting amplitude variations of the laser output,we may write the field as $E_o \exp[i\Omega t + \phi(t)]$. The instantaneous frequency of such a wave will be $\Omega^* = \Omega + (d\phi/dt)$. Our servo-induced variations $\delta\Omega = \Omega^* - \Omega = (d\phi/dt)$ should have a flat spectrum since they were introduced by the comparison of a shot-noise signal with a frequency-discrimination servo system. We can write

$$\phi(t) = \int_o^t \delta\Omega\, dt \qquad (5)$$

with $\delta\Omega$ being given by Eq. (4). Since the variations $\delta\Omega$ represent a random process, the phase variable $\phi(t)$ in Eq. (5) is also random. To calculate the spectrum of an electric field with a random phase we will have to calculate the autocorrelation function. This problem has recently been clearly treated by

Elliott et al. [12] for the case that the random noise was deliberately being imposed by the experimenter. In a way, that's what is happening here. The result of Elliott's analysis is a general expression for the spectral line width of the source, and two very simple limiting formulas. For our case where the deviations are very small in magnitude but occur with a very fast rate, the result is a Lorentzian line shape. Elliott et al. also give an expression for the opposite limit -- large excursions at a rather slow rate. This case gives a Gaussian profile. Written in the present notation Elliott et al. give the result

$$\Delta_{FWHM} = \pi(\delta\Omega)^2/B \tag{6a}$$

$$= \pi(\Delta\Omega_c \cdot \sqrt{e/I})^2 \tag{6b}$$

where the second result follows from Eq. (4). Equation (6b) shows that the stabilized laser line widths under high feedback conditions is simply π times the spectral noise density, independent of the bandwidth B (B assumed >> rms excursions). Numerically for these assumed conditions (B = 2 MHz, $\delta\Omega/2\pi$ = 225 Hz) we project a line width of the Lorentzian spectral peak to be 80 mHz.

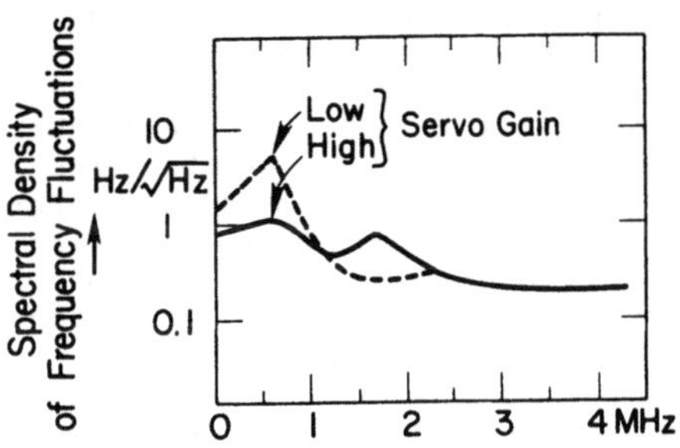

Fig. 5. Spectral density of residual noise at the frequency servo error point. Data from Ref. [11]. Note that the "low gain" is not quite sufficient to suppress the laser's intrinsic noise at low frequencies, while the "high gain" position reveals excessive phase shift in the discriminator/amplifier system. The noise for frequencies below ~1 MHz is always a few dB above the shot-noise level represented by the high-frequency asymptote, whereas if the gain were "enough" in the sense described in the text, even the apparent shot noise would be suppressed at low frequencies by the servo's imposing a (negative) image noise on the laser's frequency.

It may be interesting to confront these ideas with experimental facts. In Fig. 5 we show the measured noise spectral density obtained by Helmcke et al. [11] with a fringe-side servo system. The reader will see that they are having still some residual servo problems and were unable to completely suppress the laser's intrinsic noise. The reasons are explained in that report -- basically problems of compromising the need for a sharp reference cavity in order to have a high slope and a good signal/noise ratio with the resultant time delay and poor transient performance when the cavity line width is small. Still, a line width due to random noise of only 3 Hz represents a reasonably attractive level.

After this work was finished, a discussion with R. W. P. Drever led to a really optimal type of discriminator [8] which is free of the unpleasant coupling of reference cavity sharpness with a time delay. The basic idea is to set up an optical system which can measure the phase shift of the optical field by looking at the signal reflected from the cavity. For quick changes one can imagine that it would be possible to make homodyne or heterodyne phase measurements of the newly-changed laser field, since the fields inside the reference cavity cannot change rapidly. They can provide the phase reference for the measurement. We note that this principle, emphasizing the importance of optical phase measurements, was already appreciated in the 1960's by P. H. Lee and M. L. Skolnick [13]. FM spectroscopy was also independently invented in the same time frame as Drever and Hall's work [8] by G. C. Bjorklund [14] who desired to use the tunability of fm sidebands effectively to tune some laser light to a new spectral location, offset from the carrier. Of course, it ultimately turned out that, as usual, the essence of the "new" technique had already been worked out and applied in the NMR field [15]. As an optical frequency discriminator, the technique has a number of very important advantages, mainly technical in nature. However, the need to use a little of the light to provide a "local oscillator" wave for heterodyne detection does degrade the signal-to-random-noise performance by about 3.4 dB compared with the fringe-side servo discussed previously.

To illustrate the narrow line width possibilities of even dye lasers when they are servo-stabilized using the "rf pound stabilizer (or the Pound/Drever discriminator)," Hough et al. [16] set up two independent dye lasers. Each was stabilized to near the noise level of the discriminator. This was not as difficult to achieve as in the work with Helmcke using the fringe-side system. Figure 6 shows the spectrum of the beat between these two independently-stabilized dye lasers. The first thing to note is that the beat of the two lasers reaches the shot-noise floor for Fourier frequencies above 300 or 400 Hz. The line width due to random noise processes in the discriminator will be rather small, since we have a noise density below 1 $Hz/\sqrt{Hz}$. The result using Eq. (6b) is 1.3 Hz FWHM! (We found that the stabilizing beam initially bouncing but eventually absorbed inside the reference cavity gave local heating of the mirror: it was an attempt to avoid the long-term heating curve that prompted us to use such a low power that the random noise was near 0.6 $Hz/\sqrt{Hz}$.) Unfortunately it is also all too clear in Fig. 6 that other environmental factors are causing problems at the lower Fourier frequencies. Some effort was invested in reducing the coupling of building vibrations into the spacer of the more sensitive reference cavity by hanging it with a pendulum suspension. A typical drift rate for the two independently-stabilized lasers was a few hundred Hz per second, although when the temperature drift-sign turned around at night we found periods of 5 min during which the beat frequency was constant to within a few kiloHertz.

To conclude this discussion I want to show beat waveforms obtained by D. Hils and C. Salomon [17] when heterodyning two HeNe lasers together. The two independent lasers have been stabilized to adjacent axial orders of a high-finesse Fabry-Perot cavity. The resulting difference frequency ($c/2l \simeq 240$ MHz) has been subtracted from the optical beat signal by rf heterodyne with a highly stable frequency synthesizer. Figure 7 shows the beat wave heterodyned down into the audio (really subaudio) range. The range of time on the graph covers 10 s. One can see

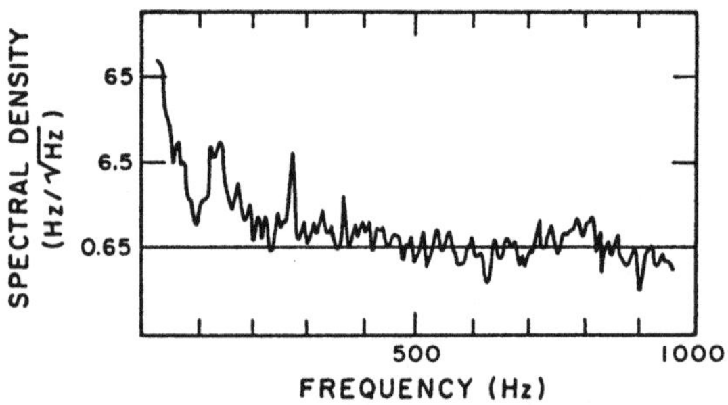

Fig. 6. Fourier spectrum of frequency noise in beat between two independent dye lasers, located in different rooms. Data from Ref. [16]. Both systems used the rf sideband discriminators of Ref. [8]. This figure shows that ultimately it is possible to closely approach the shot-noise limit. The increases of the noise density at low frequencies is due to room vibrations, in spite of our isolation efforts.

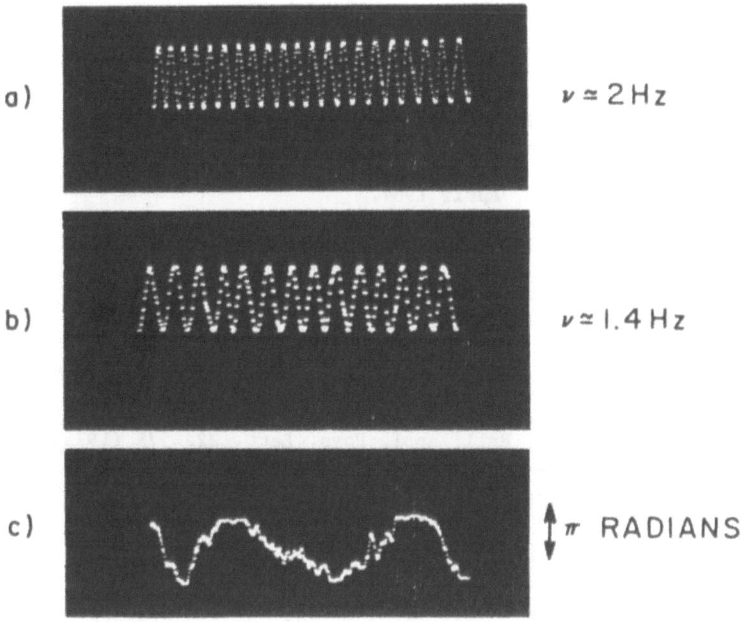

Fig. 7. Beat frequency between two HeNe lasers independently locked to a single reference cavity. Data from Ref. [17]. By locking with a unit difference in interference orders we eliminate coupling between the two systems. Heterodyne with a stable rf synthesizer displaces the laser beat to near zero frequency, a), b), c) representing different rf offsets. From the stability of the beat over a good fraction of the 10 s display, it is clear that the lasers are locked to the cavity with sub-Hertz precision.

that the two independent servo systems are able to lock the lasers rather well: If for example laser A moved forward by one radian in its 5×10^{-14} output frequency, that would show up as a 1 radian shift of this audio beat. Thus it is clear from the figure that no such rapid phase jumps are occurring. Rather the phase is stable to within a radian for a few seconds. So these lasers are certainly independently stabilized and are certainly stabilized on the sub-Hertz level. This result confirms our expectations about the servo performance and the random noise term $S(\omega)$ which feeds noise into the frequency discriminator box in Fig. 2. Experiments are now in progress to assess the nature of the cavity stability -- the $D(\omega)$ term in Fig. 2 -- by heterodyne experiments with an iodine-stabilized laser. Its short-term performance will be drastically worse, because of the lower signal/noise available from I_2, but it is expected to be the more stable for times beyond just a minute or a few minutes at most.

This presentation integrates the efforts of a large number of colleagues who have collaborated in different phases of the work over the years. Particularly, I wish to acknowledge the contributions made by R. L. Barger, A. Brillet, J. Helmcke, S. A. Lee, J. Hough, L. Hollberg, J. C. Bergquist, D. Hils, M. D. Rayman, and C. Salomon. Discussions with P. L. Bender over many years have aided my understanding immeasurably, and his project with J. E. Faller, D. Hils, and myself to put a gravity wave antenna in space may provide a secure future employment prospect for yet unborn generations of laser stabilization specialists. The work at JILA is supported in part by the National Bureau of Standards in support of its basic mission in precision measurement science, and in part by the Office of Naval Research and the National Science Foundation.

REFERENCES

1. R. E. Slusher, L. W. Hollberg, B. Yurke, J. C. Mertz, J. F. Valley, Phys. Rev. Lett. 55, 2409 (1985).
2. C. Caves, Phys. Rev. D 23, 1693 (1981).
3. H. W. Bode, Network Analysis and Feedback Amplifier Design (Van Nostrand, New York, 1945).
4. T. W. Hänsch, private communication, April 1984.
5. See Ref. 3, and other texts on servo theory.
6. Unfortunately, no adequate publication was ever produced about this work. But it did earn an IR-100 award in 1975!
7. Y. Yamamoto, in U. S. Japan Seminar on Coherence, Incoherence and Chaos, Nara Japan, September 1984; also in a preprint.
8. R. W. P. Drever, J. L. Hall, F. V. Kowalski, J. Hough, G. M. Ford, A. J. Munley, M. Ward, Appl. Phys. B 31, 97 (1983). The experiment was performed in September 1979.
9. T. W. Hänsch and B. Couillard, Opt. Commun. 35, 441 (1980).
10. R. L. Barger, M. S. Sorem and J. L. Hall, Appl. Phys. Lett. 22, 573 (1973).
11. J. Helmcke, S. A. Lee and J. L. Hall, Appl. Opt. 21, 1686 (1982).
12. D. S. Elliott, R. Roy, S. J. Smith, Phys. Rev. A 26, 12 (1982).
13. P. H. Lee and M. L. Skolnick, IEEE J. Quant. Electr. QE-2, 784 (1966).
14. G. C. Bjorklund, Opt. Lett. 5, 15 (1980).
15. See for example B. Smaller, Phys. Rev. 83, 812 (1951); J. V. Acrivos, J. Chem. Phys. 36, 1097 (1962).
16. J. Hough, D. Hils, M. D. Rayman, Ma L.-S., L. Hollberg, J. L. Hall, Appl. Phys. B 33, 179 (1984).
17. D. Hils, C. Salomon and J. L. Hall, to be published.

Index of Contributors

MIX
Papier aus verantwortungsvollen Quellen
Paper from responsible sources
FSC® C105338

If you have any concerns about our products,
you can contact us on
ProductSafety@springernature.com

In case Publisher is established outside the EU,
the EU authorized representative is:
Springer Nature Customer Service Center GmbH
Europaplatz 3, 69115 Heidelberg, Germany

Printed by Libri Plureos GmbH
in Hamburg, Germany